Handbook of Green Chemistry

Volume 3
Biocatalysis

Volume Edited by
Robert H. Crabtree

Related Titles

Wasserscheid, P., Welton, T. (eds.)

Ionic Liquids in Synthesis

2nd Edition
2008
ISBN: 978-3-527-31239-9

Sheldon, R. A., Arends, I., Hanefeld, U.

Green Chemistry and Catalysis

2007
ISBN: 978-3-527-30715-9

Cornils, B., Herrmann, W. A., Muhler, M., Wong, C.-H. (eds.)

Catalysis from A - Z

A Concise Encyclopedia

3rd Edition
2007
ISBN: 978-3-527-31438-6

Loupy, A. (ed.)

Microwaves in Organic Synthesis

2nd Edition
2006
ISBN: 978-3-527-31452-2

Kappe, C. O., Stadler, A., Mannhold, R., Kubinyi, H., Folkers, G. (eds.)

Microwaves in Organic and Medicinal Chemistry

2005
ISBN: 978-3-527-31210-8

Handbook of Green Chemistry

Volume 3
Biocatalysis

Volume Edited by Robert H. Crabtree

**WILEY-
VCH**

WILEY-VCH Verlag GmbH & Co. KGaA

The Editor

Prof. Dr. Paul T. Anastas
Yale University
Center for Green Chemistry & Green Engineering
225 Prospect Street
New Haven, CT 06520
USA

Volume Editor

Prof. Dr. Robert H. Crabtree
Yale University
Department of Chemistry
225 Prospect St.
New Haven, CT 06520-8107
USA

Handbook of Green Chemistry – Green Catalysis
Vol. 1: Homogenous Catalysis
ISBN: 978-3-527-32496-5
Vol. 2: Heterogenous Catalysis
ISBN: 978-3-527-32497-2
Vol. 3: Biocatalysis
ISBN: 978-3-527-32498-9

Set I (3 volumes):
ISBN: 978-3-527-31577-2

Handbook of Green Chemistry
Set (12 volumes):
ISBN: 978-3-527-31404-1

Library of Congress Card No.:
applied for

British Library Cataloguing-in-Publication Data
A catalogue record for this book is available from the British Library.

Bibliographic information published by the Deutsche Nationalbibliothek
The Deutsche Nationalbibliothek lists this publication in the Deutsche Nationalbibliografie; detailed bibliographic data are available on the Internet at http://dnb.d-nb.de.

© 2009 WILEY-VCH Verlag GmbH & Co. KGaA, Weinheim

Typesetting Thomson Digital, Noida, India
Printing betz-druck GmbH, Darmstadt
Binding Litges & Dopf GmbH, Heppenheim
Cover Design Adam-Design, Weinheim

Printed in the Federal Republic of Germany
Printed on acid-free paper

ISBN: 978-3-527-32498-9

Contents

Handbook of Green Chemistry, Volume 3: Biocatalysis. Edited by Robert H. Crabtree
Copyright © 2009 WILEY-VCH Verlag GmbH & Co. KGaA, Weinheim
ISBN: 978-3-527-32498-9

About the Editors

Series Editor

Paul T. Anastas joined Yale University as Professor and serves as the Director of the Center for Green Chemistry and Green Engineering there. From 2004–2006, Paul was the Director of the Green Chemistry Institute in Washington, D.C. Until June 2004 he served as Assistant Director for Environment at the White House Office of Science and Technology Policy where his responsibilities included a wide range of environmental science issues including furthering international public-private cooperation in areas of Science for Sustainability such as Green Chemistry. In 1991, he established the industry-government-university partnership Green Chemistry Program, which was expanded to include basic research, and the Presidential Green Chemistry Challenge Awards. He has published and edited several books in the field of Green Chemistry and developed the 12 Principles of Green Chemistry.

Volume Editor

Robert Crabtree took his first degree at Oxford, did his Ph.D. at Sussex and spent four years in Paris at the CNRS. He has been at Yale since 1977. He has chaired the Inorganic Division at ACS, and won the ACS and RSC organometallic chemistry prizes. He is the author of an organometallic textbook, and is the editor-in-chief of the Encyclopedia of Inorganic Chemistry and Comprehensive Organometallic Chemistry. He has contributed to C-H activation, H_2 complexes, dihydrogen bonding, and his homogeneous tritiation and hydrogenation catalyst is in wide use. More recently, he has combined molecular recognition with CH hydroxylation to obtain high selectivity with a biomimetic strategy.

Handbook of Green Chemistry, Volume 3: Biocatalysis. Edited by Robert H. Crabtree
Copyright © 2009 WILEY-VCH Verlag GmbH & Co. KGaA, Weinheim
ISBN: 978-3-527-32498-9

List of Contributors

Vincenza Andreoni
Università degli Studi di Milano
Dipartimento di Scienze e Tecnologie
Alimentari e Microbiologiche
Via Celoria 2
20133 Milan
Italy

Uwe T. Bornscheuer
University of Greifswald
Institute of Biochemistry
Department of Biotechnology and
Enzyme Catalysis
Felix-Hausdorff-Strasse 4
17487 Greifswald
Germany

Dean Brady
CSIR Biosciences
Ardeer Road
1645 Modderfontein
South Africa

Richard Cammack
King's College London
Department of Biochemistry
150 Stamford Street
London SE1 9NH
UK

Teresa De Diego
Universidad de Murcia
Facultad de Química
Departamento de Bioquímica y Biología
Molecular 'B' e Inmunología
P.O. Box 4021
30100 Murcia
Spain

António L. De Lacey
CSIC
Instituto de Catálisis
C/ Marie Curie 2
28049 Madrid
Spain

Victor M. Fernández
CSIC
Instituto de Catálisis
C/ Marie Curie 2
28049 Madrid
Spain

Oreste Ghisalba
Novartis Pharma AG
4002 Basel
Switzerland

Handbook of Green Chemistry, Volume 3: Biocatalysis. Edited by Robert H. Crabtree
Copyright © 2009 WILEY-VCH Verlag GmbH & Co. KGaA, Weinheim
ISBN: 978-3-527-32498-9

Liliana Gianfreda
Università degli Studi di Napoli
Federico II
Dipartimento di Scienze del Suolo, della
Pianta e dell'Ambiente
Via Università 100
80055 Portici (NA)
Italy

Dörte Gocke
Heinrich-Heine-Universität Düsseldorf
Institut für Molekulare
Enzymtechnologie
52426 Jülich
Germany

Leandro Helgueira Andrade
University of São Paulo
Institute of Chemistry
Caixa Postal 26077
CEP 055139-970
São Paulo, SP
Brazil

José L. Iborra
Universidad de Murcia
Facultad de Química
Departamento de Bioquímica y Biología
Molecular 'B' e Inmunología
P.O. Box 4021
30100 Murcia
Spain

Anett Kirschner
University of Groningen
Groningen Biomolecular Science and
Biotechnology Institute
Department of Biochemistry
Nijenborgh 4
9747 AG Groningen
The Netherlands

James E. Leresche
Lonza Braine SA
Chaussée de Tubize 297
1420 Braine-l'Alleud
Belgium

Pedro Lozano
Universidad de Murcia
Facultad de Química
Departamento de Bioquímica y Biología
Molecular 'B' e Inmunología
P.O. Box 4021
30100 Murcia
Spain

Hans-Peter Meyer
Lonza Ltd
Rottenstrasse
3930 Visp
Switzerland

Michael Müller
Albert-Ludwigs-Universität Freiburg
Institut für Pharmazeutische
Wissenschaften
Stefan-Meier-Strasse 19
79194 Freiburg
Germany

Kaoru Nakamura
Kyoto University
Institute for Chemical Research
Uji
Kyoto 611-0011
Japan

Martina Pohl
Heinrich-Heine-Universität Düsseldorf
Institut für Molekulare
Enzymtechnologie
52426 Jülich
Germany

Olaf Rüdiger
CSIC
Instituto de Catálisis
C/ Marie Curie 2
28049 Madrid
Spain

Vlada B. Urlacher
University of Stuttgart
Institute of Technical Biochemistry
Allmandring 31
70569 Stuttgart
Germany

1
Catalysis with Cytochrome P450 Monooxygenases

Vlada B. Urlacher

1.1
Properties of Cytochrome P450 Monooxygenases

1.1.1
General Aspects

Biocatalytic oxyfunctionalization of non-activated hydrocarbons is considered as 'potentially the most useful of all biotransformations' [1]. Since biooxidation-based applications using cytochrome P450 monooxygenases often yield compounds that are difficult to synthesize using traditional synthetic chemistry, they have attracted considerable attention from chemists, biochemists and biotechnologists. Cytochrome P450 monooxygenases (P450s or CYPs) are heme-containing monooxygenases, which were recognized and defined as a distinct class of hemoproteins about 50 years ago [2, 3]. These enzymes got their name from their unusual properties to form reduced (ferrous) iron–carbon monoxide complexes in which the heme absorption Soret band shifts from 420 to $\sim$450 nm [4, 5]. Essential for this spectral characteristic is the axial coordination of the heme iron by a cysteine thiolate, which is common to all P450 monooxygenases [6, 7]. The phylogenetically conserved cysteinate is the proximal ligand to the heme iron, with the distal ligand generally assumed to be a weakly bound water molecule [8].

In terms of nomenclature, the root symbol CYP, denoting cytochrome P450, is followed by an Arabic number representing the particular families (generally groups of proteins with more than 40% amino acid sequence identity), a letter for the respective subfamilies (greater than 55% identity) and a number determining the specific gene; for example, CYP102A1, which represents the cytochrome P450 BM-3 from *Bacillus megaterium*. An exception is the CYP51 family, where sterol Δ^{22}-desaturases are grouped together based on their identical function and not on sequence similarity [9].

Since their discovery, the P450s have been studied in enormous detail due to their involvement in a plethora of crucial cellular roles – from carbon source assimilation,

Handbook of Green Chemistry, Volume 3: Biocatalysis. Edited by Robert H. Crabtree
Copyright © 2009 WILEY-VCH Verlag GmbH & Co. KGaA, Weinheim
ISBN: 978-3-527-32498-9

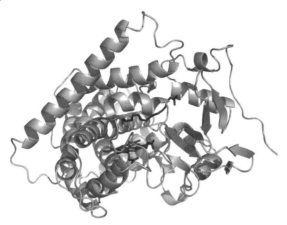

Figure 1.1 The crystal structure of the P450 BM-3 monooxygenase domain with palmitoleic acid bound (adapted from pdb 1SMJ): heme and palmitoleic acid in black.

through biosynthesis of hormones to carcinogenesis, drug activation and degradation of xenobiotics.

Cytochrome P450 monooxygenases build one of the largest gene families, with currently more than 7700 gene sequences found in all domains of life [10]. Despite less than 20% sequence identity across the gene superfamily, P450 enzymes appear to take on a similar structural fold [11] (Figure 1.1).

The number of P450 monooxygenases identified is constantly increasing due to microbial screenings and the rapid development of DNA sequencing techniques, leading to an increasing number of sequenced genomes. Functional characterization of new members of the P450 gene family thus offers a route to diverse building blocks, closely linked with the retrieval of new important compounds.

1.1.2
Chemistry of Substrate Oxidation by P450 Monooxygenases

Most P450 enzymes catalyze the reductive scission of dioxygen, while bound to the heme iron:

$$NAD(P)H + RH + O_2 + H^+ \rightarrow ROH + NAD(P)^+ + H_2O$$

The process requires the consecutive delivery of two electrons in the form of hydride ions to the heme iron. P450 monooxygenases are external monooxygenases that utilize reducing equivalents (electrons in the form of hydride ions), ultimately derived from the pyridine cofactors NADH or NADPH and transferred to the heme via redox partners [12, 13].

The classical P450 catalytic cycle as recently revised by Sligar and colleagues [14] is depicted in Scheme 1.1. Substrate binding in the active site induces the dissociation of a water molecule that is bound relatively weakly as the sixth coordinating ligand to the heme iron to the thiolate **1**. This, in turn, induces a shift of the heme iron spin

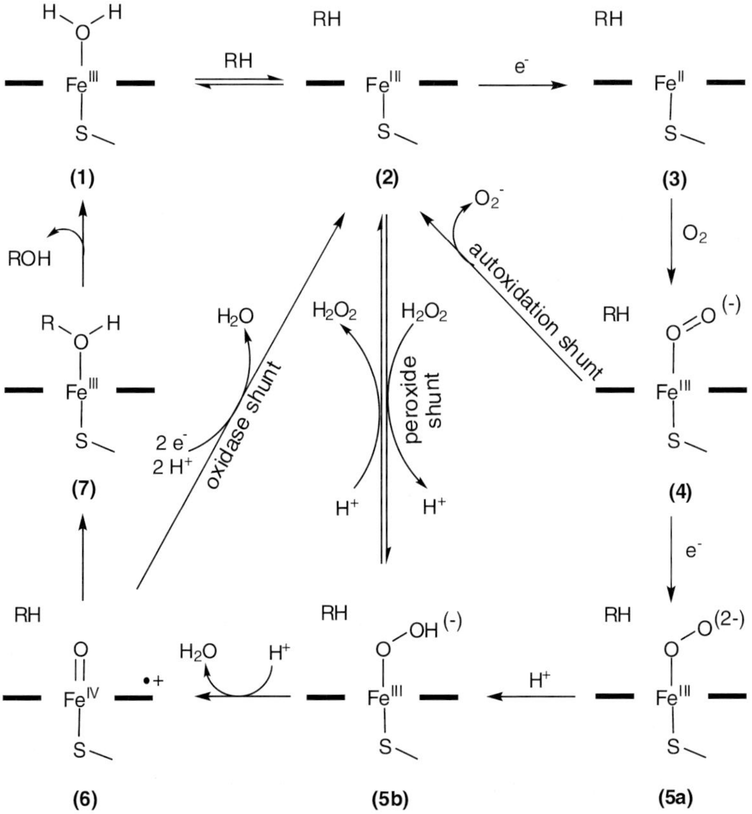

Scheme 1.1 Catalytic cycle of P450 monooxygenases.

state from low-spin ($S = 1/2$) to high-spin ($S = 5/2$) and a positive shift in heme iron reduction potential in the order of 130–140 mV [15]. The increased potential the delivery of the first electron, which reduces the heme iron from the ferric form, Fe (III) (2), to the ferrous form, Fe(II) (3). After the first electron transfer, the Fe(II) (3) binds dioxygen, resulting in a ferrous dioxygen complex (4). The timely delivery of the second electron converts this species into a ferric peroxy anion (5a). Subsequent steps in the P450 cycle are considered to be relatively fast with respect to the electron transfer. The ferric peroxy species 5a is protonated to a ferric hydroperoxy complex (5b) (compound 0) and then further protonated to a high-valent ferryl–oxo complex (6) (compound I), accompanied by the release of a water molecule through heterolytic scission of the dioxygen bond in the preceding intermediate (7). Compound I is considered to be the intermediate catalyzing the majority of P450 reactions; however, compound 0 may also be important for some P450-dependent catalytic reactions [16], for example in the epoxidation of C=C double bonds [17].

Under certain circumstances, P450 monooxygenases can enter one of three abortive cycles, also referred to as uncoupling pathways (Scheme 1.1). If the second electron is not delivered to reduce the short-lived ferrous–oxy complex 4, it can decay, forming superoxide (autoxidation shunt). The inappropriate positioning of the substrate in the active site is often the molecular reason for the two other uncoupling cycles. The ferric hydroperoxy intermediate **5b** can collapse and release hydrogen peroxide (peroxide shunt), whereas decay of **6** (compound **I**) is accompanied by the release of water (oxidase shunt). The intracellularly formed peroxide and superoxide might damage P450 monooxygenases through heme macrocycle degradation and apoprotein oxidation [18, 19]. For industrial applications, it is particularly important to note that the uncoupling pathways in all cases consume reducing equivalents from NAD(P)H without product formation. The 'peroxide shunt' can be forced into a productive direction through the addition of hydrogen peroxide or organic peroxides [20, 21].

1.1.3
Redox Partners of P450 Monooxygenases

Depending on the redox partners, traditionally two main classes of cytochrome P450 monooxygenases have been defined [22]. Class I P450 enzymes are found in mitochondria and bacteria and use a small redox 2Fe–2S iron–sulfur protein (ferredoxin) and an FAD-containing reductase (ferredoxin reductase) for transfer of electrons from NAD(P)H to the P450 component. Microsomal P450s belong to class II P450 redox systems that exploit the FAD- and FMN-containing cytochrome P450 reductase (CPR) (and sometimes cytochrome b5) for transfer of electrons from NADPH.

In recent years, the numerous genome sequencing projects have revealed many other types of electron transfer proteins, which belong neither to class I nor to class II [23]. For example, P450 BM-3 is a fusion enzyme consisting of a P450 monooxygenase at the N-terminus and a CPR-like reductase which are separated by a short linker, without any membrane anchor region for either of the two domains [24, 25]. According to an updated classification system suggested by Chapmen and colleagues (http://www.chem.ed.ac.uk/chapman/p450.html), P450 BM-3 belongs to class III. Other members of this class are CYP102A2 and CYP102A3 from *Bacillus subtilis* and several other recently discovered members from the CYP102A-subfamily [26, 27].

A further distinctive type is fused flavocytochromes that Chapmen and colleagues referred to as class IV. These enzymes belonging to the CYP116 family are found in several *Rhodococcus* strains [28], in some species of the pathogen *Burkholderia* and in *Ralstonia metallidurans* [29]. They have an N-terminal P450 domain fused to a C-terminal reductase domain very similar to the well-characterized phthalate dioxygenase reductase that binds both FMN and 2Fe–2S cofactors. P450 monooxygenases fused to their redox partners are considered to be self-sufficient, which makes them particularly interesting for biotechnological applications.

In addition to the above-mentioned electron transfer systems, many others have been found to support P450 catalysis, such as the 3Fe–4S and 4Fe–4S ferredoxins or flavodoxins [30]. Nevertheless, all of them receive reducing equivalents from NAD(P)H. The natural *S. solfataricus* electron transfer partners for the thermophilic CYP119 were recently shown to be a thermostable ferredoxin and a 2-oxo acid ferredoxin oxidoreductase. Thus, CYP119 is the first P450 monooxygenase utilizing as ultimate source of electrons not pyridine nucleotides but 2-oxo acids such as pyruvate [31].

Apart from the self-sufficient P450 monooxygenases, there are other P450 enzymes that do not require exogenous protein partners. One important group of such P450s are natural P450 peroxygenases. They do not involve the 'classical' P450 reaction cycle, but the 'peroxide shunt', which is the reverse of the uncoupling reaction that leads to the collapse of the ferric hydroperoxy species (see Section 1.1.2). This shortened catalytic cycle resembles the catalytic mechanism of peroxidases [32]. Three enzymes with a potential for biocatalytic applications are the H_2O_2-utilizing fatty acid hydroxylases SP_α from *Sphingomonas paucimobilis* (CYP152B1) [33], BS_β from *B. subtilis* (CYP152A1) [34] and a P450cla from *Clostridium acetobutylicum* (CYP152A2) [35].

Other P450s that do not rely on external redox partners are fungal P450nor enzymes, for example CYP55A1 from the rice pathogen *Fusarium oxysporum*. CYP55A1 catalyzes the reduction of two molecules of nitric oxide (NO) to dinitrogen oxide (N_2O) and reacts directly with the cofactor NADH to purchase reducing equivalents [36].

1.1.4
Major Reactions Catalyzed by P450 Monooxygenases

P450 monooxygenases are able to catalyze more than 20 different reaction types [37]. Hydroxylation of unactivated sp^3-hybridized C-atoms belongs to the classical reactions catalyzed by P450s. Examples for this reaction include the hydroxylation of saturated fatty acids (**8**) catalyzed by eukaryotic CYP4 and bacterial CYP102 (e.g. P450 BM-3) enzymes and the stereospecific hydroxylation of D-(+)-camphor (**10**) to 5-*exo*-hydroxycamphor (**11**) through P450cam (Scheme 1.2).

Epoxidation of C=C double bonds is another major reaction type catalyzed by P450s. Particularly attractive in this regard is $P450_{EpoK}$ from *Sorangium cellulosum*, which catalyzes the epoxidation of the thiazole-containing macrolactone epothilone D (**12**) to epothilone B (**13**) [38] (Scheme 1.3). Epothilones are important anti-tumor polyketides with high microtubule stabilizing activity.

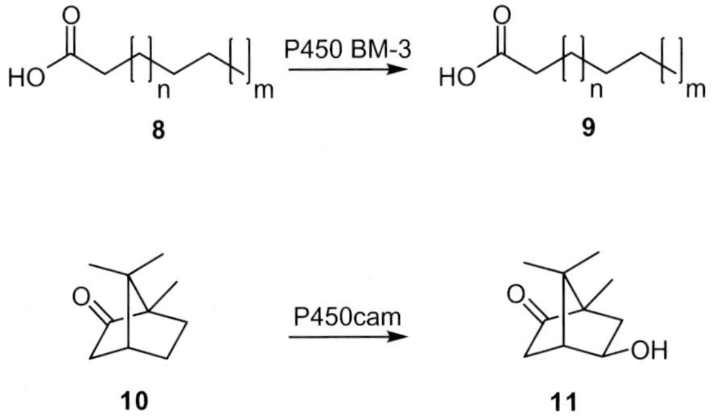

Scheme 1.2 Fatty acid hydroxylation catalyzed by P450 BM-3 and camphor hydroxylation catalyzed by P450cam.

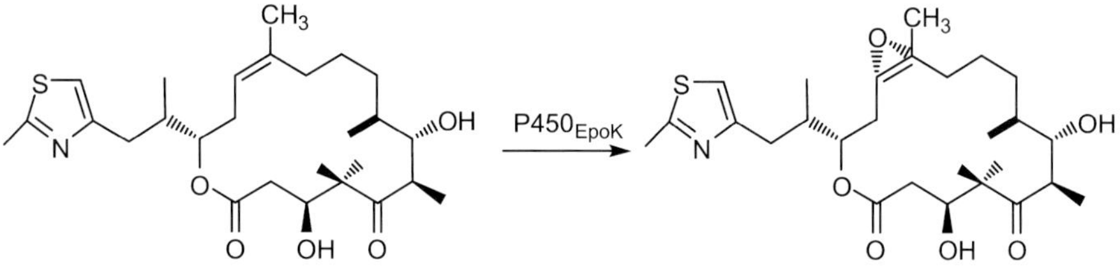

Scheme 1.3 Epoxidation of epothilon D to epothilon B through P450$_{EpoK}$.

Aromatic hydroxylation also belongs to the common P450 reactions. P450$_{NikF}$ from *Streptomyces tendae* Tü901 has been reported to catalyze the aromatic hydroxylation of the pyridyl ring of the peptidyl moiety of nikkomycin (**14**), an inhibitor of chitin synthase [39] (Scheme 1.4). Many P450 monooxygenases have been engineered towards aromatic hydroxylation, since this ability make such enzymes attractive candidates for bioremediation (examples will be described later).

Additionally, some P450 monooxygenases are also able to catalyze the cleavage of C−C bonds via multiple substrate oxidations, for example, demethylation of lanosterol (**15**) to a precursor of cholesterol, 4,4-dimethyl-5α-cholesta-8,14,24-dien-3β-ol (**18**), by a lanosterol 14α-demethylase (CYP51) [40]. The mechanism includes three steps and proceeds via initial hydroxylation of the C-14 methyl group, followed by further oxidation of the alcohol **16** to the aldehyde **17**. Finally, acyl cleavage occurs, leading to formation of a double bond in the steroid (Scheme 1.5).

A similar cascade of reactions is assumed to be catalyzed by CYP107H1 (P450$_{Biol}$) from *B. subtilis* during conversion of long chain fatty acyl CoA esters to pimeloyl CoA in the biotin biosynthesis pathway [41].

P450 enzymes not only catalyze oxidation and dealkylation at C-atoms, but also at N-, S- and O-atoms. Heteroatom dealkylation is believed to be the following step of the hydroxylation of the α-carbon atom. Examples include the *N*-oxygenation of *N,N*-dimethylalanine and *N,N*-dialkylarylamines by mammalian CYP2B1 and CYP2B4, respectively [42, 43], and *O*-demethylation of 5-methoxytryptamine by CYP2D6 [44].

Some P450 enzymes are able to catalyze oxidative phenol coupling, a reaction usually carried out by peroxidases. Three independent P450 monooxygenases with such activity have been proven to be involved into the synthesis of vancomycin-type antibiotics in *Amycolatopsis balhimycina* [45].

Oxygenation P450 reactions include also oxidative deamination, desulfurylation, dehalogenation and Baeyer–Villiger oxidation [46, 47]. Many other unusual types of

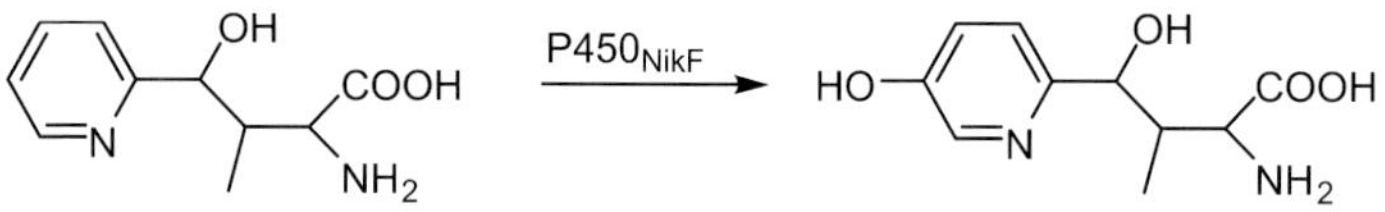

Scheme 1.4 Aromatic hydroxylation of the pyridyl ring of the peptidyl moiety catalyzed by P450$_{NikF}$ during nikkomycin biosynthesis.

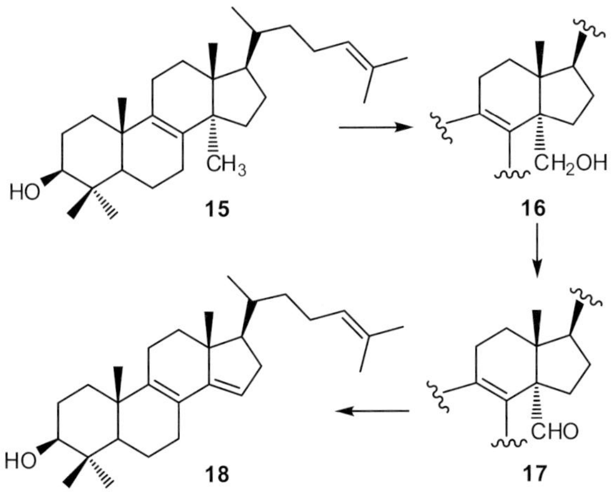

Scheme 1.5 Mechanism of sterol demethylation by CYP51.

reactions catalyzed by P450s have been described in the literature, including dehydrogenation, dehydration, reductive dehalogenation, epoxide reduction and isomerization [48].

1.2
Biotechnological Applications of P450 Monooxygenases

1.2.1
Human P450s in Drug Development

Drug development is based on the detailed characterization of metabolic pathways and their relevance for drug safety. This type of analysis often requires milligram quantities of metabolites, which are difficult to synthesize by chemical routes, especially when the metabolites result from stereoselective oxidations. An ongoing field of research in drug discovery and development is the use of recombinant human P450s. Human P450s expressed in baculovirus-infected insect cell lines have been used for some time to synthesize drug metabolites in order to assess the toxicity of potential pharmaceuticals [49–51]. Recent achievements in the expression of recombinant human P450s in *Saccharomyces cerevisiae* [52–54], *Yarrowia lipolytica* [55], *Pichia pastoris* [56] and *Escherichia coli* [57, 58] facilitate their use for the synthesis of drug metabolites. Co-expression of mammalian P450 reductase and P450s in *S. cerevisiae* and *P. pastoris* seems to be necessary to achieve a catalytically highly active monooxygenase system, since the host P450 reductase couples poorly with microsomal P450s. In the absence of human P450 reductase, the human P450s CYP3A4 and CYP2D6, expressed in *S. cerevisiae*, were catalytically inactive; however, upon expression of this ancillary factor they displayed activity towards prototypical substrates such as testosterone and bufuralol, respectively. Expression of human

CYP1A1 in *Y. lipolytica* [55] was performed with or without overproduction of *Y. lipolytica* CPR. Remarkably, co-expression of human CPR was not required in this case. Increase in *Y. lipolytica* CPR expression resulted in 6.3–12.8-fold higher activity towards ethoxyresorufin, depending on the CPR transformant [55].

E. coli is an attractive expression system because high levels of expression and growth to high cell densities can be achieved [59]. In addition, *E. coli* is easily manipulated and a wide variety of strain variants and vectors with powerful promoters are available. A limitation of this expression system is that in almost all cases mammalian cDNAs have to be modified before they can be expressed [59]. Initial attempts to improve expression involved modification of the N-terminal sequence since this region, and especially the second codon, have been proposed to be important in determining protein yield [60, 61]. This strategy was applied to express a number of mammalian P450s (e.g. CYPs 2E1 [62], 4A4 [63], 6A1 [64], 1A1 [65]). Using this approach, secondary structure formation in the transcript could be minimized, which enhanced the protein expression level in *E. coli*. Further optimization could be achieved by in-frame fusion of a bacterial signal peptide (for example, a modified ompA-leader sequence) to the human P450 cDNAs [51]. This leader is removed during P450 processing, thus releasing the native P450. Following this strategy, 14 recombinant human P450 enzymes co-expressed with NADPH–P450 reductase in *E. coli* have been used by several pharmaceutical companies both as biocatalysts for the preparation of metabolites of drug candidates and for high-throughput P450 inhibition screening. Up to 300 mg of the different metabolites could be obtained using the permeabilized recombinant *E. coli* cells expressing human P450s [66].

A different approach was pursued based on the hypothesis that the hydrophobicity of the N-terminal anchor of mammalian P450s impeded efficient expression in bacteria. Therefore, the N-terminal hydrophobic region of many P450s responsible for the interaction with the membrane was truncated to a greater or lesser extent (for example, CYPs 2E1 [67], 2D6 [68], 2C9 [69] and many others). Three human cytochrome P450s, 3A4, 2C9 and 1A2 with truncated N-termini, were each co-expressed with human CPR in *E. coli* using bicistronic expression system. Intact *E. coli* cells and membranes containing P450 and CPR were used in the preparative synthesis of drug metabolites. The optimized bioconversions conducted on a 1 L scale yielded 59 mg of 6β-hydroxytestosterone, 110 mg of 4′-hydroxydiclofenac and 88 mg of acetaminophen [58].

1.2.2
Microbial Oxidation for Synthesis of Pharmaceutical Intermediates

Microbial oxidation (using fungi, yeast, archea and bacteria) can be performed by using (1) native P450-producing strains, which can be altered by metabolic engineering if needed, or (2) recombinant whole cells, harboring microbial P450s. Microbial equivalents of human P450-catalyzed reactions are an additional alternative, especially of interest when from several hundred milligrams to multi-gram amounts of metabolites are required and also for the identification and production of non-human metabolites with new biological properties. Many pharmaceutical companies maintain collections

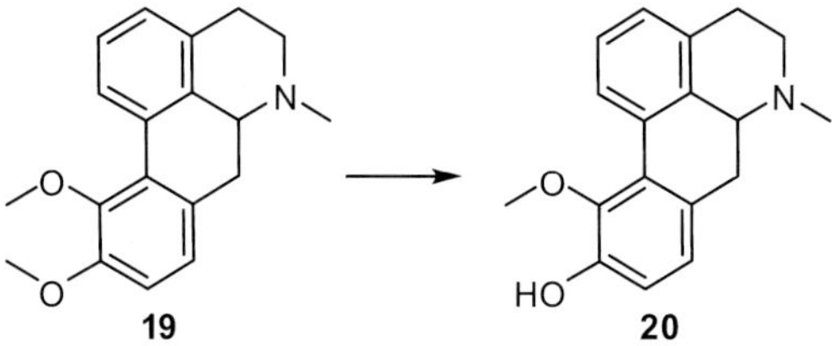

Scheme 1.6 Two examples of microbial oxidations.

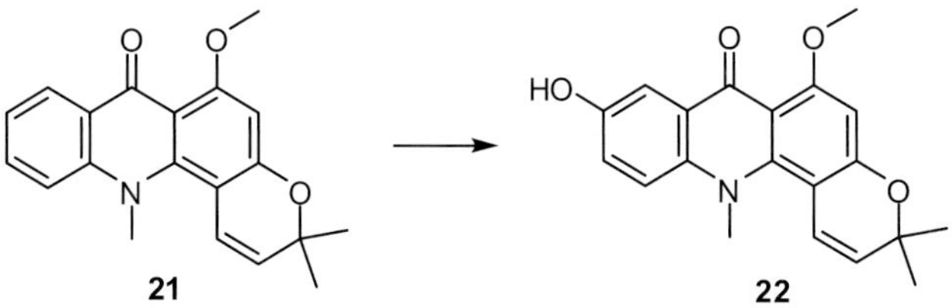

of bacteria, yeasts and fungi to screen systematically target xenobiotics with the goal of identifying additional P450s with new substrate ranges. Interesting candidates have been identified in the genera *Cunninghamella, Curvularia, Aspergillus, Rhizopus, Saccharomyces, Streptomyces* and some others. For example, *Cunninghamella elegans* catalyzes *O*-demethylation of 10,11-dimethoxyaporphine (**19**) to isoapocodeine (**20**) and *Aspergillus alleaceus* is capable of hydroxylating acronycine (**21**) to 9-hydroxyacronycine (**22**). Both reactions run on a preparative scale [70] (Scheme 1.6).

Recently, the construction and application of a novel P450 library, based on about 250 bacterial cytochrome P450 genes (about 70% from actinomycetes), co-expressed with putidaredoxin and putidaredoxin reductase in *E. coli* have been reported [71]. Screening of the P450 array with testosterone identified 24 bacterial P450s, which stereoselectively monohydroxylate testosterone at the 2α-, 2β-, 6β-, 7β-, 11β-, 12β-, 15β-, 16α- and 17-positions. Most of these hydroxylations are common for both prokaryotic and human P450s. Hence the identified bacterial candidates can be further applied for the production of drug metabolites on a preparative scale.

Microbial oxidations of steroids represent the very well-established large-scale commercial applications of P450 monooxygenases. The 11β-hydroxylation of 11-deoxycortisol (Reichstein S) to hydrocortisone using P450 of *Curvularia* sp. [72] is applied by Schering (acquired by Merck, Germany, in 2006) on an industrial scale of approximately 100 t per year [73]. Another example is the conversion of progesterone to cortisone through 11α-hydroxylation by *Rhizopus* sp., developed in the 1950s by Pharmacia and Upjohn (later acquired by Pfizer, USA) [73–75]. Both processes are one-step biotransformations, which cannot be achieved by chemical routes.

Production of the cholesterol-reducing pravastatin by oxidation of compactin, catalyzed by the P450 monooxygenase from *Mucor hiemalis* (Daiichi Sankyo, USA, and Bristol-Myers Squibb, USA) is another example of the commercial application of microbial oxidation [76, 77].

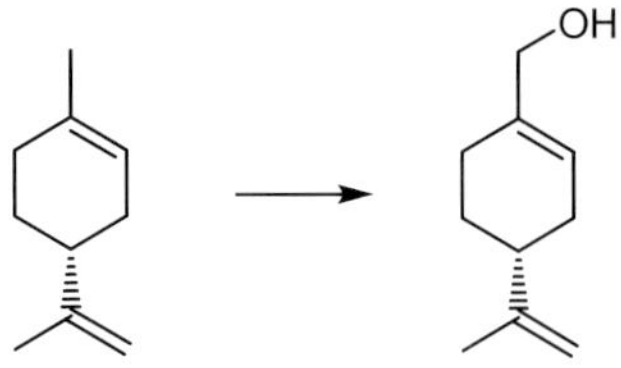

23   **24**

Scheme 1.7 Hydroxylation of L-limonene to (−)-perillyl alcohol catalyzed by CYP153 from *Mycobacterium* sp. HXH-1500.

Diverse activities of microbial P450 monooxygenases have potential applications in the synthesis of new antibiotics, especially with the issue of widespread bacterial antibiotic resistance. P450$_{eryF}$ (CYP107A1) from *Saccharopolyspora erythraea*, involved in the biosynthesis of the macrolide antibiotic erythromycin, catalyzes a hydroxylation step leading to the functional molecule [78]. Another example is OxyA, OxyB and OxyC (CYP165A1, CYP165B1 and CYP165C1) from *Amycolatopsis orientalis*, involved in the biosynthesis of a glycopeptide antibiotic [45]. Crystal structures of OxyC and OxyB [79, 80], and also that of P450$_{eryF}$ [81], have been published. A structural knowledge of these P450 monooxygenases opens up a way to enzyme optimization and engineering, leading to new antibacterial properties.

A screening among 1800 bacterial strains identified the *Mycobacterium* sp. strain HXH-1500, catalyzing regio- and stereospecific hydroxylation of L-limonene (**23**) at position 7 to yield the anticancer drug (−)-perillyl alcohol (**24**) (Scheme 1.7). The biocatalytic production of (−)-perillyl alcohol from L-limonene was performed using the *Mycobacterium* sp. P450 alkane hydroxylase (CYP153 family) recombinantly expressed in *Pseudomonas putida* cells [82]. The whole-cell process was performed in a two-phase system resulting in 6.8 g L^{-1} product in the organic phase.

1.2.3
Plant P450s and Transgenic Plants

Of all organisms, plants express the largest number of different P450 enzymes. For example, the genome sequence of the model plant *Arabidopsis thaliana* contains more than 230 P450 and P450-like gene sequences and the genome of *Oryza sativa* codes for more than 300 P450s. Unlike mammalian P450 enzymes, most plant P450 enzymes are involved in the synthesis of secondary metabolites. Cloning and characterization of an increasing number of recombinant plant P450 genes and the construction of transgenic plants has allowed their application in the synthesis of important pharmaceutical precursors and for engineering herbicide tolerance and bioremediation.

Recently, functional expression of eight taxol biosynthetic genes from yew, *Taxus cuspidate*, including the taxoid 10β-hydroxylase (CYP725A1) and a corresponding cytochrome P450 reductase, in yeast has been reported. This *in vivo* catalytic system yields baccatin III, an intermediate of taxol biosynthesis and a useful precursor for the semisynthesis of anti-cancer drugs [83].

An interesting industrial application of plant P450s exploits their role in anthocyanin biosynthesis that results in the delphinidin-derived pigments found in blue and violet flowers. The types of anthocyanins synthesized in plants are controlled by two P450 enzymes, flavonoid 3'-hydroxylase and flavonoid 3',5'-hydroxylase [84]. Suntory, Japan, and Calgene Pacific (now Florigene), Australia, have used the P450s from blue pansy and dihydroflavonol reductase from petunia in parallel with suppression of the rose dihydroflavonol reductase to develop genetically engineered transgenic 'blue roses', which cannot be obtained by classical plant hybridization methods [85]. Although the color of 'blue roses' is far from being 'pansy blue', because of the pH in vacuoles, the approach has been applied successfully for the production of carnations with new colors, which are currently on the market [86].

An ongoing field of research is phytoremediation using transgenic plants. The effectiveness of phytoremediation can be greatly enhanced by introducing genes known to be involved in the metabolism of pollutants in other organisms. For example, the nitroaromatic explosive hexahydro-1,3,5-trinitro-1,3,5-triazine (royal demolition explosive, RDX) are toxic not only to mammals, but also to plants and therefore cannot effectively be degraded by using conventional phytoremediation. Recently, the identification and characterization of a natural fusion enzyme between a soluble P450 monooxygenase and a flavodoxin (together named xplA/B) from *Rhodococcus rhodocrous* strain Y-11, which is capable of degradation of RDX, has been reported [87]. At present, the physiological role of this P450 enzyme is not clear, but the degradative reaction can be catalyzed not only through oxidation, but also through reduction, both pathways resulting in ring cleavage and release of nitrite. However, the reaction occurs more efficiently at low environmental oxygen concentrations. Expression of this bacterial P450 gene in *A. thaliana* considerably improved the tolerance and uptake of RDX by these plants from soil and water. Furthermore, the transgenic plants used nitrogen derived from RDX degradation for growth [88].

The existing strategies for the remediation of small volatile hydrocarbons are still insufficient; their harmful effects on health are, however, very serious. Introduction of the rabbit CYP2E1 gene into tobacco plants resulted in transgenic lines with improved metabolism of trichloroethylene [89]. Further developments focused on the construction of hybrid poplar plants that efficiently express CYP2E1. These hybrids were not only capable of degradation of trichloroethylene, vinyl chloride, carbon tetrachloride, benzene and chloroform from hydroponic solution, but also demonstrated removal of gaseous trichloroethylene, chloroform and benzene from the air [90].

Phytoremediation of herbicides can be enhanced by using transgenic plants expressing P450s. For example, transgenic rice plants, expressing human CYP1A1, demonstrated a broad cross-resistance towards various herbicides and were able to metabolize rapidly chlorotoluron and norflurazon [91]. Transgenic rice plants, expressing human P450 genes CYP1A1, CYP2B6 and CYP2C19, could decrease the amounts of atrazine and metolachlor in contaminated water and soil [92].

1.3
Optimization of P450 Monooxygenase-Based Catalytic Systems

The development of technical applications of P450s faces numerous specific challenges that do not occur with other industrially applied enzymes such as hydrolases, lyases and transferases. Nearly all P450s require costly cofactors NADPH or NADH, rendering industrial applications impossible if the cofactor has to be added in a stoichiometric amount. Therefore, industrial applications of P450s have so far been restricted to whole-cell systems, which mostly solve the problem of cofactor regeneration. In such instances, however, physiological effects such as limited substrate uptake, toxicity of substrate or product, product degradation and elaborate downstream processing must be taken into account. The use of isolated P450 monooxygenases allows the physiological limitations of whole-cell biocatalysts to be overcome, but still faces problems of cofactor regeneration and low operational stability. Another factor important for efficient biocatalysis with P450s is the yield of product based on NAD(P)H consumed or the coupling efficiency. In addition to reducing the efficiency of cofactor usage, uncoupling between NADPH oxidation and product formation, resulting in reactive oxygen species (ROS) such as superoxide anions and hydrogen peroxide, may cause oxidative destruction of the heme and oxidative damage of the protein.

Although the substrate spectra and reactions catalyzed by mammalian P450s are often more interesting for the pharmaceutical and chemical industry, the catalytic activity and stability of these enzymes are far too low for their technical implementation. Bacterial P450 monooxygenases are the most promising candidates for biotransformations as they are water-soluble cytosolic stable enzymes. However, bacterial P450s often have narrow substrate specificity or low regioselectivity.

This means that optimization strategies for P450 monooxygenases target divers areas such as the substitution or regeneration of the cofactor NAD(P)H, the extension of substrate spectra, enzyme activity, coupling efficiency and selectivity and the enhancement of enzyme stability (Figure 1.2).

1.3.1
Replacement or Regeneration of the Cofactor NAD(P)H

The limited use of P450-catalyzed reactions in industry stems from the high cost of NAD(P)H cofactors. An appealing strategy for overcoming this limitation on both the

Figure 1.2 Engineering of P450 enzymes for potential biotechnological application.

laboratory and industrial scales is the application of whole-cell biocatalysts. Provided that living cells are provided with all required nutrients, all endogenous cofactor recycling systems are functional and there is no need to supplement cofactors. Whole-cell processes, however, have other limitations mentioned above [73]. Moreover, when concentrations of recombinant P450 biocatalysts within the cell reach a certain level, cofactor concentration may again become a bottleneck for the overall process. The catalytic efficiency of P450cam expressed in *E. coli* could be increased up to 25-fold by the coupling of its native electron transfer system (putidaredoxin and putidaredoxin reductase) to enzymatic NADH regeneration catalyzed by a recombinantly expressed glycerol dehydrogenase. This whole-cell system was applied to camphor hydroxylation in aqueous solution and in a biphasic system with isooctane [93].

P450 monooxygenase *in vitro* applications require cofactor substitution or regeneration systems. Several approaches have been developed to avoid or minimize the use of natural expensive cofactors. Direct chemical reduction of the heme iron by the strong reducing agent sodium dithionite has been shown to support the hydroxylation reactions catalyzed by P450 BM-3. However, in this case the reduction rate of the heme iron was approximately 8000 times slower than with NADPH [94].

Other attempts to substitute or regenerate NAD(P)H by non-enzymatic methods have focused on the use of organometallic complexes such as cobalt(III) sepulchrate (CoIIIsep) [95, 96]. Usually, the mediators work in two ways – coupled with electrochemical reduction on electrodes and with electron-donating chemical compounds. Limitations of this approach include low system efficacy, low sensitivity of mediators to molecular oxygen and the production of reactive oxygen species or mediator aggregation.

Direct electrochemical heme reduction has been studied in detail for more than 15 years [97, 98]. Heme reduction on a cathode is complicated and often insufficient for biocatalysis. This can be partly attributed to the deeply buried heme iron and partly to the instability of the enzyme upon interaction with the electrode surface. Improvements to this approach include the modification of the electrode surface [99, 100].

One of the most common approaches for overcoming the stoichiometric need for the costly NAD(P)H cofactor involves enzymatic regeneration systems [101]. Strategies for the regeneration of NADPH-dependent P450 monooxygenases are based on D/L-isocitrate dehydrogenase [102], an engineered formate dehydrogenase from *Pseudomonas* sp. [101, 103], glucose-6-phosphate dehydrogenase [104] and alcohol dehydrogenase from *Thermoanaerobium brockii* [105]. Although these systems work reasonably well, further reducing the costs of enzymatic oxyfunctionalization can be achieved by engineering of NADPH-dependent P450s to accept NADH [106, 107]. NADH has the advantage that it is about 10 times less expensive and more stable than NADPH and there are more NAD^+-dependent enzymes available for cofactor regeneration at lower prices. For instance, using homology modeling and site-directed mutagenesis, P450 BM-3 mutants have been constructed showing altered cofactor specificity from NADPH to NADH. The best mutant, W1046S/S965D, demonstrated the highest turnover rates during cytochrome c reduction and the lowest K_M for NADH ($k_{cat} = 14\,600$ min^{-1}, $K_M = 17.6$ μM). It had equal affinity to both NADH and NADPH and displayed a 2.6-times higher activity when NADH was used [108].

The use of cheaper chemicals to replace NAD(P)H directly is another approach to facilitate P450 enzyme-based applications. Peroxides that directly convert the heme iron of P450s to a ferric hydroperoxy complex by the 'peroxide shunt' (e.g. hydrogen peroxide, cumene peroxide, *tert*-butyl oxide) might be useful for hydroxylating various substrates. Among mammalian P450s, human 2D6 and 3A4 have been reported to perform efficiently in the presence of a low concentration of cumene hydroperoxide [109, 110]. As mentioned earlier, the essential problem in utilizing the 'peroxide shunt' for the P450 biocatalysis seem to lie in the time-dependent degradation of the heme and in oxidation of the protein [18, 19]. Methods of directed evolution were applied to evolve P450s to enhance the efficiency of the 'peroxide shunt' pathway. By random mutagenesis and subsequent DNA shuffling of P450cam, its H_2O_2 hydroxylation activity towards naphthalene was increased 20-fold [111]. Using random and site-specific mutagenesis, a P450 BM-3 mutant was obtained, which support the H_2O_2-driven hydroxylation of alkanes and epoxidation of styrene without a rapid loss of activity [21]. Furthermore, a double mutant of mammalian CYP3A4 has been developed with an 11-fold improvement in catalytic efficiency during debenzylation of 7-benzylquinoline supported by hydrogen peroxide [112].

1.3.2
Engineering of New Substrate Specificities of P450s

Various techniques of protein engineering have been applied to change or optimize features of P450 monooxygenases, including site-directed and random mutagenesis, DNA recombination or combinations of these methods. The strategies applied have their advantages and drawbacks, which have been discussed in excellent reviews and books [113–116]. Therefore, in the following section we will not focus on methods of enzyme engineering or on assays for mutant screening, but only on properties of P450 monooxygenases that were engineered.

1.3.2.1 Engineering of Bacterial P450s
Although bacterial P450 monooxygenases exhibit high stability, activity and expression levels in recombinant hosts, their substrate spectra are often fairly narrow. Therefore, bacterial P450s are ideal targets for engineering of new substrate specificities.

Among bacterial P450s, P450cam from *P. putida* [117] and P450 BM-3 from *B. megaterium* [118, 119] are the best characterized. The structure, catalytic mechanism and biochemistry of these two enzymes have been studied in detail [120]. Mutagenesis studies on these P450s have led to significant improvements in our understanding of general aspects of P450 catalytic function.

In the presence of its redox partners, putidaredoxin reductase and putidaredoxin, P450cam catalyzes the stereospecific oxidation of D-(+)-camphor to 5-*exo*-hydroxycamphor (Scheme 1.2). It was demonstrated that mutations of the active site residue Y96 to more hydrophobic ones considerably increased the activity of P450cam towards the oxidation of hydrophobic molecules smaller than camphor, such as styrene and alkanes [121, 122], or larger than camphor, such as naphthalene and pyrene [123, 124]. The Y96F mutant was capable of hydroxylation of tetralin stereo- and regiospecifically to the 1-(*R*)-alcohol [125]. Using saturation mutagenesis at positions Y96 and F87, several P450cam mutants were constructed with activity towards indole and diphenylmethane [126]. Some P450cam mutants have been engineered for biodegradation of environmentally harmful pollutants, such as di-, tri-, tetra-, penta- and hexachlorobenzene [127, 128]. Amazing results have been achieved in engineering P450cam to an alkane hydroxylase. The authors used step-by-step adaptation of the enzyme to smaller *n*-alkanes beginning with hexane [129], then to butane and propane [130] and finally to ethane [131]. The best mutant with eight substitutions oxidized propane at 500 min^{-1} with 86% coupling, which was comparable to that of the wild-type enzyme towards (+)-camphor, the natural substrate. The NADH turnover rate for ethane was even higher ($\sim$800 min^{-1}), but the coupling for ethanol formation was low (10.5%) [131].

Successful attempts to engineer P450cam mutants for selective oxidation of cheap terpenes to expensive oxidized derivatives have also been accomplished. Rational mutants of P450cam, although exhibiting low activity towards the sesquiterpene (+)-valencene, produced >85% of (+)-*trans*-nootkatol and (+)-nootkatone, both grapefruit flavors [132, 133]. A cheap waste product of the wood industry, the monoterpene (+)-α-pinene, was oxidized by P450cam mutants to 86% (+)-*cis*-verbenol or to a mixture of (+)-verbenone and (+)-*cis*-verbenol [134]. Both verbenol and verbenone are active pheromones against various beetle species.

P450 BM-3 is an obvious target enzyme for developing cell-free biotechnological applications, since it is a self-sufficient single-component protein with a high catalytic activity. The turnover rates of P450 BM-3 towards fatty acids are among the highest activity values reported for P450 monooxygenases [135]. P450 BM-3 has been engineered to accept substrates interesting for fine chemistry synthesis. For instance, the mutant A74G/F87V/L188Q, designed by saturation mutagenesis of multiple sites, was shown to oxidize indole, *n*-octane, highly branched fatty acids and fatty alcohols, polychlorinated dibenzo-*p*-dioxins, polyaromatic hydrocarbons, styrene and many other chemical compounds [136–140]. The monoterpene geranylacetone was converted by the P450 BM-3 R47L/Y51F/F87V with high activity (2080 min^{-1}) and stereoselectivity (97% *ee*) to a single product, 9,10-epoxygeranylacetone [141]. The mutant A74E/F87V/P386S exhibited an 80-fold improved activity towards β-ionone (270 min^{-1}) compared with the wild-type enzyme, producing the flavorant (*R*)-4-hydroxy-β-ionone (40% *ee*) as the only product [142].

Using a combination of directed evolution and site-directed mutagenesis, Arnold and co-workers altered the selectivity of P450 BM-3 from hydroxylation of dodecane (C$_{12}$) first to octane (C$_8$) and hexane (C$_6$) and further on to gaseous propane (C$_3$) and ethane (C$_2$) [143–146]. Some mutants were found with high stereoselectivity, which led to either the (*R*)- or (*S*)-2-octanol enantiomer products of alkane hydroxylation [147].

A recent area of research interest covers the design and development of novel bacterial P450s for the production of pharmaceuticals and human metabolites. Although wild-type P450 BM-3 is not able to metabolize any drug-like compounds, it has been turned by protein design and directed evolution into an enzyme which oxidizes human drugs. The R47L/F87V/L188Q mutant was shown to metabolize testosterone, amodiaquine, dextromethorphan, acetaminophen and 3,4-methylene-dioxymethylamphetamine [148]. Several mutants of P450 BM-3 were obtained by means of directed evolution, which are able to accept and convert propranolol, a multi-function β-adrenergic blocker [149]. Another mutant, 9-10A-F87A, was capable of stereo- and regioselective hydroxylation of the peptide group of buspirone in the α-position. (*R*)-6-Hydroxybuspirone was the sole product formed with >99.5% *ee* [150]. Buspirone and 6-hydroxybuspirone are anti-anxiety agents. Recently, a new approach for DNA recombination (the so-called SCHEMA algorithm) was applied to obtain chimeras based on P450 BM-3 and its homologues CYP102A2 and CYP102A3, sharing only 63% amino acid identity [151]. A survey of the activities of new chimeras demonstrated that this approach created completely new functions, absent in the wild-type enzymes, including the ability to accept drugs (e.g. propranolol, tolbutamide, zoxazolamine [152], verapamil and astemizole [153]). Although in most cases the yields of human metabolites obtained with P450 BM-3 mutants were comparable to those produced by recombinant human P450s in bacterial or baculovirus systems, the catalytic activities, coupling efficiencies and total turnover numbers achieved were much lower than those obtained with natural substrates of P450 BM-3, suggesting that further mutagenesis is required.

1.3.2.2 Engineering of Mammalian P450s

Whereas the understanding of the determinants for function of bacterial P450s is generally satisfactory, the structure–function relationships are far less clear for eukaryotic P450s. Therefore, most protein engineering efforts conducted on mammalian P450s have been used for the identification of key residues involved in substrate binding. Nevertheless, as more crystal structures of membrane-bound P450s are becoming available, the structural knowledge can be used for the construction of new enzyme properties.

Homology modeling and rational design performed on CYP2B1 allowed switching of the regioselectivity of progesterone 16α-hydroxylation to 21-hydroxylation [154]. CYP2B1 has also been engineered by a combination of random and site-directed mutagenesis to accept the anti-cancer prodrugs cyclophosphamide and ifosfamide [155]. Using random mutagenesis to the substrate recognition sequences (SRS) of the indole-hydroxylating CYP2A6 enzyme, mutants have been constructed exhibiting higher regioselectivity towards indole and high activity towards several indole analogs [156]. The CYP2D6 mutants were subsequently applied to the generation of a library of biologically active indigoids for the inhibition of cyclin-dependent kinases [157]. Later, the substrate spectrum of human CYP2D6 was expanded towards bulky indole derivatives, such as 5-hydroxybenzylindole, by random mutagenesis and site-directed mutagenesis [158].

The ability of human CYP1A2 to catalyze *O*-demethylation of 7-methoxyresorufin was improved by three rounds of mutagenesis and screening. The triple mutant E163K/V193M/K170Q exhibited turnover rates more than five times faster than wild-type CYP1A2 [159].

To create a chimeric library of CYP1A1 and CYP1A2, Gillam's group used restriction enzyme-mediated DNA family shuffling [160]. Different activity profiles towards various luciferin derivatives were observed among active clones, including improved specific activity, novel activities and broadening of substrate range [160].

1.3.3
Stability of P450s

1.3.3.1 Thermostability of P450s
Although considerable progress has been made in tailoring P450 enzymes so that they can effectively convert new substrates, their poor stability often remains a significant obstacle for applications in the synthesis of chemicals. Cytochrome P450s are generally labile and highly temperature sensitive. The application of thermostable P450s could solve this problem to at least some extent [161]. At present, only four thermostable cytochrome P450s have been reported. CYP119 was identified in archaebacterium *Sulfolobus solfataricus* and subsequently characterized in detail [162]. However, natural substrates are still unknown. A similar thermostable P450 mono-oxygenase related to CYP119 was recently identified in another *Sulfolobus* strain [163], but has not yet been further investigated. A thermostable cytochrome P450 steroid hydroxylase was found in a thermophilic *Bacillus* strain [164] and a thermostable P450 enzyme, CYP175A1, was identified in the genome of *Thermus thermophilus* HB27 strain and characterized [165]. However, natural electron transfer proteins have not yet been identified. Thus, detailed characterization and enzyme engineering on thermostable P450s is still a field of ongoing research.

Directed evolution and DNA recombination has also been used to good effect in enhancing thermostability of mesophilic P450s. After several rounds of random mutagenesis, a variant of the P450 BM-3 heme domain was found with enhanced thermal stability, showing a 50-fold increased half-life at 57.5 °C [166]. A family of 44 novel thermostable P450s with half-lives of inactivation at 57 °C up to 108 times that of the most stable parent has been constructed suing the SCHEMA algorithm (see Section 1.3.2.1) [153].

Since the thermal inactivation of the P450 BM-3 reductase under reaction conditions might be responsible for the low overall operational stability of the fusion enzyme, the reductase domain has been exchanged with that of CYP102A3. Maximum activity of the chimeric enzyme was obtained at 51 °C and the half-life at 50 °C, 100 min, was more than 10 times longer than that of P450 BM-3 [167].

1.3.3.2 Process Stability of P450s
Low water solubility of P450 substrates is currently a significant obstacle, hindering the realization of biooxidation processes on a large scale. Therefore, the addition of organic (co)-solvents is often necessary for effective catalysis. On the other hand, organic solvents, particularly, water-miscible ones, might be harmful for enzyme performance. To enhance the stability of P450 monooxygenases, different enzyme stabilization techniques, including protein engineering, immobilization and reaction engineering, have been applied. For instance, a quadruple mutant of mammalian

CYP2B1 obtained by random mutagenesis has improved tolerance to elevated temperatures and dimethyl sulfoxide (DMSO) [168]. Several mutants of P450 BM-3 were constructed which demonstrated high activity even in 25% DMSO, whereas the activity of the wild-type enzyme under these conditions decreased to 50% [169].

Furthermore, if the product of the monooxygenase-catalyzed reaction is toxic to the biocatalyst, *in situ* product recovery might be necessary. This can be realized by using biphasic organic–aqueous phase systems. Biphasic systems, where a pure liquid substrate represents the organic phase, were particularly suitable for the biooxidation of small volatile substrates such as cyclohexane and *n*-octane. The P450 BM-3 mutants retained its activity for over 100 h and the enzyme's total turnover numbers (TTNs) reached 10 000–12 000 [108].

Recently, Auclair and colleagues demonstrated that even typically unstable mammalian enzymes CYP2D6 [170] and CYP3A4 [171] are compatible with biphasic systems and can be lyophilized, facilitating their adaptation to process requirements [172].

Another approach for the stabilization of P450s is their immobilization on solid supports. The first example dates back to 1988 when Wiseman and co-workers [173] immobilized purified P450s from *S. cerevisiae* along with the corresponding reductases by entrapment in calcium alginate or in polyacrylamide. A decade later, the plant CYP71B1, fused to a P450 reductase, was immobilized on colloidal liquid aphrons [174]. Kelly and co-workers reported the co-immobilization of prokaryotic CYP105D1 with a ferredoxin on the ionic exchange resin DE52-72 [175]. Maurer *et al.* demonstrated efficient immobilization of P450 BM-3 in sol–gel [103]. This approach yielded a biocatalyst with high long-term storage stability and high tolerance towards DMSO. The combination of P450 BM-3 and a formate dehydrogenase for cofactor regeneration, co-immobilized in sol–gel, represents a solid P450 biocatalyst, useful for biocatalysis.

1.4
Outlook

Although many interesting P450-mediated reactions have been described in the literature, examples of process implementation and scale-up to pilot or industrial scales are comparatively rare. Obviously, it needs a long time and excellent knowledge of the corresponding P450 systems to develop industrial processes due to the complexity of these enzymes.

Progress has been made in engineering effective electron supply systems to support P450 catalysis. However, the shortcoming of these systems is low coupling efficiency and inefficient catalytic activity. Furthermore, to reduce the complexity of the P450 systems, attempts have been made to construct artificial self-sufficient systems, which, instead of using the expensive cofactor NAD(P)H, rely on peroxides. In this respect, a problem which remains to be solved is heme degradation in the presence of peroxides.

The invention of new methods of protein engineering in the past decade has led to the construction of an abundance of P450 mutants with new tailored properties. By extending the substrate spectra, stabilizing the enzymes and finding alternatives for electron supply, we have come one step closer to the technical application of cytochrome P450 monooxygenases without the need for dangerous or toxic reagents.

The number of P450 monooxygenases identified is constantly increasing through the sequencing of genomes and microbial screening. Recently, several novel P450 enzymes have been identified in bacterial strains and plants associated with unknown pathways of secondary metabolism. Functional characterization and utilization of unusual cytochrome P450s and P450 mutants in drug development, fine chemical synthesis, gene therapy, bioremediation, biosensors and plant improvement represent a new exciting perspective.

Abbreviations

DMSO   dimethyl sulfoxide
FAD   flavin adenine dinucleotide
FMN   flavin mononucleotide
NAD(P)H   nicotinamide adenine dinucleotide (phosphate), reduced form

Acknowledgments

The author thanks Dr Sabine Eiben and Katja Koschorreck for critical reading of the manuscript and Matthias Dietrich for help in preparing Figure 1.1.

References

1 Davies, H.G., Green, R.H., Kelly, D.R. and Roberts, S.M. (1989) *Biotransformations in Preparative Organic Chemistry: the Use of Isolated Enzymes and Whole-cell Systems in Synthesis*, Academic Press, London.

2 Garfinkel, D. (1958) *Archives of Biochemistry and Biophysics*, **77**, 493–509.

3 Klingenberg, M. (1958) *Archives of Biochemistry and Biophysics*, **75**, 376–386.

4 Omura, T. and Sato, R. (1964) *The Journal of Biological Chemistry*, **239**, 2370–2378.

5 Raag, R. and Poulos, T.L. (1989) *Biochemistry*, **28**, 7586–7592.

6 E. Bayer, Hill, H.O.A., Röder, A. and Williams, R.J.P. (1969) *Chemical Communications*, 109–116.

7 Hill, H.A.O., Roder, A. and Williams, R.J. (1969) *The Biochemical Journal*, **115**, 59P–60P.

8 Poulos, T.L., Finzel, B.C. and Howard, A.J. (1986) *Biochemistry*, **25**, 5314–5322.

9 Kelly, S.L., Lamb, D.C. and Kelly, D.E. (2006) *Biochemical Society Transactions*, **34**, 1159–1160.

10 Nelson, D.R. (2006) *Methods in Molecular Biology*, **320**, 1–10.

11 Graham, S.E. and Peterson, J.A. (1999) *Archives of Biochemistry and Biophysics*, **369**, 24–29.

12 Bernhardt, R. (2004) *Chemistry & Biology*, **11**, 287–288.

13 Hannemann, F., Bichet, A., Ewen, K.M. and Bernhardt, R. (2007) *Biochimica et Biophysica Acta*, **1770**, 330–344.

14 Denisov, I.G., Makris, T.M., Sligar, S.G. and Schlichting, I. (2005) *Chemical Reviews*, **105**, 2253–2277.

15 Sligar, S.G. (1976) Biochemistry, 15, 5399–5406.

16 Sligar, S.G., Makris, T.M. and Denisov, I.G. (2005) Biochemical and Biophysical Research Communications, 338, 346–354.

17 Jin, S., Bryson, T.A. and Dawson, J.H. (2004) Journal of Biological Inorganic Chemistry, 9, 644–653.

18 Gutteridge, J.M. (1989) Acta Paediatrica Scandinavia Supplement, 361, 78–85.

19 Gutteridge, J.M. (1990) Free Radical Research Communications, 9, 119–125.

20 Joo, H., Lin, Z. and Arnold, F.H. (1999) Nature, 399, 670–673.

21 Cirino, P.C. and Arnold, F.H. (2003) Angewandte Chemie-International Edition, 42, 3299–3301.

22 Bernhardt, R. (1996) Reviews of Physiology, Biochemistry and Pharmacology, 127, 137–221.

23 Munro, A.W., Girvan, H.M. and McLean, K.J. (2007) Biochimica et Biophysica Acta, 1770, 345–359.

24 Narhi, L.O. and Fulco, A.J. (1987) The Journal of Biological Chemistry, 262, 6683–6690.

25 Munro, A.W., Daff, S., Coggins, J.R., Lindsay, J.G. and Chapman, S.K. (1996) European Journal of Biochemistry/FEBS, 239, 403–409.

26 Gustafsson, M.C., Roitel, O., Marshall, K.R., Noble, M.A., Chapman, S.K., Pessegueiro, A., Fulco, A.J., Cheesman, M.R., von Wachenfeldt, C. and Munro, A.W. (2004) Biochemistry, 43, 5474–5487.

27 Girvan, H.M., Waltham, T.N., Neeli, R., Collins, H.F., McLean, K.J., Scrutton, N.S., Leys, D. and Munro, A.W. (2006) Biochemical Society Transactions, 34, 1173–1177.

28 Roberts, G.A., Celik, A., Hunter, D.J., Ost, T.W., White, J.H., Chapman, S.K., Turner, N.J. and Flitsch, S.L. (2003) The Journal of Biological Chemistry, 278, 48914–48920.

29 Mot, R.D. and Parret, A.H. (2002) Trends in Microbiology, 10, 502–508.

30 McLean, K.J., Sabri, M., Marshall, K.R., Lawson, R.J., Lewis, D.G., Clift, D., Balding, P.R., Dunford, A.J., Warman,

A.J., McVey, J.P., Quinn, A.M., Sutcliffe, M.J., Scrutton, N.S. and Munro, A.W. (2005) Biochemical Society Transactions, 33, 796–801.

31 Puchkaev, A.V. and Ortiz de Montellano, P.R. (2005) Archives of Biochemistry and Biophysics, 434, 169–177.

32 Matsunaga, I. and Shiro, Y. (2004) Current Opinion in Chemical Biology, 8, 127–132.

33 Matsunaga, I., Yokotani, N., Gotoh, O., Kusunose, E., Yamada, M. and Ichihara, K. (1997) The Journal of Biological Chemistry, 272, 23592–23596.

34 Matsunaga, I., Ueda, A., Fujiwara, N., Sumimoto, T. and Ichihara, K. (1999) Lipids, 34, 841–846.

35 Girhard, M., Schuster, S., Dietrich, M., Durre, P. and Urlacher, V.B. (2007) Biochemical and Biophysical Research Communications, 362, 114–119.

36 Nakahara, K., Tanimoto, T., Hatano, K., Usuda, K. and Shoun, H. (1993) The Journal of Biological Chemistry, 268, 8350–8355.

37 Sono, M., Roach, M.P., Coulter, E.D. and Dawson, J.H. (1996) Chemical Reviews, 96, 2841–2888.

38 Tang, L., Shah, S., Chung, L., Carney, J., Katz, L., Khosla, C. and Julien, B. (2000) Science, 287, 640–642.

39 Bruntner, C., Lauer, B., Schwarz, W., Mohrle, V. and Bormann, C. (1999) Molecular & General Genetics, 262, 102–114.

40 Shyadehi, A.Z., Lamb, D.C., Kelly, S.L., Kelly, D.E., Schunck, W.H., Wright, J.N., Corina, D. and Akhtar, M. (1996) The Journal of Biological Chemistry, 271, 12445–12450.

41 Bower, S., Perkins, J.B., Yocum, R.R., Howitt, C.L., Rahaim, P. and Pero, J. (1996) Journal of Bacteriology, 178, 4122–4130.

42 Hlavica, P. and Kunzel-Mulas, U. (1993) Biochimica et Biophysica Acta, 1158, 83–90.

43 Seto, Y. and Guengerich, F.P. (1993) The Journal of Biological Chemistry, 268, 9986–9997.

44 Yu, A.M., Idle, J.R., Herraiz, T., Kupfer, A. and Gonzalez, F.J. (2003) Pharmacogenetics, 13, 307–319.

45 Bischoff, D., Bister, B., Bertazzo, M., Pfeifer, V., Stegmann, E., Nicholson, G.J., Keller, S., Pelzer, S., Wohlleben, W. and Sussmuth, R.D. (2005) *Chembiochem*, **6**, 267–272.

46 Cryle, M.J., Stok, J.E. and De Voss, J.J. (2003) *Australian Journal of Chemistry*, **56**, 749–762.

47 Isin, E.M. and Guengerich, F.P. (2007) *Biochimica et Biophysica Acta*, **1770**, 314–329.

48 Munro, A.W., Girvan, H.M. and McLean, K.J. (2007) *Natural Product Reports*, **24**, 585–609.

49 Miners, J.O. (2002) *Clinical and Experimental Pharmacology & Physiology*, **29**, 1040–1044.

50 Gillam, E.M., Aguinaldo, A.M., Notley, L.M., Kim, D., Mundkowski, R.G., Volkov, A.A., Arnold, F.H., Soucek, P., DeVoss, J.J. and Guengerich, F.P. (1999) *Biochemical and Biophysical Research Communications*, **265**, 469–472.

51 Pritchard, M.P., McLaughlin, L. and Friedberg, T. (2006) *Methods in Molecular Biology*, **320**, 19–29.

52 Hanioka, N., Okumura, Y., Saito, Y., Hichiya, H., Soyama, A., Saito, K., Ueno, K., Sawada, J. and Narimatsu, S. (2006) *Biochemical Pharmacology*, **71**, 1386–1395.

53 Yabusaki, Y. (1998) *Methods in Molecular Biology*, **107**, 195–202.

54 Pompon, D., Louerat, B., Bronine, A. and Urban, P. (1996) *Methods in Enzymology*, **272**, 51–64.

55 Nthangeni, M.B., Urban, P., Pompon, D., Smit, M.S. and Nicaud, J.M. (2004) *Yeast*, **21**, 583–592.

56 Dietrich, M., Grundmann, L., Kurr, K., Valinotto, L., Saussele, T., Schmid, R.D. and Lange, S. (2005) *Chembiochem*, **6**, 2014–2022.

57 Breinholt, V.M., Rasmussen, S.E., Brosen, K. and Friedberg, T.H. (2003) *Pharmacology & Toxicology*, **93**, 14–22.

58 Vail, R.B., Homann, M.J., Hanna, I. and Zaks, A. (2005) *Journal of Industrial Microbiology & Biotechnology*, **32**, 67–74.

59 Gold, L. (1990) *Methods in Enzymology*, **185**, 11–37.

60 Barnes, H.J., Arlotto, M.P. and Waterman, M.R. (1991) *Proceedings of the National Academy of Sciences of the United States of America*, **88**, 5597–5601.

61 Chen, G.F. and Inouye, M. (1990) *Nucleic Acids Research*, **18**, 1465–1473.

62 Winters, D.K. and Cederbaum, A.I. (1992) *Biochimica et Biophysica Acta*, **1156**, 43–49.

63 Nishimoto, M., Clark, J.E. and Masters, B.S. (1993) *Biochemistry*, **32**, 8863–8870.

64 Andersen, J.F., Utermohlen, J.G. and Feyereisen, R. (1994) *Biochemistry*, **33**, 2171–2177.

65 Guo, Z., Gillam, E.M., Ohmori, S., Tukey, R.H. and Guengerich, F.P. (1994) *Archives of Biochemistry and Biophysics*, **312**, 436–446.

66 Hanlon, S.P., Friedberg, T., Wolf, C.R., Ghisalba, O. and Kittelmann, M., (2007) in *Modern Biooxidation* (eds R.D. Schmid and V.B. Urlacher), Wiley-VCH Verlag GmbH, Weinheim, pp. 233–252.

67 Gillam, E.M., Guo, Z. and Guengerich, F.P. (1994) *Archives of Biochemistry and Biophysics*, **312**, 59–66.

68 Gillam, E.M., Guo, Z., Martin, M.V., Jenkins, C.M. and Guengerich, F.P. (1995) *Archives of Biochemistry and Biophysics*, **319**, 540–550.

69 Sandhu, P., Baba, T. and Guengerich, F.P. (1993) *Archives of Biochemistry and Biophysics*, **306**, 443–450.

70 Ghisalba, O. and Kittelmann, M. (2007) in *Modern biooxidation* (eds R.D. Schmid and V.B. Urlacher), Wiley-VCH Verlag GmbH, Weinheim, pp. 211–232.

71 Agematu, H., Matsumoto, N., Fujii, Y., Kabumoto, H., Doi, S., Machida, K., Ishikawa, J. and Arisawa, A. (2006) *Bioscience, Biotechnology, and Biochemistry*, **70**, 307–311.

72 Petzoldt, K., Annen, K., Laurent, H. and Wiechert, R. (1982) US Patent 4353985.

73 van Beilen, J.B., Duetz, W.A., Schmid, A. and Witholt, B. (2003) *Trends in Biotechnology*, **21**, 170–177.

74 Peterson, D.H., Murray, H.C., Eppstein, S.H., Reineke, L.M., Weintraub, A., Meister, P.D. and Leigh, H.M. (1952) *Journal of the American Chemical Society*, **74**, 5933–5936.

75 Hogg, J.A. (1992) *Steroids*, **57**, 593–616.

76 Serizawa, N. (1997) *Journal of Synthetic Organic Chemistry Japan*, **55**, 334–338.

77 Serizawa, N., Nakagawa, K., Hamano, K., Tsujita, Y., Terahara, A. and Kuwano, H. (1983) *The Journal of Antibiotics Japan*, **36**, 604–607.

78 Andersen, J.F., Tatsuta, K., Gunji, H., Ishiyama, T. and Hutchinson, C.R. (1993) *Biochemistry*, **32**, 1905–1913.

79 Zerbe, K., Pylypenko, O., Vitali, F., Zhang, W., Rouset, S., Heck, M., Vrijbloed, J.W., Bischoff, D., Bister, B., Sussmuth, R.D., Pelzer, S., Wohlleben, W., Robinson, J.A. and Schlichting, I. (2002) *The Journal of Biological Chemistry*, **277**, 47476–47485.

80 Pylypenko, O., Vitali, F., Zerbe, K., Robinson, J.A. and Schlichting, I. (2003) *The Journal of Biological Chemistry*, **278**, 46727–46733.

81 Nagano, S., Cupp-Vickery, J.R. and Poulos, T.L. (2005) *The Journal of Biological Chemistry*, **280**, 22102–22107.

82 van Beilen, J.B., Holtackers, R., Luscher, D., Bauer, U., Witholt, B. and Duetz, W.A. (2005) *Applied and Environmental Microbiology*, **71**, 1737–1744.

83 Dejong, J.M., Liu, Y., Bollon, A.P., Long, R.M., Jennewein, S., Williams, D. and Croteau, R.B. (2006) *Biotechnology and Bioengineering*, **93**, 212–224.

84 Forkmann, G. and Martens, S. (2001) *Current Opinion in Biotechnology*, **12**, 155–160.

85 Ogata, J., Kanno, Y., Itoh, Y., Tsugawa, H. and Suzuki, M. (2005) *Nature*, **435**, 757–758.

86 Fukui, Y., Tanaka, Y., Kusumi, T., Iwashita, T. and Nomoto, K. (2003) *Phytochemistry*, **63**, 15–23.

87 Jackson, R.G., Rylott, E.L., Fournier, D., Hawari, J. and Bruce, N.C. (2007) *Proceedings of the National Academy of Sciences of the United States of America*, **104**, 16822–16827.

88 Rylott, E.L., Jackson, R.G., Edwards, J., Womack, G.L., Seth-Smith, H.M., Rathbone, D.A., Strand, S.E. and Bruce, N.C. (2006) *Nature Biotechnology*, **24**, 216–219.

89 S.L. Doty, Shang, T.Q., Wilson, A.M., Tangen, J., Westergreen, A.D., Newman, L.A., Strand, S.E. and Gordon, M.P. (2000) *Proceedings of the National Academy of Sciences of the United States of America*, **97**, 6287–6291.

90 Doty, S.L., James, C.A., Moore, A.L., Vajzovic, A., Singleton, G.L., Ma, C., Khan, Z., Xin, G., Kang, J.W., Park, J.Y., Meilan, R., Strauss, S.H., Wilkerson, J., Farin, F. and Strand, S.E. (2007) *Proceedings of the National Academy of Sciences of the United States of America*, **104**, 16816–16821.

91 Kawahigashi, H., Hirose, S., Ohkawa, H. and Ohkawa, Y. (2007) *Biotechnology Advances*, **25**, 75–84.

92 Kawahigashi, H., Hirose, S., Ohkawa, H. and Ohkawa, Y. (2006) *Journal of Agricultural and Food Chemistry*, **54**, 2985–2991.

93 Mouri, T., Michizoe, J., Ichinose, H., Kamiya, N. and Goto, M. (2006) *Applied Microbiology and Biotechnology*, **2**, 514–520.

94 Fang, X. and Halpert, J.R. (1996) *Drug Metabolism and Disposition*, **24**, 1282–1285.

95 Faulkner, K.M., Shet, M.S., Fisher, C.W. and Estabrook, R.W. (1995) *Proceedings of the National Academy of Sciences of the United States of America*, **92**, 7705–7709.

96 Schwaneberg, U., Appel, D., Schmitt, J. and Schmid, R.D. (2000) *Journal of Biotechnology*, **84**, 249–257.

97 Kazlauskaite, J., Westlake, A.C.G., Wong, L.-L. and Hill, H.A.O. (1996) *Chemical Communications*, 2189–2190.

98 Scheller, F., Renneberg, R., Schwarze, W., Strnad, G., Pommerening, K., Prumke, H.J. and Mohr, P. (1979) *Acta Biologica et Medica Germanica*, **38**, 503–509.

99 Fleming, B.D., Tian, Y., Bell, S.G., Wong, L.L., Urlacher, V. and Hill, H.A. (2003) *European Journal of Biochemistry/FEBS*, **270**, 4082–4088.

100 Shumyantseva, V.V., Ivanov, Y.D., Bistolas, N., Scheller, F.W., Archakov, A.I. and Wollenberger, U. (2004) *Analytical Chemistry*, **76**, 6046–6052.

101 Faber, K. (2004) *Biotransformations in Organic Chemistry*, 5th. edn, Springer, Berlin.

102 Schwaneberg, U., Otey, C., Cirino, P.C., Farinas, E. and Arnold, F.H. (2001) *Journal of Biomolecular Screening*, **6**, 111–117.

103 Maurer, S.C., Schulze, H., Schmid, R.D. and Urlacher, V. (2003) *Advanced Synthesis and Catalysis*, **345**, 802–810.

104 Falck, J.R., Reddy, Y.K., Haines, D.C., Reddy, K.M., Krishna, U.M., Graham, S., Murry, B. and Peterson, J.A. (2001) *Tetrahedron Letters*, **42**, 4131–4133.

105 Kubo, T., Peters, M.W., Meinhold, P. and Arnold, F.H. (2006) *Chemistry*, **12**, 1216–1220.

106 Dohr, O., Paine, M.J., Friedberg, T., Roberts, G.C. and Wolf, C.R. (2001) *Proceedings of the National Academy of Sciences of the United States of America*, **98**, 81–86.

107 Neeli, R., Roitel, O., Scrutton, N.S. and Munro, A.W. (2005) *The Journal of Biological Chemistry*, **280**, 17634–17644.

108 Maurer, S.C., Kuhnel, K., Kaysser, L.A., Eiben, S., Schmid, R.D. and Urlacher, V.B. (2005) *Advanced Synthesis and Catalysis*, **347**, 1090–1098.

109 Chefson, A., Zhao, J. and Auclair, K. (2006) *Chembiochem*, **7**, 916–919.

110 Zanger, U.M., Vilbois, F., Hardwick, J.P. and Meyer, U.A. (1988) *Biochemistry*, **27**, 5447–5454.

111 Joo, H., Lin, Z. and Arnold, F.H. (1999) *Nature*, **399**, 670–673.

112 Kumar, S., Liu, H. and Halpert, J.R. (2006) *Drug Metabolism and Disposition*, **34**, 1958–1965.

113 Cirino, P.C. and Arnold, F.H. (2002) *Current Opinion in Chemical Biology*, **6**, 130–135.

114 Gillam, E.M. (2007) *Chemical Research in Toxicology*, **21**, 220–231.

115 Miles, C.S., Ost, T.W., Noble, M.A., Munro, A.W. and Chapman, S.K. (2000) *Biochimica et Biophysica Acta*, **1543**, 383–407.

116 Tee, K.L. and Schwaneberg, U. (2007) *Combinatorial Chemistry & High Throughput Screening*, **10**, 197–217.

117 Gunsalus, I.C. and Wagner, G.C. (1978) *Methods in Enzymology*, **52**, 166–188.

118 Miura, Y. and Fulco, A.J. (1975) *Biochimica et Biophysica Acta*, **388**, 305–317.

119 Fulco, A.J. (1991) *Annual Review of Pharmacology and Toxicology*, **31**, 177–203.

120 P.R. Ortiz de Montellano (ed.) (2005) *Cytochrome P450: Structure, Mechanism, and Biochemistry*, 3rd edn, Kluwer Academic/Plenum Press, New York.

121 Nickerson, D.P., Harford-Cross, C.F., Fulcher, S.R. and Wong, L.L. (1997) *FEBS Letters*, **405**, 153–156.

122 Stevenson, J.-A., Bearpark, J.K. and Wong, L.-L. (1998) *New Journal of Chemistry*, 551–552.

123 England, P.A., Harford-Cross, C.F., Stevenson, J.A., Rouch, D.A. and Wong, L.L. (1998) *FEBS Letters*, **424**, 271–274.

124 Harford-Cross, C.F., Carmichael, A.B., Allan, F.K., England, P.A., Rouch, D.A. and Wong, L.-L. (2000) *Protein Engineering*, **13**, 121–128.

125 Grayson, D.A., Tewari, Y.B., Mayhew, M.P., Vilker, V.L. and Goldberg, R.N. (1996) *Archives of Biochemistry and Biophysics*, **332**, 239–247.

126 Celik, A., Speight, R.E. and Turner, N.J. (2005) *Chemical Communications*, 3652–3644.

127 Chen, X., Christopher, A., Jones, J.P., Bell, S.G., Guo, Q., Xu, F., Rao, Z. and Wong, L.L. (2002) *The Journal of Biological Chemistry*, **277**, 37519–37526.

128 Jones, J.P., O'Hare, E.J. and Wong, L.L. (2001) *European Journal of Biochemistry/FEBS*, **268**, 1460–1467.

129 Stevenson, J.-A., Westlake, A.C.G., Whittock, C. and Wong, L.-L. (1996) *Journal of the American Chemical Society*, **118**, 12846–12847.

130 Bell, S.G., Stevenson, J.A., Boyd, H.D., Campbell, S., Riddle, A.D., Orton, E.L. and Wong, L.L. (2002) *Chemical Communications*, 490–491.

131 Xu, F., Bell, S.G., Lednik, J., Insley, A., Rao, Z. and Wong, L.L. (2005) *Angewandte Chemie-International Edition*, **44**, 4029–4032.

132 Wong, L.L., Bell, S.G. and Carmichael, A.B. (2000) British Patent WO 00/31273.

133 Sowden, R.J., Yasmin, S., Rees, N.H., Bell, S.G. and Wong, L.L. (2005) *Organic and Biomolecular Chemistry*, **3**, 57–64.

134 Bell, S.G., Chen, X., Sowden, R.J., Xu, F., Williams, J.N., Wong, L.L. and Rao, Z. (2003) *Journal of the American Chemical Society*, **125**, 705–714.

135 Munro, A.W., Leys, D.G., McLean, K.J., Marshall, K.R., Ost, T.W., Daff, S., Miles, C.S., Chapman, S.K., Lysek, D.A., Moser, C.C., Page, C.C. and Dutton, P.L. (2002) *Trends in Biochemical Sciences*, **27**, 250–257.

136 Appel, D., Lutz-Wahl, S., Fischer, P., Schwaneberg, U. and Schmid, R.D. (2001) *Journal of Biotechnology*, **88**, 167–171.

137 Budde, M., Morr, M., Schmid, R.D. and Urlacher, V.B. (2006) *Chembiochem: A European Journal of Chemical Biology*, **7**, 789–794.

138 Li, Q.S., Ogawa, J., Schmid, R.D. and Shimizu, S. (2001) *Applied and Environmental Microbiology*, **67**, 5735–5739.

139 Li, Q.S., Schwaneberg, U., Fischer, P. and Schmid, R.D. (2000) *Chemistry - A European Journal*, **6**, 1531–1536.

140 Sulistyaningdyah, W.T., Ogawa, J., Li, Q.S., Shinkyo, R., Sakaki, T., Inouye, K., Schmid, R.D. and Shimizu, S. (2004) *Biotechnology Letters*, **26**, 1857–1860.

141 Watanabe, Y., Laschat, S., Budde, M., Affolter, O., Shimada, Y. and Urlacher, V.B. (2007) *Tetrahedron*, **63**, 9413–9422.

142 Urlacher, V.B., Makhsumkhanov, A. and Schmid, R.D. (2006) *Applied Microbiology and Biotechnology*, **70**, 53–59.

143 Fasan, R., Chen, M.M., Crook, N.C. and Arnold, F.H. (2007) *Angewandte Chemie-International Edition*, **46**, 8414–8418.

144 Farinas, E.T., Schwaneberg, U., Glieder, A. and Arnold, F.H. (2001) *Advanced Synthesis and Catalysis*, **343**, 601–606.

145 Glieder, A., Farinas, E.T. and Arnold, F.H. (2002) *Nature Biotechnology*, **20**, 1135–1139.

146 Meinhold, P., Peters, M.W., Chen, M.M.Y., Takahashi, K. and Arnold, F.H. (2005) *Chembiochem: A European Journal of Chemical Biology*, **6**, 1765–1768.

147 Peters, M.W., Meinhold, P., Glieder, A. and Arnold, F.H. (2003) *Journal of the American Chemical Society*, **125**, 13442–13450.

148 van Vugt-Lussenburg, B.M., Damsten, M.C., Maasdijk, D.M., Vermeulen, N.P. and Commandeur, J.N. (2006) *Biochemical and Biophysical Research Communications*, **346**, 810–818.

149 Otey, C.R., Bandara, G., Lalonde, J., Takahashi, K. and Arnold, F.H. (2006) *Biotechnology and Bioengineering*, **93**, 494–499.

150 Landwehr, M., Hochrein, L., Otey, C.R., Kasrayan, A., Backvall, J.E. and Arnold, F.H. (2006) *Journal of the American Chemical Society*, **128**, 6058–6059.

151 Otey, C.R., Landwehr, M., Endelman, J.B., Hiraga, K., Bloom, J.D. and Arnold, F.H. (2006) *PLoS Biology*, **4**, e112.

152 Landwehr, M., Carbone, M., Otey, C.R., Li, Y. and Arnold, F.H. (2007) *Chemistry & Biology*, **14**, 269–278.

153 Li, Y., Drummond, D.A., Sawayama, A.M., Snow, C.D., Bloom, J.D. and Arnold, F.H. (2007) *Nature Biotechnology*, **25**, 1051–1056.

154 Kumar, S., Scott, E.E., Liu, H. and Halpert, J.R. (2003) *The Journal of Biological Chemistry*, **278**, 17178–17184.

155 Kumar, S., Chen, C.S., Waxman, D.J. and Halpert, J.R. (2005) *The Journal of Biological Chemistry*, **280**, 19569–19575.

156 Nakamura, K., Martin, M.V. and Guengerich, F.P. (2001) *Archives of Biochemistry and Biophysics*, **395**, 25–31.

157 Wu, Z.L., Aryal, P., Lozach, O., Meijer, L. and Guengerich, F.P. (2005) *Chemistry & Biodiversity*, **2**, 51–65.

158 Wu, Z.L., Podust, L.M. and Guengerich, F.P. (2005) *The Journal of Biological Chemistry*, **280**, 41090–41100.

159 Kim, D. and Guengerich, F.P. (2004) *Archives of Biochemistry and Biophysics*, **432**, 102–108.

160 Johnston, W.A., Huang, W., Voss, J.J.D., Hayes, M.A. and Gillam, E.M. (2007) *Drug Metabolism and Disposition*, **35**, 2177–2185.

161 Nishida, C.R. and Ortiz de Montellano, P.R. (2005) *Biochemical and Biophysical Research Communications*, **338**, 437–445.

162 Wright, R.L., Harris, K., Solow, B., White, R.H. and Kennelly, P.J. (1996) *FEBS Letters*, **384**, 235–239.

163 Oku, Y., Ohtaki, A., Kamitori, S., Nakamura, N., Yohda, M., Ohno, H. and Kawarabayashi, Y. (2004) *Journal of Inorganic Biochemistry*, **98**, 1194–1199.

164 Sideso, O., Smith, K.E., Welch, S.G. and Williams, R.A. (1997) *Biochemical Society Transactions*, **25**, 17S.

165 Yano, J.K., Blasco, F., Li, H., Schmid, R.D., Henne, A. and Poulos, T.L. (2003) *The Journal of Biological Chemistry*, **278**, 608–616.

166 Salazar, O., Cirino, P.C. and Arnold, F.H. (2003) *Chembiochem: A European Journal of Chemical Biology*, **4**, 891–893.

167 Eiben, S., Bartelmas, H. and Urlacher, V.B. (2007) *Applied Microbiology and Biotechnology*, **75**, 1055–1061.

168 Kumar, S., Sun, L., Liu, H., Muralidhara, B.K. and Halpert, J.R. (2006) *Protein Engineering, Design and Selection*, **19**, 547–554.

169 Wong, T.S., Arnold, F.H. and Schwaneberg, U. (2004) *Biotechnology and Bioengineering*, **85**, 351–358.

170 Zhao, J., Tan, E., Ferras, J. and Auclair, K. (2007) *Biotechnology and Bioengineering*, **98**, 508–513.

171 Chefson, A. and Auclair, K. (2007) *Chembiochem: A European Journal of Chemical Biology*, **8**, 1189–1197.

172 Chefson, A., Zhao, J. and Auclair, K. (2007) *Journal of Biotechnology*, **130**, 436–440.

173 King, D.L., Azari, M.R. and Wiseman, A. (1988) *Methods in Enzymology*, **137**, 675–686.

174 Lamb, S.B., Lamb, D.C., Kelly, S.L. and Stuckey, D.C. (1998) *FEBS Letters*, **431**, 343–346.

175 Taylor, M., Lamb, D.C., Cannell, R.J., Dawson, M.J. and Kelly, S.L. (2000) *Biochemical and Biophysical Research Communications*, **279**, 708–711.

2
Biocatalytic Hydrolysis of Nitriles

Dean Brady

2.1
The Problem with Nitrile Hydrolysis

Nitrile functional groups are simple to synthesize and are found in compounds used as solvents, extractants, polymers (acrylonitrile and adiponitrile are required for the production of polyacrylonitrile and nylon 66), synthetic rubber, pharmaceuticals, chiral intermediates, herbicides and pesticides (such as dichlobenil, bromoxynil, ioxynil and buctril). They are also important intermediates in the organic synthesis of other functional groups, including amines, amides, lactams, amidines, carboxylic acids (and hence carboxylic esters and lactones), aldehydes and ketones and heterocyclic compounds [1].

Cyanide addition provides for an effective means of single carbon homologation of the carbon backbone [2]. Organonitriles can be synthesized by various means, such as the addition of a cyanide ion to alkyl halides, the Sandmeyer reaction and the reaction of aryl halides with copper cyanide [3]. α,β-Unsaturated nitriles can also be synthesized using Wittig-type reactions for aldehydes or by Knoevenagel reaction from cyanoacetates and aldehydes or ketones [1c, 4]. As cyanide is one of the few water-stable carbanions, cyanide provides a C_1-synthon that can be used for reactions in aqueous solutions [2]. The synthesis of α-hydroxynitriles is spontaneous in water from the cyanide and corresponding aldehyde whereas the Strecker reaction provides α-aminonitriles [5]. β-Hydroxynitriles can be generated by opening of epoxides by cyanide [5, 6]

Nitriles are usually synthetically more accessible than carboxylic acids [7] and hence are often used as the entry point to both carboxylic acids and amides. However, the subsequent hydrolysis of the nitrile into amides or carboxylic acids is an environmental problem due to the requirement for harsh reaction conditions involving either strong mineral acids (e.g. hydrochloric or phosphoric acid) or bases (e.g. potassium or sodium hydroxide) and relatively high reaction temperatures (Scheme 2.1). These reactions sometimes give low yields due to unwanted byproduct formation and generate concentrated contaminating waste salt streams (e.g. 6 mol L^{-1}) when the acid or base is pH neutralized prior to disposal [1b, 2, 7]. Indeed, acid catalysis is one of

Handbook of Green Chemistry, Volume 3: Biocatalysis. Edited by Robert H. Crabtree
Copyright © 2009 WILEY-VCH Verlag GmbH & Co. KGaA, Weinheim
ISBN: 978-3-527-32498-9

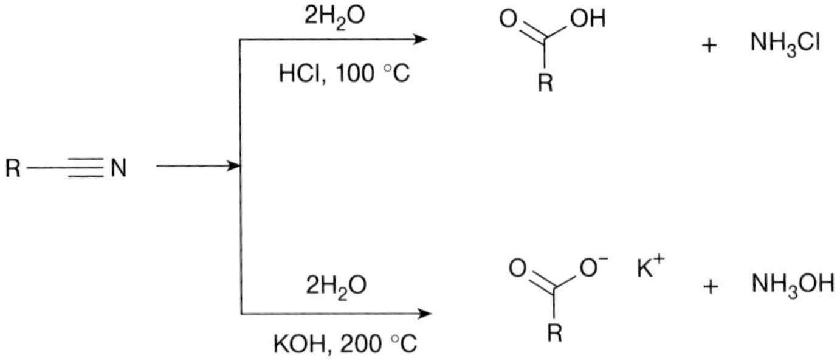

Scheme 2.1

the two largest generic areas of chemistry and as such creates vast quantities of inorganic waste [8].

This type of process is counter to the principles of green chemistry, where catalytic agents are better than stoichiometric agents and it is considered to be better to prevent waste than to treat or clean up after reaction [9].

Moreover, acid or base hydrolysis presents practical difficulties in reaction control, as halting the reaction at the amide stage in high yield is not necessarily simple. Further, the reaction conditions may be incompatible with retention of other functional groups (such as esters) within the same molecule that are hydrolyzable or acid or base sensitive. Additionally, regioselective hydrolysis of one nitrile in a polynitrile compound is problematic. Lastly, enantioselective hydrolysis of a nitrile racemate is impossible under the usual conditions of acid or base hydrolysis.

There has been some success in replacing acid or base hydrolysis of nitriles by using metals under neutral conditions, but it often requires using expensive reagent systems such as platinum, palladium and cobalt complexes and usually retains the disadvantage of elevated reaction temperatures [3].

2.2
Biocatalysis as a Green Solution

Enzymes provide many aspects that epitomize green chemistry [10]:

• They function under mild reaction conditions, physiological pH and temperature (but can work beyond this range).

• They function in a non-polluting solvent (water).

• They are non-corrosive.

• They have a relatively small environmental footprint as they are biodegradable and are generated by environmentally benign processes (microbial cultivation).

• Furthermore, due to their inherent high activities and chemo-, regio- and stereo-selectivity in reactions of multifunctionalized molecules, they can circumvent the protection and deprotection steps required in traditional organic syntheses. This is possible due to the structure of the active site, which often allows for selective conversion of certain functional groups, based on charge, polarity, size and chirality.

2.3
Nitrile Biocatalysts

In the hydrolysis of nitriles, we are fortunate to have two different enzyme systems to work with (Scheme 2.2). Nitrilases are hydrolases (EC 3.5.5.1) which convert nitriles directly to carboxylic acids. The second system is a combination of a nitrile hydratase (EC 4.2.1.84), a lyase, which converts the nitrile to carboxamide through a hydrolysis step, and an amidase (EC 3.5.1.4) that subsequently hydrolyzes the carboxamide to the carboxylic acid. The catalytic mechanisms of the two nitrile converting enzymes are distinct [2].

2.3.1
Nitrilase

Nitrilases hydrolyze nitriles to the corresponding carboxylic acid and ammonia. Nitrilases are a branch of the 12-branch nitrilase superfamily that also includes amidases and N-acyltransferases [11]. The nitrilases were found to belong to one of six clades based on amino acid sequence [12]. The nitrilase protein structure is a novel α–β–β–α sandwich protein fold, with a triad of residues Glu–Lys–Cys that is essential for catalysis in all members of the superfamily [11], the importance of the cysteine residues having been previously determined [13]. Nitrilases have no metal cofactor or prosthetic group. The catalytic mechanism of nitrilases probably involves a nucleophilic attack by the thiol group of the catalytic cysteine residue on the nitrile carbon atom with concomitant protonation of nitrogen to form a tetrahedral thioimidate intermediate. Addition of water leads to protonation of the nitrogen atom and hydroxylation of C-1, which rearranges to release ammonia; subsequently the resultant carboxylic acid is released and the enzyme regenerated by addition of a second water molecule (Scheme 2.3) [14].

Some nitrilases preferentially hydrolyze either aliphatic nitriles or arylacetonitriles, whereas others hydrolyze aromatic or heterocyclic nitriles to their respective carboxylic acids [5, 15, 16], as summarized by Banerjee et al. [1a].

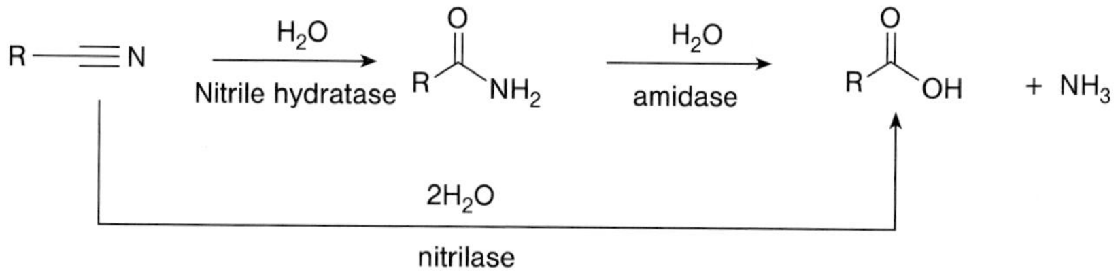

Scheme 2.2

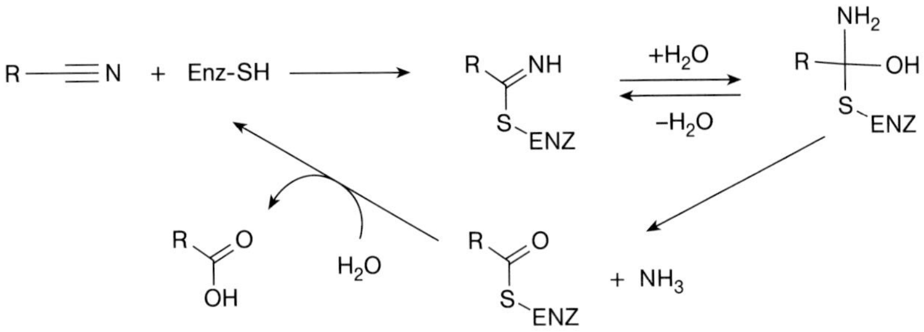

Scheme 2.3

The 47 kDa monomers of *Nocardia* sp. NCIB 11216 nitrilase where found to oligomerize to form a 560 kDa dodecamer in the presence of benzonitrile [16]. Moreover, Nagasawa *et al.* [17] determined that the activity of a nitrilase from *Rhodococcus rhodochrous* J1 was only active towards acrylonitrile with subunit oligomerization initiated by preincubation with benzonitrile. In this process, the enzyme undergoes post-translational cleavage, releasing 39 C-terminal amino acids and resulting in the formation of active, helical homo-oligomers of indefinite length [18].

Although nitrilases cannot be crystallized, the 3D structure of these helical homo-oligomers was elegantly determined by Sewell's group using electron microscopy, wherein the image of the stain envelope was fitted with a protein homology model based on other members of the nitrilase superfamily [18].

2.3.2
Nitrile Hydratase

Nitrile hydratases are typically α,β-heterodimeric metalloenzymes (Figure 2.1) composed of subunits ranging in size from 26 to 35 kDa. There are two types of nitrile hydratases that are classified according to their catalytically essential metal cation. The ferric nitrile hydratases contain a non-heme Fe(III), whereas the cobalt nitrile hydratases contain a non-corrinoid Co(III) [19]. The metal of the α-subunit in the active site is a catalyst for –CN hydration and is required for the folding of the enzyme [1a, 20]. The enzymes have a highly conserved motif CXLCSC, where X represents either a T (threonine) or S (serine) residue for cobalt- or iron-containing nitrile hydratase, respectively. The mechanism has recently been elucidated in detail [21] (Scheme 2.4).

The nitrile hydratase active site is interesting in that it is light activated at the non-heme iron center on the α-subunit and activation occurs by displacement of nitric oxide bound to the non-heme iron [22].

The two genes for the nitrile hydratase subunits are usually found as part of a gene cluster which also includes an amidase and regulatory protein-encoding

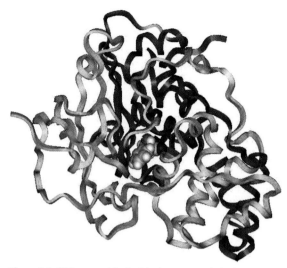

Figure 2.1 Ribbon model of a *Rhodococcus* nitrile hydratase α,β-heterodimer depicting a substrate 3-hydroxy-3-phenylpropionitrile in the active site.

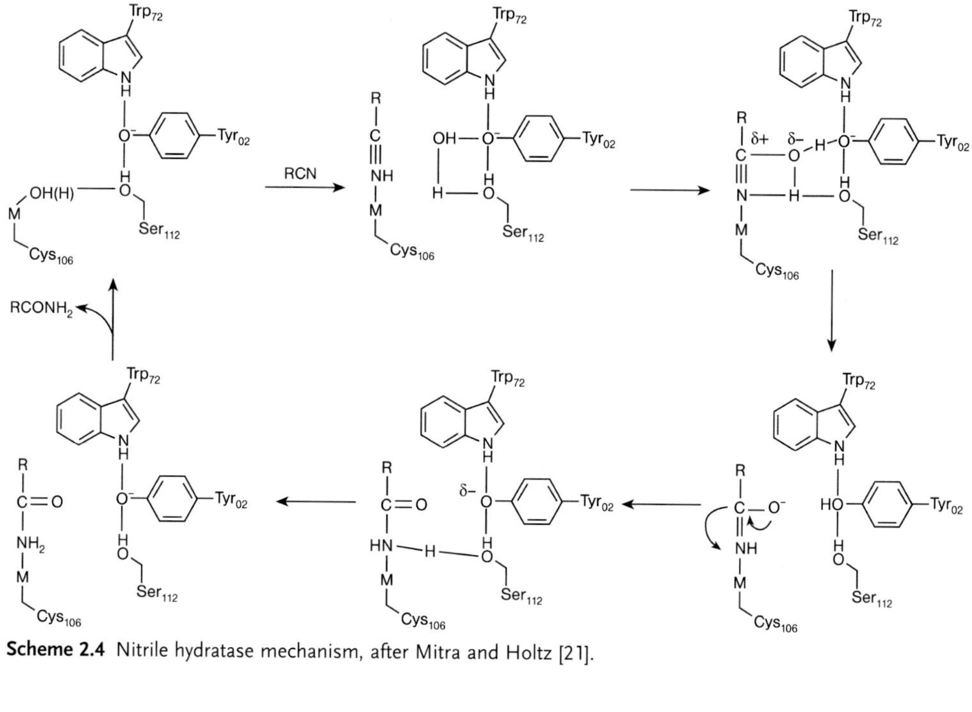

Scheme 2.4 Nitrile hydratase mechanism, after Mitra and Holtz [21].

genes and are generally induced by their amide products rather than their nitrile substrates [20, 23].

Similarly to nitrilases, nitrile hydratases differ in their substrate preferences [1a,d, 5, 24]. Although some nitrile hydratases are stereoselective [25], this does not appear to be a common trait.

2.3.3
Amidase

The carboxamide generated by a nitrile hydratase can be catalyzed by an amidase with suitable substrate specificity. In the case of the nitrile hydratase and amidase system, it is often the amidase that provides the enantioselectivity, whereas the nitrile hydratase is usually non-enantioselective for most substrates [1d, 26].

The amidase is found in the same superfamily as the nitrilases (branch 2) and has a similar catalytic mechanism. The active site contains a Cys–Glu–Lys catalytic triad [27]. The carbonyl group of amide undergoes a nucleophilic attack, resulting in the formation of a tetrahedral intermediate, which is subsequently converted to acyl-enzyme with the removal of ammonia and hydrolyzed to the corresponding carboxylic acid [1a]. The enzyme is usually found as a homodimer or homotetramer of 39–63 kDa subunits [1d], but larger oligomers have been discovered, such as the homo-hexamer of 38 kDa subunits from *Geobacillus pallidus* RAPc8 [28].

Nitrilases sometimes fail to add the second water molecule to the nitrile and the amide rather than the acid is released, an event that is fairly common with some substrates, while the amidase of *R. rhodochrous* J1 can occasionally hydrolyze nitriles [14]. This cross-over of function is probably due to the similar mechanism of the two enzymes.

2.4
Synthetic Utility

Enzymatic hydrolysis of various nitriles and primary amides was recently reviewed in detail by Martínková and Křen [1d] and covered by Schulze [29]. The broad concepts are described below.

2.4.1
Chemoselectivity

The availability of two separate systems for nitrile biocatalysis allows for a choice of amide or carboxylic acid product in high purity, which would be unattainable using alkali or acid hydrolysis. Using a nitrile hydratase amidase system, one can very specifically halt the reaction at the amide using purified or over-expressed nitrile hydratase or by using amidase inhibitors [5]. Nitrile biocatalysts have been shown not to hydrolyze acid or alkali labile groups on the same molecule, such as esters [1d, 5].

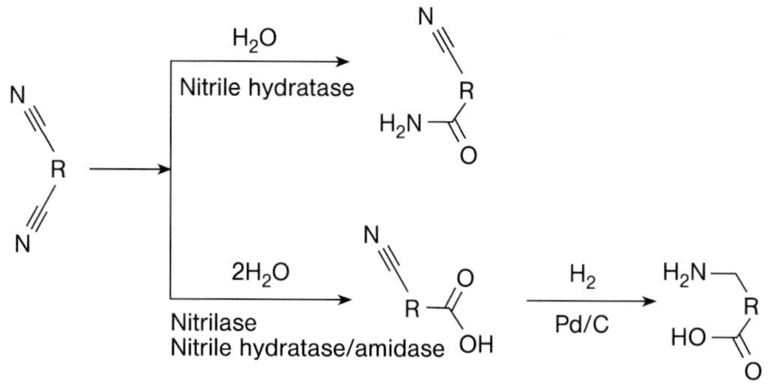

Scheme 2.5

2.4.2
Regioselectivity

Nitrile biocatalysts show specificity depending on charge, branching position, substituents and aliphatic or aromatic structure [1d, 30] which would allow for selective hydrolysis in mixtures of nitrile compounds or on polynitriles [4, 25c].

Selective hydrolysis of a single nitrile in a poly- or dinitrile affords the opportunity to retain the other nitrile group or to convert it to an amine through hydrogenation [3, 5, 14, 31] (Scheme 2.5). Biocatalysis tends to yield only one product as the monohydrolysis of a dinitrile produces a cyanocarboxylic acid or cyanoamide that is more polar than the substrate and not transformed further by the enzyme [2].

2.4.3
Enantioselectivity

It is expected that up to 70% of pharmaceutical intermediates will be chiral by 2010, a doubling from the previous decade with a trend towards inclusion of multiple stereogenic centers [32]. The commercial synthetic process generally requires a minimum of 98% enantiomeric excess (*ee*). Nitrile biocatalysts are superb at meeting this challenge [2]. For example, Benz *et al.* [1b] recently applied a nitrilase from *Rhodococcus* R312 to hydrolyze enantioselectively (*R,S*)-cyano-1,4-benzodioxane to (*R*)-1,4-benzodioxane-2-carboxylic acid and (*R,S*)-cyano-6-formyl-1,4-benzodioxane to (*R*)-6-formyl-1,4-benzodioxane-2-carboxylic acid with 98 and 99% *ee*, respectively. With careful selection of biocatalytic systems, it is possible to provide enantiomeric amides, carboxylic acids and nitriles with high enantiopurity (Table 2.1).

Diastereoselectivity   α,β-Unsaturated nitriles can be resolved using biocatalysts. Effenberger and Osswald [4] used the Diels–Alder reaction to prepare the four diastereomers of 3-(2-cyanocyclohex-3-enyl)propenenitrile from 1-cyano-1,3-

Table 2.1 Product options using enantioselective nitrile biocatalysts.

Biocatalyst type	Substrate class		
	Nitrile	Amide	Carboxylic acid
Selective nitrilase	Single enantiomer in high yield	None	Single enantiomer
Non-selective NHase, selective amidase	Racemate in low yield	Single enantiomer	Single enantiomer
Selective NHase and amidase	Single enantiomer	Single enantiomer in low yield	Single enantiomer
Selective NHase and amidase with opposite selection	Single enantiomer in high yield	Single enantiomer in high enantiopurity	Single enantiomer in low yield

butadiene. The recombinant nitrilase AtNIT1 from *Arabidopsis thaliana* hydrolyzed the *E*-isomers exclusively and showed a high level of *cis/trans* selectivity to give predominantly the corresponding (*E*)-*cis*-3-(2-cyanocyclohex-3-enyl)propenoic acid (97%), with only small amounts of the *trans* product being seen (Scheme 2.6).

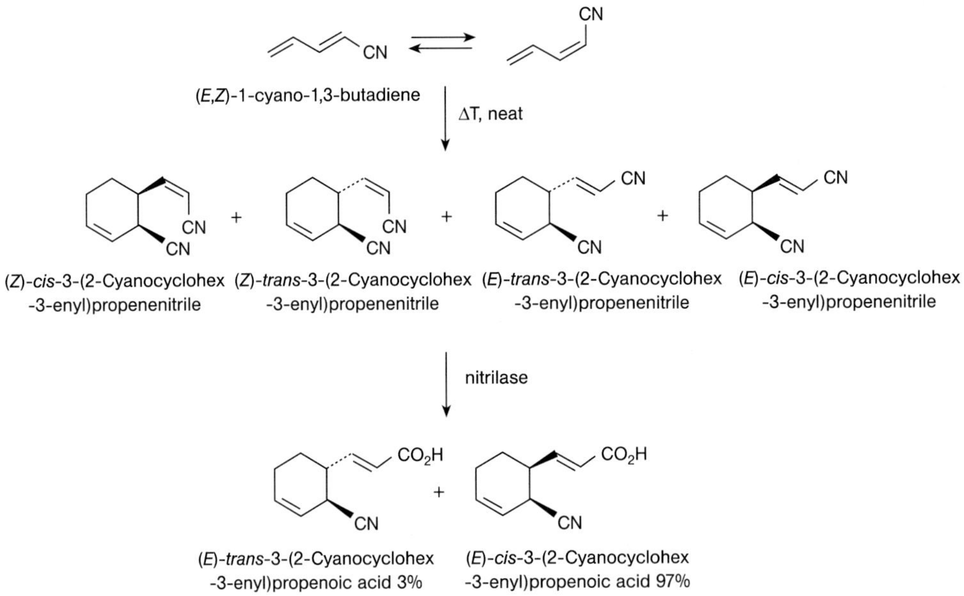

(*E,Z*)-1-cyano-1,3-butadiene

ΔT, neat

(*Z*)-*cis*-3-(2-Cyanocyclohex -3-enyl)propenenitrile + (*Z*)-*trans*-3-(2-Cyanocyclohex -3-enyl)propenenitrile + (*E*)-*trans*-3-(2-Cyanocyclohex -3-enyl)propenenitrile + (*E*)-*cis*-3-(2-Cyanocyclohex -3-enyl)propenenitrile

nitrilase

(*E*)-*trans*-3-(2-Cyanocyclohex -3-enyl)propenoic acid 3%   (*E*)-*cis*-3-(2-Cyanocyclohex -3-enyl)propenoic acid 97%

Scheme 2.6

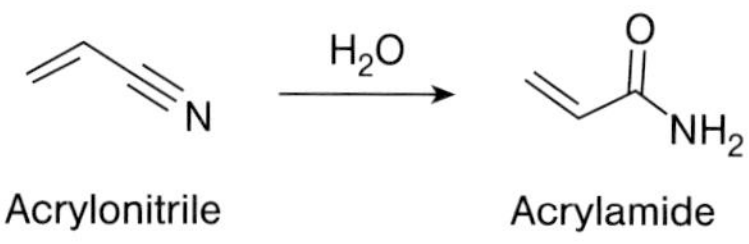

Acrylonitrile   Acrylamide

Scheme 2.7

2.5
Commercial Examples

2.5.1
Chemical Synthesis

2.5.1.1 Acrylamide

The earliest commercial success in nitrile biocatalysis was also the first industrial bulk chemical biotransformation for biocatalysis as a whole and hence has been extensively reviewed [2, 33]. The commercial production of acrylamide (propenamide) is performed by hydration of acrylonitrile using immobilized whole bacterial cells of *Rhodococcus rhodochrous* J1 with high nitrile hydratase activity (Scheme 2.7). This biotransformation is performed on an industrial scale by Mitsubishi Rayon at 65 000 t per annum. The high yield (>99.9%) is possible as the formation of acrylic acid byproducts that form using traditional methodology does not occur with biocatalysis (conversion >99.99%, selectivity >99.99%) [33]. Moreover no polymerization inhibitor is required as the reaction occurs under mild conditions (5°C). The technology has been updated and the present production organism *R. rhodochrous* J1 represents a third-generation biocatalyst and is immobilized in polyacrylamide gel. Acrylamide productivity exceeds 400 g L^{-1} h^{-1} in this fed-batch reaction and the product concentration can reach 600–700 g L^{-1} as it is insoluble and precipitates out of solution [34].

2.5.1.2 Nicotinamide and Nicotinic Acid [35]

Nicotinamide and nicotinic acid (niacin) are B-complex vitamins, with an annual global consumption of 8000 and 14 000 t, respectively. Nicotinamide can be synthesized from 3-cyanopyridine by chemical hydrolysis, but the hydrolysis reaction is difficult to halt at the amide and some of the material is converted to the carboxylic acid and hence nicotinate salt byproduct is generated.

The application of a suitable nitrile hydratase could yield nicotinamide from 3-cyanopyridine (Scheme 2.8). Nitto (now Mitsubishi Rayon) in Japan, BASF in Germany and Lonza in Switzerland have developed this reaction. Using the same nitrile hydratase expressing biocatalyst as for acrylamide production, *R. rhodochrous* J1, the same nitrile hydratase biocatalyst as applied in acrylamide production, is being used by Lonza in China for the annual production of 3000 t of nicotinamide from 3-cyanopyridine with 100% yield on starting material owing to the elimination of byproduct formation and a volumetric yield of 1465 g L^{-1} [29, 35]. These high yields are achieved as the biocatalyst may be added to a concentrated solution of

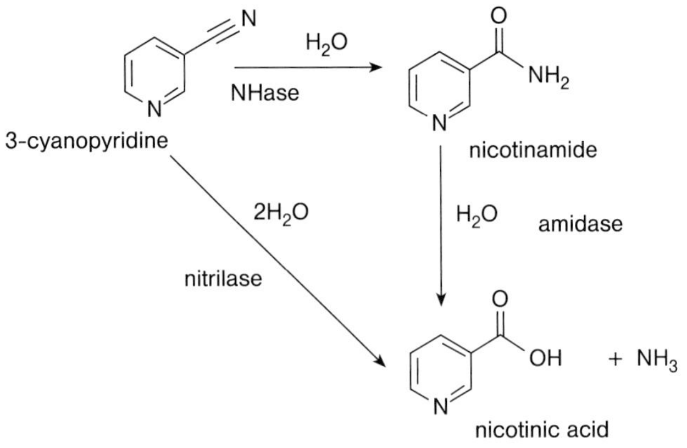

Scheme 2.8

3-cyanopyridine, providing a nicotinamide product of pharmaceutical quality that is subsequently spray-dried.

Similarly, the production of nicotinic acid can be achieved by nitrilase-catalyzed hydrolysis of 3-cyanopyridine to nicotinic acid. *Nocardia globerula* NHB-2, which expresses nitrilase activity, converted 98.6% of 40 mM 3-cyanopyridine in 9 h at 26 mmol h^{-1} g^{-1} dry cell weight (dcw) on the bench scale without detectable formation of nicotinamide, which is a byproduct of other processes [36].

2.5.1.3 Atorvastatin

Statins are a class of HMG-CoA reductase inhibitors that are administered in the treatment of hypocholesterolemia and atherosclerosis. The statin atorvastatin (Lipitor) with sales of $12 billion in 2004 is currently the global leading pharmaceutical. One synthetic option is the desymmetrization of 3-hydroxyglutaronitrile, which is readily synthesized by reaction of racemic epichlorohydrin with cyanide.

Researchers at Diversa generated a mutant nitrilase enzyme that had been identified from an environmental screen (see below). The mutated nitrilase catalyzed desymmetrization of 3-hydroxyglutaronitrile at high substrate concentration (3 M) in a laboratory-scale reaction. The advantage of the introduction of the stereogenic center by desymmetrization, as opposed to kinetic resolution, is that it permitted a yield of 100% on the initial nitrile. After a 15 h reaction at 20°C, (*R*)-4-cyano-3-hydroxybutyric acid was isolated in 96% yield, with an excellent *ee* of 98.5% and a volumetric productivity of 619 g L^{-1} d^{-1} [37].

Subsequently Dow Chirotech, a subsidiary of Dow Chemical Company, developed the Diversa nitrilase further into a biocatalysis process [38] and using the Pfenex expression system (a *Pseudomonas fluorescens*-based host expression system) to overproduce the enzyme. Synthesized 3-hydroxyglutaronitrile is desymmetrized by the nitrilase. Optimal reaction conditions were 3 M (330 g L^{-1}) 3-hydroxyglutaronitrile, pH 7.5, 27°C, reacted for 16 h. A conversion of 100% and 99% product *ee* were obtained. (*R*)-4-Cyano-3-hydroxybutyric acid was subsequently esterified to give the

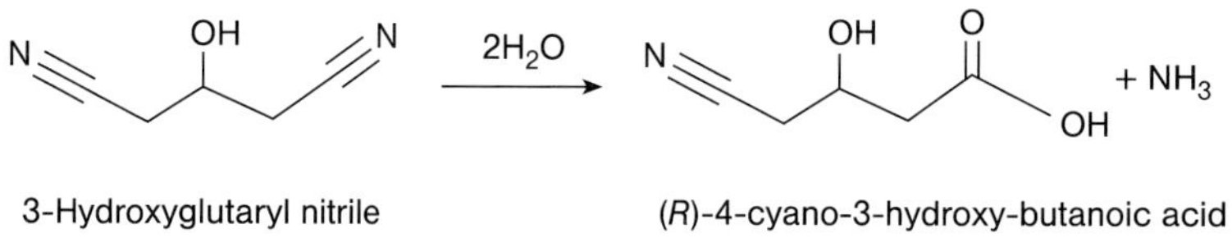

3-Hydroxyglutaryl nitrile

(R)-4-cyano-3-hydroxy-butanoic acid

Scheme 2.9

desired intermediate, ethyl (R)-4-cyano-3-hydroxybutyrate. Overall, a highly efficient three-stage synthesis of ethyl (R)-4-cyano-3-hydroxybutyrate from low-cost epichlorohydrin was achieved with an overall yield of 23%, 98.8% *ee* and 97% purity (Scheme 2.9) [38].

2.5.1.4 5-Cyanovaleramide

DuPont replaced inefficient industrial chemical hydration of adiponitrile to 5-cyanovaleramide (MnO_2 catalyst, $130°C$), which resulted in a significant amount of byproducts, with a biocatalytic regioselective hydration of the dinitrile (Scheme 2.10) [33b]. 5-Cyanovaleramide is an early intermediate in the synthesis of the herbicide azafenidin. The nitrile hydratase biocatalytic process is more efficient and results in fewer byproducts, allowing for 93% yield and 96% selectivity (97% conversion). The process uses whole cells of *Pseudomonas chlororaphis* B23 immobilized in calcium alginate beads. By recycling the biocatalyst in multiple batch reactions, over 12 t of adiponitrile can be transformed to 5-cyanovaleramide with a catalytic efficiency of $3150 \ kg \ kg^{-1}$ catalyst [39].

2.5.1.5 Mandelic Acid

(R)-(−)-Mandelic acid is an intermediate in the production of semisynthetic cephalosporins and penicillins [40]. Racemic mandelonitrile, an α-hydroxynitrile, can be transformed to (R)-(−)-mandelic acid using a nitrilase as a catalyst. A considerable advantage of this reaction is that cyanohydrins racemize in aqueous solution through a spontaneous reversible reaction between aldehyde and hydrogen cyanide and the α-hydroxynitrile. Therefore, if one enantiomer of the α-hydroxynitrile is enantioselectively hydrolyzed, the enantiomeric equilibrium of the substrate is re-established (Scheme 2.11). Hence the the R-enantioselective nitrilase from *Alcaligenes faecalis* ATCC 8750 can provide a 100% yield of (R)-(−)-mandelic acid with an *ee* of >99% on the basis of the racemic mandelonitrile [41]. BASF currently uses a nitrilase to make (R)-mandelic acid on a multi-ton scale [33a, 42].

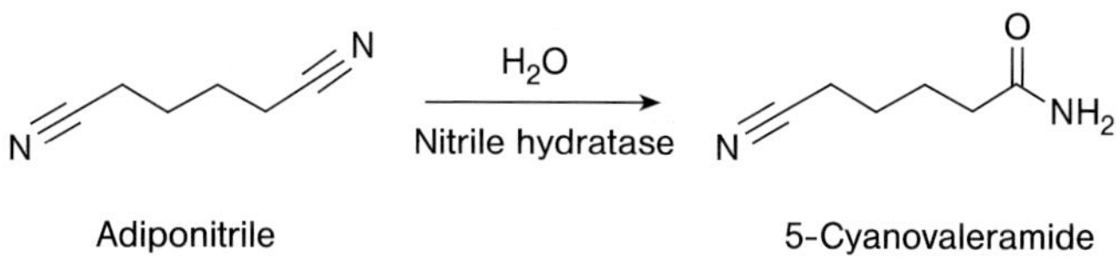

Adiponitrile

5-Cyanovaleramide

Scheme 2.10

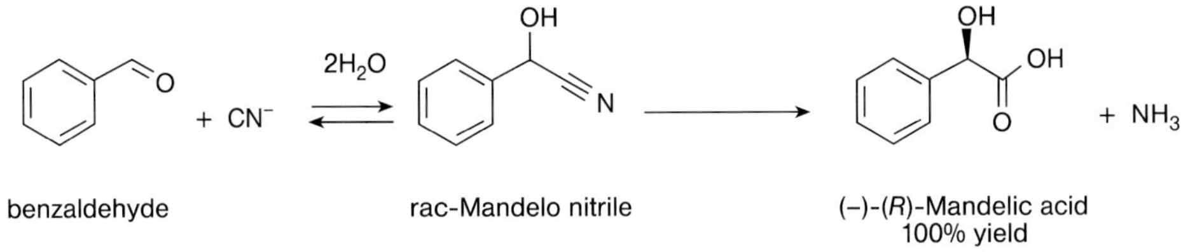

benzaldehyde   rac-Mandelo nitrile   (–)-(R)-Mandelic acid
100% yield

Scheme 2.11

2.5.1.6 Pyrazinecarboxylic Acid

Lonza has a biocatalytic route to antituberculosis drugs under development. Bioconversion of 2-cyanopyrazine to pyrazinecarboxylic acid using whole cells of *Agrobacterium* DSM 6336 expressing a nitrilase is followed by regioselective hydroxylation to 5-hydroxypyrazine-2-carboxylic acid by a carboxylic acid hydroxylase (Scheme 2.12). The product concentration can be as high as 40 g L^{-1}. The nitrilase and carboxylic acid hydroxylase are induced through growth on 3-cyanopyridine as a carbon and energy source. The lower isolated yield of 80% compared with the analytical yield of 95% is due to repeated precipitation during the process [33a, c]

2.5.1.7 (E)-2-Methyl-2-Butenoic Acid [43]

(E)-2-Methyl-2-butenoic acid (tiglic acid) and (Z)-2-methyl-2-butenoic acid (angelic acid) are useful starting materials in the synthesis of pharmaceutical intermediates, flavors and fragrances. A mixture of (E/Z)-2-methyl-2-butenenitrile can be obtained by isomerizing 2-methyl-3-butenenitrile, a commercially available byproduct of adiponitrile manufacture. However, the geometric isomers are so similar in physical properties that their separation is a problem.

Researchers at DuPont discovered that the nitrilase activity of immobilized *Acidovorax facilis* 72W regioselectively hydrolyzed the E-isomer to (E)-2-methyl-2-butenoic acid without conversion of (Z)-2-methyl-2-butenenitrile, permitting facile separation of the acid salt and the nitrile with high yield and purity (Scheme 2.13).

2.5.1.8 1,5-Dimethyl-2-Piperidone, a Lactam [44]

DuPont has developed a water-soluble biodegradable and non-flammable solvent, 1,5-dimethyl-2-piperidone (Xolvone), with applications in cleaning electronic parts, such as computer circuit boards. Initially, commercial production of Xolvone

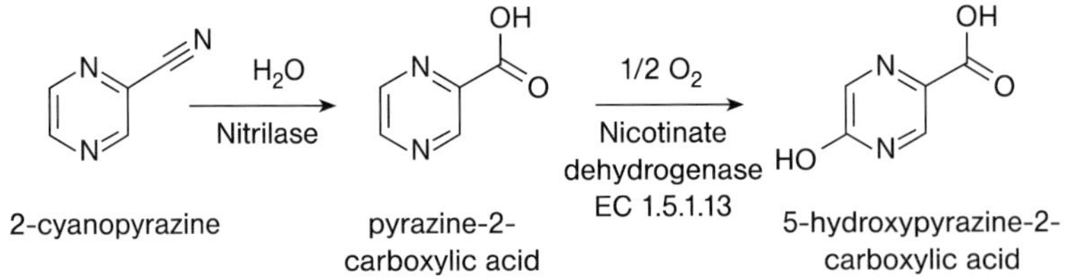

2-cyanopyrazine   pyrazine-2-
carboxylic acid   5-hydroxypyrazine-2-
carboxylic acid

Scheme 2.12

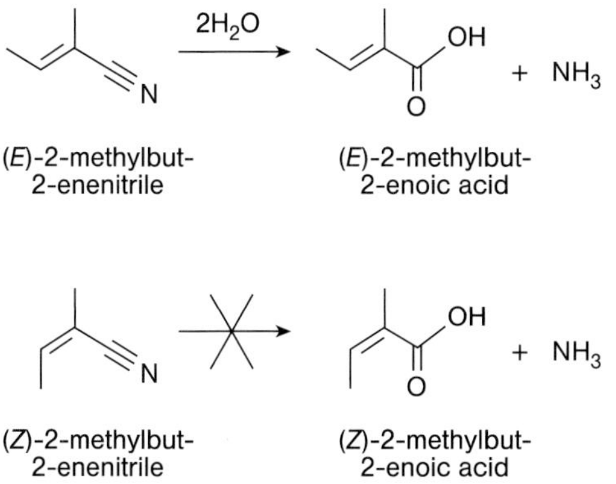

Scheme 2.13

(E)-2-methylbut-
2-enenitrile

$2H_2O$

(E)-2-methylbut-
2-enoic acid

+ NH_3

(Z)-2-methylbut-
2-enenitrile

(Z)-2-methylbut-
2-enoic acid

+ NH_3

involved direct hydrogenation of 2-methylglutaronitrile in the presence of methyl-amine and produced a mixture of geometric isomers of 1,3- and 1,5-dimethyl-2-piperidones. 2-Methylglutaronitrile is a byproduct in the manufacture of adiponitrile and is generated at 105 000 t per annum.

To improve the yield of 1,5-dimethyl-2-piperidone, a novel chemoenzymatic process was subsequently developed as a replacement. Immobilized *Acidovorax facilis* 72W (ATCC 55746) cells were used to convert 2-methylglutaronitrile with a substrate loading of approximately 200 g L^{-1} to the ammonium salt of 4-cyanopentanoic acid at 100% conversion and with >98% regioselectivity (Scheme 2.14). Volumetric productivity of 4-cyanopentanoic acid in the wild-type organism (up to 49 g L^{-1} h^{-1}) could be increased to 79 g L^{-1} h^{-1} by expressing the *A. facilis* 72W nitrilase in *E. coli*. The alginate-immobilized biocatalyst could be re-used repeatedly in a 500 L batch reaction, providing 3.5 kg product g^{-1} catalyst.

Without the need for isolation, the aqueous solution of 4-cyanopentanoic acid, after concentration through distillation, is hydrogenated in the presence of methylamine to yield 88% of 1,5-dimethyl-2-piperidone. Compared with the previous inefficient process, the chemoenzymatic process produces only the single geometric isomer of 1,5-dimethyl-2-piperidone, less waste and byproducts are produced and product recovery is improved.

2.5.1.9 3-Hydroxyvaleric Acid [45]
3-Hydroxyalkanoic acids are used in the synthesis of co-polyesters for fibers and other applications. Reaction conditions for the chemical hydrolysis of hydroxyalkaneni-

2-Methylglutaronitrile

$2H_2O$
nitrilase or
nitrile hydratase
and amidase

4-Cyanopentanoic
acid

+ NH_4^+

H_2
Pd/C
CH_3NH_2

1,5-dimethyl-2-piperidone
(Xolvone)

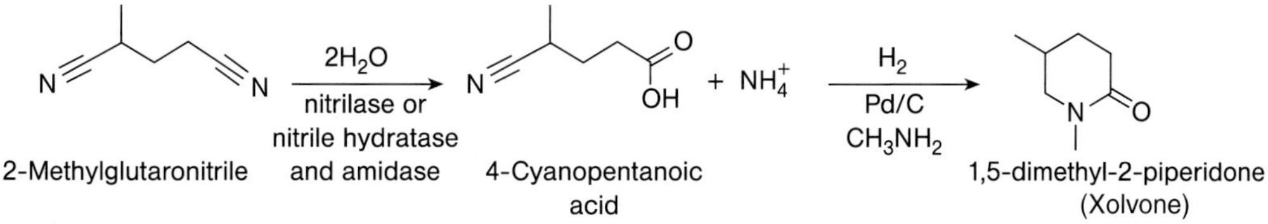

Scheme 2.14

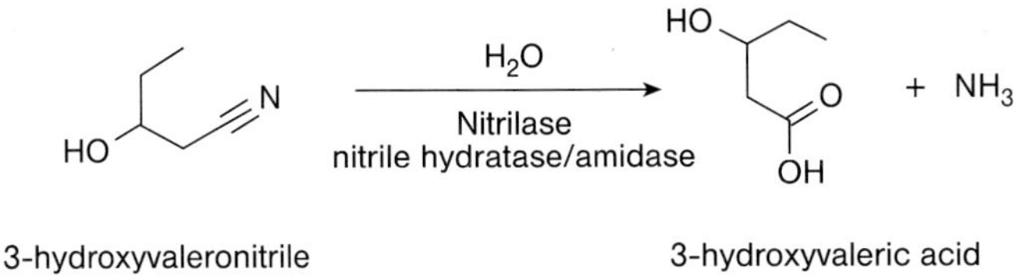

3-hydroxyvaleronitrile   3-hydroxyvaleric acid

Scheme 2.15

triles results in a degree of elimination of hydroxy groups to produce undesirable alkenylnitrile byproducts that must be removed prior to the polymerization reaction to prevent color formation.

To avoid co-product formation, 3-hydroxyvaleronitrile can be converted to 3-hydroxyvaleric acid using biocatalysts (Scheme 2.15). *Comamonas testosteroni* 5-MGAM-4D expressed a nitrile hydratase and amidase activity that provided 1.0 M 3-hydroxyvaleric acid (as the ammonium salt) in nearly 100% yield. GA/PEI-cross-linked cells immobilized in alginate could be recycled for 100 reactions while still retaining useful levels of activity and could therefore achieve overall catalytic productivities of 670 g g^{-1} dcw of 3-hydroxyvaleric acid and an initial volumetric productivity of 44 g L^{-1} h^{-1} of 3-hydroxyvaleric acid. DuPont has produced 100 kg quantities of 3-hydroxyvaleric acid using this process.

2.5.2
Surface Modification of Polymers [46]

Approximately 2.7×10^6 t per annum of polyacrylonitrile (PAN) for the textile industry are produced by radical polymerization of acrylonitrile. PAN is highly hydrophobic, but a degree of hydrophilicity can have advantages in PAN-based textiles by improving wettability, dyeability, fastness of special finishes and the feel of the product. Higher hydrophilicity would similarly facilitate specialty applications such as coating of PAN for the production of technical textiles, the performance of filtration devices and polymers, medical devices and electronics. Alkaline treatments can lead to increased hydrophilicity by hydrolyzing a percentage of the nitrile groups on the polymer, but this crude method leads to some physical deterioration in strength of the fibers and irreversible yellowing. Application of enzymes as biocatalysts would be more selective, specifically hydrolyzing only surface nitrile groups, and therefore would be less destructive.

However, the nature of the substrate places some constraints on the use of biocatalysts. First, the biocatalyst would have to be cell free as an intracellular enzyme would not have contact with the polymer. Second, the active site would have to be near the enzyme surface as the polymer would cause steric hindrance.

Hence only a few enzymes from *Brevibacterium imperiale*, *Corynebacterium nitrilophilus*, *Arthrobacter* sp., *Rhodococcus rhodochrous* NCIMB 11216 and *Agrobacterium tumefaciens* (BST05) have been effective.

The nitrile hydratase of *Rhodococcus rhodochrous* NCIMB 11216 converted 16% of surface nitrile groups of PAN granules to amides. *Agrobacterium tumefaciens* (BST05) also expresses an effective combination of nitrile hydratase and amidase activity after cultivation in a medium containing acetonitrile. *In vitro* the enzymes were able to convert 1.1% of the nitrile groups of powdered PAN to carboxylic acids, but the majority of conversions stopped at the amide stage.

2.5.3
Bioremediation

Nitriles tend to be toxic, carcinogenic and mutagenic in nature. Green chemistry is aimed at pollution prevention and not remediation. However, due to release of chemical process wastewater, synthetic nitrile compounds are widespread in the environment. As environmentally unfriendly processes will take time to phase out, there is a current need for application of nitrile biocatalysts for bioremediation. Microbial systems expressing nitrile-hydrolyzing enzymes have been used to trans-form environmental nitriles to the less harmful corresponding amides and acids. For example, aqueous latex waste streams were decontaminated by hydrolysis of acrylo-nitrile by nitrile hydratase activity [46e]. Acetonitrile in nylon process wastewater was efficiently degraded to acetic acid using *Rhodococcus pyridinivorans* S85-2 [47].

2.6
Challenges

2.6.1
Biocatalyst Stability

Schmid *et al.* [33a] noted that biocatalysis can be applied broadly in organic syn-thesis. In the case of nitrile-hydrolyzing biocatalysts, there are a few considerations. Unlike many of the other hydrolases currently used in industry (lipases, proteases, amylases), nitrile hydratases and nitrilases are multimeric intracellular enzymes which tend to be less robust. Therefore, there are a number of factors that need to be tackled for nitrile biocatalysis to reach its full potential:

- *Thermostability:* nitrile-hydrolyzing enzymes are often affected by temperature (the biocatalytic hydrolysis of acrylamide by *R. rhodochrous* J1 is performed near 5°C) and not many thermostable enzymes have been identified. Moreover, enantios-electivity tends to decrease with increase in reaction temperature. However, thermostable nitrile-hydrolyzing enzymes do exist, such as the nitrile hydratase discovered by Cramp and Cowan [48], and further effort should be expended in searching for them as thermostability is usually an indicator of overall stability. Alternatively, genetic engineering may be used to improve the stability of the enzymes (see below).

- *Substrate/product inhibition:* as intracellular proteins, nitrile-hydrolyzing enzymes would usually be exposed to millimolar levels of substrate. However, to be comparable to chemical processes, the biocatalytic reactions must achieve product concentrations of at least 50–100 g L^{-1} [32]. However, although there seem to be frequent examples of inhibition by reactants in the literature, some of the examples above show that the biocatalysts can function at 3 *M* reactant concentrations.

- *Organic solvents:* application of nitrilases and nitrile hydratases in organic solvents has not been extensively investigated, although they can withstand dilute alcohols that are used to solubilize hydrophobic substrates (around 15% m/m). Studies have also shown that *Pseudomonas* sp. DSM 11387 and nitrile hydratase from *Rhodococcus* sp. DSM 11397 retained activity in some organic solvents [49].

- *Oxidation (chemical denaturation):* nitrilases are well known to be oxygen-sensitive, which is commonly ascribed to autoxidation of the cysteine residues, and they typically lose most or all of their activity upon exposure to oxygen within times that would be required for a single reaction or a few catalyst recycles. This inactivation can be avoided by purging oxygen from the reaction, although this would have implications for reactor design [50]. The antioxidants mercaptoethanol and dithiothreitol may be used, but these may have unwanted side-effects [51].

  Mateo *et al.* [17a] have tried a different approach, in which they reduced the local oxygen concentration using a salting-out effect of polycationic polymers. They prepared co-aggregates of the nitrilases and high molecular weight polyethylenimine (PEI) by precipitation. The resulting PEI co-aggregates of the nitrilases were considerably more oxygen tolerant than the freely dissolved enzymes, with the nitrilase from *Pseudomonas fluorescens* EBC 191 retaining complete activity on exposure to oxygen for 40 h.

Physical Robustness   *Whole cell biocatalysts* are hardier (and cheaper) than purified enzymes and are preferred unless there are other considerations, such as membrane barrier problems. In particular, where multiple enzyme processes are concerned, it is usually better to express and retain both enzymes within the cell and this obviously applies to the nitrile hydratase–amidase system. However, late-stage biocatalytic processes using whole cells can pose a problem in that for pharmaceutical products unwanted impurities (such as nucleic acids) need to be removed [32]. Moving the whole cell biocatalytic reaction earlier can also occasionally pose problems as biological material may poison chemical catalysts used in later stages, and water has to be removed if the next synthetic step is anhydrous.

  Immobilization provides additional stability for biocatalysts and facilitates recycling, thereby reducing overall biocatalyst cost and waste in the long term if the appropriate immobilization system has been selected [40, 52].

  In the examples above, it is apparent that the DuPont researchers have opted for alginate immobilization of whole cells, whereas the Mitsubishi acrylamide process uses whole cells appropriately immobilized in polyacrylamide. Newer techniques are under investigation. Martínková's group immobilized whole cells of the nitrile biocatalyst *Rhodococcus equi* A4 using the LentiKats procedure using a copolymer of poly(vinyl alcohol) and poly(ethylene glycol) [52a]. The resultant lens-shaped hydrogel particles were successfully applied to the biotransformation of benzonitrile,

3-cyanopyridine, (R,S)-3-hydroxy-2-methylenebutanenitrile and (R,S)-3-hydroxy-2-methylene-3-phenylpropanenitrile.

Immobilization of enzymes requires preprocessing to rupture cells and at least partial purification of the enzymes, making them more expensive. However, an advantage for commercial processes is that the unit activity for enzymes can greatly exceed that of whole-cell systems [32].

CLEA technology (an enzyme self-immobilization technology that requires no physical support) has been successfully applied to the immobilization of nitrilases [50a, 52b]

2.6.2
Availability

An analysis of the literature [12, 53] shows that less than 20 nitrilases have been described. Although a few of these are commercial nitrilase preparations and have been made available (notably by Biocatalytics, now part of Codexis), they are currently too expensive for commercial manufacturing. A crude immobilized *Rhodococcus* sp. whole cell preparation (a nitrile hydratase–amidase biocatalyst) was available from Novozymes for some time (SP 409 and SP 361, based on the immobilization material), but was discontinued [2]. Diversa generated a nitrilase toolbox, but the company has since merged to form Verenium and changed its research focus.

Hence the lack of commercial preparations has prompted the discovery of new nitrile-hydrolyzing organisms [2, 40]. Microorganisms having nitrilase activity can be enriched from Nature by using nitriles as the only nitrogen and/or carbon source in the growth medium [54], and specific types of activity, based on reaction requirements, can be selected for by a wide range of screening methods [55].

Nitrile-hydrolyzing activity is commonly found in *Alcaligenes, Bacillus, Pseudomonas* and particularly *Rhodococcus* species [56, 57] and broadly found in other life forms, for example plants such as *Arabidopsis thaliana* [7], fungi [58] and yeast [56, 59]. Catalytic diversity, based on substrate preference, is high [1d, 57, 60], even within species [56, 61].

Large fractions (85–99%) of the organisms existing in Nature appear to be refractory to cultivation and hence one should not be limited by culturing techniques. Metagenomics is a good option, as it involves isolation of DNA directly from the environment and bypasses the need for microbial cultivation [20]. Researchers at Diversa applied metagenomics and robotic high-throughput selection techniques based on catalytic function to screen the clone libraries, generating 137 novel nitrilase sequences [12, 53]. The different clades had distinct preferences for specific substrates and enantiomers.

Nitrile hydratases are harder to isolate by this means as they require the presence of a suitable amidase to allow for substrate enrichment growth. However, the genes can be isolated from the environmental DNA sample by molecular biology techniques. Precigou *et al.* [20] used the polymerase chain reaction (PCR) for specific identification of nitrile hydratase (NHase)-encoding genes extracted from soil samples, using a primer design based on the highly conserved sequences coding for the metal-binding site of the α-subunit gene. Isolated genes were expressed in *E. coli*.

Finally, one need not be limited even to the wide selection available from Nature. In order to improve the enantioselectivity of their lead biocatalyst for 3-hydroxyglutar-ylnitrile at high molar concentrations, researchers at Diversa used a molecular-directed evolution technique which involved a systematic saturation mutagenesis method (GSSM) wherein each amino acid of a protein is replaced with each of the other 19 naturally occurring amino acids, providing the broadest range of structural permuta-tions. A total of 31 584 clones were evaluated for enantioselectivity. Whereas the wild-type nitrilase obtained by metagenomics gave an *ee* value of 87.8%, some mutants displayed improved *ee* values between 95.4 and 98.1% towards 3-hydroxyglutarylnitrile to afford (*R*)-4-cyano-3-hydroxybutyric acid, providing a basis for a commercially relevant process [53a] (Section 2.5.1.3). The 10% improvement in *ee* at 3 *M* concentra-tion substrate was achieved through substitution of an Ala codon for a His codon.

The genes, natural or mutated, for selected nitrile- and amide-hydrolyzing enzymes can be isolated and over-expressed in a suitable protein production host. The molecular biology revolution of the last couple of decades has made it possible to produce high-purity enzymes at low cost, to the point where the biocatalyst is no longer a major contributor to the techno-economics of the process.

2.7
Conclusion

Nitrile biocatalysts can provide environmentally attractive processes in three ways: by negating the need for a high-salt waste stream, by reducing the number of overall process steps and increasing the atom efficiency of the process by removing the need for temporary protecting groups and practically eliminating byproduct formation. This is achieved primarily through their selectivity. Their high activity can also decrease waste by increasing volumetric productivity.

The technology opens up the opportunity for following new synthetic routes for existing compounds that were not possible previously. Schulze [29] depicted the reduction in four steps required for the synthesis of acrylamide by replacement of the typical reaction with a biocatalyst, whereas the enantioselective synthesis of β-proline by commercially available nitrilases could be considerably reduced to only four steps [62].

Moreover, due to their chemoselectivity, there is more flexibility as to where the target reaction occurs in the synthetic sequence [32]. They can also provide the opportunity to synthesize complex molecules that were previously economically or technically impractical to manufacture, particularly those with multiple stereogenic centers [32].

The enzymes are increasingly available and at ever-decreasing prices as protein expression technology and fermentation protocols improve. The selection is also increasing through metagenomics and directed evolution, allowing the synthesis of 'designer' catalysts fit for purpose for a particular reaction.

Nitrile biocatalysis provides a significant percentage of all commercial bio-catalytic reactions [2]. However, in spite of the great synthetic potential of nitrilases, their utilization as a versatile biocatalyst is underexploited. Now that the biocatalytic toolbox is being expanded beyond amylase, esterase and protease to include nitrile hydrolysis for green chemistry [63], we can expect to see a broader application of these enzymes. The greatest opportunity for these nitrile biocatalysts resides in the pharmaceutical industry, where the *E* factor ranges from 25 to >100 with $10–10^3$ kg of waste generated per kilogram of product [10]. It will be here that the regio- and stereoselective properties of biocatalysts will allow the circumvention of multiple protection and deprotection steps typical of difficult syntheses [32] and no doubt thereby provide the basis for many new processes.

Abbreviations

dcw	dry cell weight
ee	enantiomeric excess.
GA/PEI	glutaraldehyde/polyethylenimine (crosslinking reagent)
NHase	nitrile hydratase
PAN	polyacrylonitrile
PCR	polymerase chain reaction

Acknowledgments

The author would like to thank Joni Frederick, Varsha Chhiba, Dr Henok Kinfe, Kgama Mathiba, Alison Beeton, Nosisa Dube, Dr Mapitso Molefe, Dr Neeresh Rohitlall and Dr Paul Steenkamp at CSIR Biosciences, and also Professor Roger Sheldon, Dr Fred van Rantwijk and Dr Luuk van Langen at TU Delft. Dr Colin Kenyon and Uli Horn (CSIR Biosciences) kindly rendered the nitrile hydratase structure using Accelrys software. ISIS DRAW (MDL) was used to for the chemical structures.

References

1 (a) Banerjee, A., Sharma, R. and Banerjee, U.C. (2002) *Applied Microbiology and Biotechnology*, **60**, 33–44; (b) Benz, P., Muntwyler, R. and Wohlgemuth, R. (2007) *Journal of Chemical Technology and Biotechnology*, **82**, 1087–1098; (c) March, J. (ed.) (1992) *Advanced Organic Chemistry*, 4th edn. Wiley–Interscience, New York; (d) Martínková, L. and Křen, V. (2002) *Biocatalysis and Biotransformation*, **20**, 73–93; (e) Pollock, J.A., Clark, K.M., Martynowicz, B.J., Pridgeon, M.G., Rycenga, M.J., Stolle, K.E. and Taylor, S.K. (2007) *Tetrahedron: Asymmetry*, **18**, 1888–1892.

2 Faber, K. (1997) *Pure and Applied Chemistry*, **69**, 1613–1632.

3 Rey, P., Rossi, J.-C., Taillades, J., Gros, G. and Nore, O. (2004) *Journal of Agricultural and Food Chemistry*, **52**, 8155–8162.

4 Effenberger, F. and Osswald, S. (2001) *Tetrahedron: Asymmetry*, **12**, 2581–2587

5 Brady, D., Beeton, A., Kgaje, C., Zeevaart, J., van Rantwijk, F. and Sheldon, R.A. (2004) *Applied Microbiology and Biotechnology*, **64**, 76–85.

6 Kamal, A. and Khanna, G.B. R. (2001) *Tetrahedron Asymmetry*, **12**, 405–410.

7 (a) Osswald, S., Wajant, H. and Effenberger, F. (2002) *European Journal of Biochemistry/FEBS*, **269**, 680–687; (b) Pace, H.C. and Brenner, C. (2001) *Genome Biology*, **2**, 1.1–1.9.

8 Clark, J.H. (1999) *Green Chemistry*, **1**, 1–8.

9 Anastas, P.T. and Kirchhoff, M.M. (2002) *Accounts of Chemical Research*, **35**, 686–694.

10 Sheldon, R.A. (2007) *Green Chemistry*, **9**, 1261–1384.

11 Brenner, C. (2002) *Current Opinion in Structural Biology*, **12**, 775–782

12 Robertson, D.E., Chaplin, J.A., DeSantis, G., Podar, M., Madden, M., Chi, E., Richardson, T., Milan, A., Miller, M., Weiner, D.P., Wong, K., McQuaid, J., Farwell, B., Preston, L.A., Tan, X., Snead, M.A., Keller, M., Mathur, E., Kretz, P.L., Burk, M.J. and Short, J.M. (2004) *Applied and Environmental Microbiology*, **70**, 2429–2436.

13 (a) Kobayashi, M., Komeda, H., Yanaka, N., Nagasawa, T. and Yamada, H. (1992) *The Journal of Biological Chemistry*, **267**, 20746–20751; (b) Kobayashi, M., Yanaka, N., Nagasawa, T. and Yamada, H. (1992) *Biochemistry*, **31**, 9000–9007.

14 Kobayashi, M., Goda, M. and Shimizu, S. (1998) *Biochemical and Biophysical Research Communications*, **253**, 662–666.

15 (a) Kobayashi, M., Yanaka, N., Nagasawa, T. and Yamada, H. (1990) *Journal of Bacteriology*, **172**, 4807–4815; (b) Bhalla, T.C., Miura, A., Wakamoto, A., Ohba, Y. and Furuhashi, K. (1992) *Applied Microbiology and Biotechnology*, **37**, 184–190

16 Harper, D.B. (1977) *The Biochemical Journal*, **165**, 309–319.

17 Nagasawa, T., Wieser, M., Nakamura, T., Iwahara, H., Yoshida, T. and Gekko, K. (2000) *European Journal of Biochemistry/FEBS*, **267**, 138–144.

18 Thuku, R.N., Weber, B.W., Varsani, A. and Sewell, B.T. (2007) *FEBS Journal*, **274**, 2099–2108.

19 (a) Kobayashi, M. and Shimizu, S. (1998) *Nature Biotechnology*, **16**, 733–736; (b) Miyanaga, A., Fushinobu, S., Ito, K., Shoun, H. and Wakagi, T. (2004) *European Journal of Biochemistry/FEBS*, **271**, 429–438; (c) Okamoto, S. and Eltis, L.D. (2007) *Molecular Microbiology*, **65**, 828–838. (d) Scarrow, R.C., Brennan, B.A., Cummings, J.G., Jin, H., Duong, D.J., Kindt, J.T. and Nelson, M.J. (1996) *Biochemistry*, **35**, 10078–10088. (e) Nelson, M.J., Jin, H., Turner, I.M.J., Grove, G., Scarrow, R.C., Brennan, B.A. and Que, L. (1991) *Journal of the American Chemical Society*, **113**, 7072–7073; (f) Sugiura, Y., Kuwahara, J., Nagasawa, T. and Yamada, H. (1997) *Journal of the American Chemical Society*, **109**, 5848–5850; (g) Jin, H., Turner, I.M. J., Nelson, M.J., Gurbiel, R.J., Doan, P.E. and Hoffman, B. (1993) *Journal of the American Chemical Society*, **115**, 5290–5291.

20 Precigou, S., Goulas, P. and Duran, R. (2001) *FEMS Microbiology Letters*, **204**, 155–161.

21 Mitra, S. and and Holtz, R.C. (2007) *The Journal of Biological Chemistry*, **282**, 7397–7404.

22 (a) Okada, M., Noguchi, T., Nagashima, T., Yohda, M., Yabuki, S., Hoshino, M., Inoue, Y. and Endo, I. (1996) *Biochemical and Biophysical Research Communications*, **221**, 146–150; (b) Endo, I. (1994) *Biochemistry*, **33**, 3577–3583; (c) Endo, I. and Odaka, M. (2000) *Journal of Molecular Catalysis B-ENZYMATIC*, **10**, 81–86; (d) Endo, I., Odaka, M. and Yohda, M. (1999) *Trends in Biotechnology*, **17**, 244–249; (e) Popescu, V.-C., Münck, E., Fox, B.G., Sanakis, Y., Cummings, J.G., Turner, I.M. J. and Nelson, M.J. (2001) *Biochemistry*, **40**, 7984–7991.

23 (a) Komeda, H., Kobayashi, M. and Shimizu, S. (1996) *The Journal of Biological*

Chemistry, **271**, 15796–15802; (b) Komeda, H., Kobayashi, M. and Shimizu, S. (1996) *Proceedings of the National Academy of Sciences of the United States of America*, **93**, 4267–4272; (c) Lu, J., Zheng, Y., Yamagishi, H., Odaka, M., Tsujimura, M., Maeda, M. and Endo, I. (2003) *FEBS Letters*, **553**, 391–396; (d) Komeda, H., Kobayashi, M. and Shimizu, S. (1996) *The Journal of Biological Chemistry*, **271**, 15796–15802.

24 Beard, T.M. and Page, M.I. (1998) *Antonie Van Leeuwenhoek*, **74**, 99–106

25 (a) Payne, M.S., Wu, S., Fallon, R.D., Tudor, G., Stieglitz, B., Turner Jr., I.M. and Nelson, M.J. (1997) *Biochemistry*, **36**, 5447–5454; (b) Wegman, M.A., Heinemann, U., Stolz, A., van Rantwijk, F. and Sheldon, R.A. (2000) *Organic Process Research & Development*, **4**, 318–322; (c) Blakey, A.J., Colby, J., Williams, E. and O'Reilly, C. (1995) *FEMS Microbiology Letters*, **129**, 57–61.

26 Yokoyama, M., Sugai, T. and Ohta, H. (1993) *Tetrahedron: Asymmetry*, **4**, 1081–1084.

27 Novo, C., Farnaud, S., Tata, R., Clemente, A. and Brown, P.R. (2002) *The Biochemical Journal*, **365**, 731–738.

28 Makhongela, H.S., Glowacka, A.E., Agarkar, V.B., Sewell, B.T., Weber, B., Cameron, R.A., Cowan, D.A. and Burton, S.G. (2007) *Applied Microbiology and Biotechnology*, **75**, 801–811.

29 Schulze, B. (2002) in *Enzyme Catalysis in Organic Synthesis* (ed. K. Drauz and H. Waldmann), 2nd edn, Wiley-VCH Verlag GmbH, Vol. 2, pp. 699–715.

30 Wieser, M. and Nagasawa, T. (2000) in *Stereoselective Biocatalysis* (ed. R.N. Patel), Marcel Dekker Inc., New York, pp. 467–486.

31 Vejvoda, V., Kaplan, O., Přikrylová, V., Elišáková, V., Himl, M., Kubáč, D., Pelantová, H., Kuzma, M., Křen, V. and Martínková, L. (2007) *Biotechnology Letters*, **29**, 1119–1124.

32 (a) Pollard, D.J. and Woodley, J.M. (2007) *Trends in Biotechnology*, **25**, 66–73; (b) Rouchi, M.A. (2002) *Chemical & Engineering NEWS*, **80**, 43–50.

33 (a) Schmid, A., Dordick, J.S., Hauer, B., Kiener, A., Wubbolts, M. and Witholt, B. (2001) *Nature*, **409**, 258–268; (b) Zaks, A. (2001) *Current Opinion in Chemical Biology*, **2**, 130–136; (c) Liese, A., Seelbach, K. and Wandrey, C. (2000) *Industrial Biotransformations*, Wiley-VCH Verlag GmbH, Weinheim; (d) Yamada, H. and Kobayashi, M. (1996) *Bioscience, Biotechnology, and Biochemistry*, **60**, 1391–1400; (e) Kobayashi, M., Nagasawa, T. and Yamada, H. (1992) *Trends in Biotechnology*, **10**, 402–408.

34 Nagasawa, T. and Yamada, H. (1990) *Pure and Applied Chemistry*, **62**, 1441–1444.

35 (a) Chuck, R. (2005) *Applied Catalysis A-General*, **280**, 75–82; (b) Petersen, M. and Kiener, A. (1999) *Green Chemistry*, **4**, 99–106.

36 Sharma, N.N., Sharma, M., Kumar, H. and Bhalla, T.C. (2006) *Proc. Biochem.*, **41**, 2078–2081.

37 Müller, M. (2005) *Angewandte Chemie (International Edition in English)*, **44**, 362–365.

38 Bergeron, S., Chaplin, D.A., Edwards, J.H., Ellis, B.S. W., Hill, C.L., Holt-Tiffin, K., Knight, J.R., Mahoney, T., Osborne, A.P. and Ruecroft, G. (2006) *Organic Process Research & Development*, **10**, 661–665.

39 Hann, E.C., Eisenberg, A., Fager, S.K., Perkins, N.E., Gallagher, F.G., Cooper, S.M., Gavagan, J.E., Stieglitz, B., Hennessey, S.M. and DiCosimo, R. (1999) *Bioorganic and Medicinal Chemistry*, **7**, 2239–2245.

40 Kaul, P., Banerjee, A. and Banerjee, U.C. (2006) *Biomacromolecules*, **7**, 1536–1541.

41 Yamamoto, K., Oishi, K., Fujimatsu, I. and Komatsu, K.-I. (1992) *Applied and Environmental Microbiology*, **57**, 3028–3032.

42 Gröger, H. (2001) *Advanced Synthesis and Catalysis*, **343**, 6–7.

43 Hann, E.C., Sigmund, A.E., Fager, S.K., Cooling, F.B., Gavagan, J.E., Bramucci, M.G., Chauhan, S., Payne, M.S. and DiCosimo, R. (2004) *Tetrahedron*, **60**, 577–581.

44 (a) Hann, E.C., Sigmund, A.E., Hennessey, S.M., Gavagan, J.E., Short, D.R., Ben-Bassat, A., Chauhan, S., Fallon, R.D., Payne, M.S. and DiCosimo, R. (2002) *Organic Process Research & Development*, **6**, 492–496; (b) Gavagan, J.E., Fager, S.K., Fallon, R.D., Folsom, P.W., Herkes, F.E., Eisenberg, A., Hann, E.C. and DiCosimo, R. (1998) *The Journal of Organic Chemistry*, **63**, 4792–4801; (c) Fallon, R.D., DiCosimo, R., Gavagan, J.E. and Herkes, F.E. (2004). European Patent 1449830; (d) Cooling, F.B., Fager, S.K., Fallon, R.D., Folsom, P.W., Gallagher, F.G., Gavagan, J.E., Hann, E.C., Herkes, F.E., Phillips, R.L., Sigmund, A., Wagner, L.W., Wu, W. and DiCosimo, R. (2001) *Journal of Molecular Catalysis B-Enzymatic*, **11**, 295–306; (e) Gavagan, Hann, E.C., Herkes, F.E., Phillips, R.L., Sigmund, A., Wagner, L.W., Wu, W. and DiCosimo, R. (2001) *Journal of Molecular Catalysis B-Enzymatic*, **11**, 295–306.

45 Hann, E.C., Sigmund, A.E., Fager, S.K., Cooling, F.B., Gavagan, J.E., Ben-Bassat, A., Chauhan, S., Payne, M.S., Hennessey, S.M. and DiCosimo, R. (2003) *Advanced Synthesis and Catalysis*, **345**, 775–782.

46 (a) Fischer-Colbrie, G., Herrmann, M., Heumann, S., Puolakka, A., Wirth, A., Cavaco-Paulo, A. and Guebitz, G.M. (2006) *Biocatalysis and Biotransformation*, **24**, 419–425; (b) Gübitz, G.M. and Cavaco Paulo, A. (2003) *Current Opinion in Biotechnology*, **14**, 577–582; (c) Wang, M.-X., Feng, G.-Q. and Zheng, Q.-Y. (2004) *Tetrahedron: Asymmetry*, **15**, 347–354; (d) Tauber, M.M., Cavaco-Paulo, A., Robra, K.-H. and Gübitz, G.M. (2000) *Applied and Environmental Microbiology*, **66**, 1634–1638; (e) Battistel, E., Morra, M. and Marinetti, M. (2001) *Applied Surface Science*, **177**, 32–41.

47 Hjort, C.M., Godtferedson, S.E. and Emborg, C. (1990) *Journal of Chemical Technology and Biotechnology*, **48**, 217–226.

48 Cramp, A. and Cowan, D.A. (1999) *Biochimica et Biophysica Acta*, **1431**, 249–260.

49 Lyah, N. and Willets, A. (1998) *Biotechnology Letters*, **20**, 329–331.

50 (a) Mateo, C., Fernandes, B., van Rantwijk, F., Stolz, A. and Sheldon, R.A. (2006) *Journal of Molecular Catalysis B-Enzymatic*, **38**, 154–157; (b) Mustacchi, R., Knowles, C.J., Li, H., Dalrymple, I., Sunderland, G., Skibar, W. and Jackman, S.A. (2005) *Biotechnology and Bioengineering*, **91**, 436–440.

51 (a) Winkler, M., Glieder, A. and Klempier, N. (2006) *Chemical Communications* 1298–1300; (b) Nagasawa, T., Shimizu, S. and Yamada, H. (1993) *Applied Microbiology and Biotechnology*, **40**, 189–195.

52 (a) Kubáč, D., Čejková, A., Masák, J., Jirků, V., Lemaire, M., Gallienne, E., Bolte, J., Stloukal, R. and Martínková, L. (2006) *Journal of Molecular Catalysis B-Enzymatic*, **39**, 59–61; (b) Sheldon, R.A., Schoevaart, R. and van Langen, L.M. (2005), *Biocatalysis and Biotransformation*, **23**, 141–147.

53 (a) DeSantis, G., Wong, K., Farwell, B., Chatman, K., Zhu, Z., Tomlinson, G., Huang, H., Tan, X., Bibbs, L., Chen, P., Kretz, K. and Burk, M.J. (2003) *Journal of the American Chemical Society*, **125**, 11476–11477; (b) DeSantis, G., Zhu, Z., Greenberg, W.A., Wong, K., Chaplin, J., Hanson, S.R., Farwell, B., Nicholson, L.W., Rand, C.L., Weiner, D.P., Robertson, D.E. and Burk, M.J. (2002) *Journal of the American Chemical Society*, **124**, 9024–9025.

54 Layh, N., Hirrlinger, B., Stolz, A. and Knackmuss, H.-J. (1997) *Applied Microbiology and Biotechnology*, **47**, 668–674.

55 Martínková, L., Vejvoda, V. and Křen, V., (2008) *Journal of Biotechnology*, **133**, 318–326.

56 Brady, D., Dube, N. and Petersen, R. (2006) *South African Journal of Science*, **102**, 339–344.

57 Demakov, V.A., Maksimov, A.Y., Kuznetsova, M.V., Ovechkina, G.V., Remezovskaya, N.B. and Maksimova, Y.G.

(2007) *Russian Journal of Ecology*, **38**, 168–173.

58 Vejvoda, V., Kaplan, O., Kubáč, D., Křen, V. and Martínková, L. (2006) *Biocatalysis and Biotransformation*, **24**, 414–418.

59 Rezende, R.P., Dias, J.C. T., Rosa, C.A., Carazza, F. and Linardi, V.R. (1999) *The Journal of General and Applied Microbiology*, **45**, 185–192.

60 Heald, S.C., Brandao, P.F., Hardicre, R. and Bull, A.T. (2001) *Antonie Van Leeuwenhoek*, **80**, 169–183.

61 Brandão, P.F.B., Clapp, J.P. and Bull, A.T. (2002) *Environmental Microbiology*, **4**, 262–276.

62 Winkler, M., Meischler, D. and Klempier, N. (2007) *Advanced Synthesis and Catalysis*, **349**, 1475–1480.

63 (a) Singh, R., Sharma, R., Tewari, N., Geetanjali and Rawat, D. (2006) *Chemistry & Biodiversity*, **3**, 1279–1287; (b) Straathof, A.J.J., Panke, S. and Schmid, A. (2002) *Current Opinion in Biotechnology*, **13**, 548–556.

3
Biocatalytic Processes Using Ionic Liquids and Supercritical Carbon Dioxide

Pedro Lozano, Teresa De Diego, and José L. Iborra

3.1
Introduction

The use of safer solvents and reaction conditions and the prevention of waste are two of the 12 principles of green chemistry, which encourage the use of environmentally benign non-aqueous solvents and efficient catalysts for chemical reactions and/or processes [1]. Currently, ionic liquids (ILs) and supercritical fluids (SCFs) are the non-aqueous green solvents receiving most attention worldwide. Ionic liquids are a new class of liquid solvents that have led to a new green chemical revolution, because of their unique array of physico-chemical properties which make them suitable for numerous industrial applications. SCFs, especially supercritical carbon dioxide ($scCO_2$), are another class of useful solvents with unique properties that can be applied in green reaction, extraction and fractionation processes [2].

However, the goal of green chemistry is much more than simply replacing hazardous solvents with environmentally benign ones. The efficiency of catalyzed processes in reducing undesired reactions and/or by-products and in facilitating the recovery of products is just as important. Enzymes are catalytic proteins obtained from living systems; they show a high level of activity and selectivity (stereo-, chemo- and regio-) towards catalyzed reactions using a wide variety of substrates/chemicals and their application in chemical processes greatly improves the efficiency. Enzymatic catalysis in non-aqueous environments significantly broadens the potential of conventional aqueous-based biocatalysis. Water is a poor solvent for nearly all applications in industrial chemistry since most organic compounds of commercial interest are poorly soluble and are sometimes unstable in aqueous solution. Enzyme catalysis in non-aqueous reaction media offers several advantages, such as higher solubility for hydrophobic substrates, the enzyme insolubility that easily allows their reuse, the shifting of thermodynamic equilibrium to favor synthetic reactions over hydrolysis and the elimination of microbial contamination in reactors [3]. In this context of non-aqueous environments, the use of both IL and $scCO_2$ neoteric solvents has enhanced the potential of enzymes because of the improvements to their catalytic

Handbook of Green Chemistry, Volume 3: Biocatalysis. Edited by Robert H. Crabtree
Copyright © 2009 WILEY-VCH Verlag GmbH & Co. KGaA, Weinheim
ISBN: 978-3-527-32498-9

properties and operational stability [4]. The combination of these neoteric solvents with enzymes may represent the most important arsenal of green tools to develop integral clean chemical processes of industrial interest in the near future.

3.2
Biocatalytic Processes in Ionic Liquids

Since 2000, ionic liquids have emerged as new and attractive non-aqueous reaction media for biocatalysis because of their unique solvent properties (headed by their negligible vapor pressure) and their exceptional ability to maintain enzymes in active and stable conformations, which sometimes leads to remarkable results as regards catalytic performance; pioneering studies on the use of ILs in enzymatic catalysis [5] and reviews on biocatalysis [6] have been published.

3.2.1
Solvent Properties of ILs for Biocatalysis

The term ionic liquids (ILs), also named room temperature ionic liquids (RTILs), molten salts, liquid organic salts or fused salts, is applied to a class of substances that are completely composed of ions and which are liquid at temperatures close to room temperature. Typical ILs are based on organic cations, for example 1-alkyl-3-methylimidazolium, *N*-alkylpyridinium and tetraalkylammonium, paired with a variety of anions that have a strongly delocalized negative charge (for example $[BF_4]$, $[PF_6]$, $[NTf_2]$) (Figure 3.1), resulting in colorless, low-viscosity and easily handled materials with very interesting solvent properties. ILs have several advantages over conventional molecular organic solvents, which makes them suitable for use as environmentally benign non-aqueous solvents. These advantages include a very low vapor pressure, meaning that they do not evaporate, high thermal stability (up to 300 °C), a great ability to dissolve many different organic and inorganic substances, including gases (for example H_2, CO_2), non-flammable nature, high conductivity and a large electrochemical window [7]. Additionally, the phase behavior of ILs with classical molecular solvents, such as organic solvents and water, is interesting because ILs can be fully miscible, partially miscible or non-miscible with them as a function of the ions involved (Table 3.1).

Since ILs are usually composed of poorly coordinating ions, which makes them highly polar but non-coordinating solvents, they are non-miscible with most hydrophobic organic solvents (for example hexane, benzene, octane), thus providing a non-aqueous and polar medium to develop two-phase systems. Also, they may be fully or partially miscible with polar organic solvents, such as ethanol, acetonitrile and acetone, depending on the alkyl chain length of the cation. In the same way, most ILs are non-miscible with water and so can be used to develop biphasic systems with polar characteristics; there are databases of ILs that provide information about molecular formula, structure, melting point, density, viscosity, miscibility and so on [8]. The use of liquid biphasic systems

CATIONS:

(a)

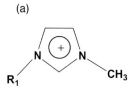

$R_1 = C_2$ to C_8

1-Alkyl-3-methylimidazolium

(b)

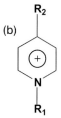

$R_1 = C_2$ to C_4
$R_2 = H$ or C_1

1,4-Dialkylpyridinium

(c)

$R_1, R_2, R_3, R_4 = C_1$ to C_8

Tetraalkylammonium

R_1	Abbrev.
C_2	[EMim]
C_3	[PMim]
C_4	[BMim]
C_6	[HMim]
C_8	[OMim]

R_1	R_2	Abbrev.
C_2	H	[EPy]
C_3	H	[PPy]
C_4	H	[BPy]
C_4	C_1	[BMPy]

R_1	R_2	R_3	R_4	Abbrev.
C_2	C_2	C_2	C_1	[E$_3$MeN]
C_4	C_1	C_1	C_1	[BMe$_3$N]
C_6	C_1	C_1	C_1	[HMe$_3$N]
C_8	C_8	C_8	C_1	[O$_3$MeN]

ANIONS:

Anion	Full name	Abbreviation
PF_6^-	Hexafluorophosphate	[PF$_6$]
$(CF_3SO_2)_2N^-$	Bis[(trifluoromethyl)sulfonyl]imide	[NTf$_2$]
BF_4^-	Tetrafluoroborate	[BF$_4$]
CH_3SO_4	Methylsulfate	[MeSO$_4$]
$CH_3CO_2^-$	Acetate	[Ac]
NO_3^-	Nitrate	[NO$_3$]
$CF_3CO_2^-$	Trifluoroacetate	[TFA]
$CF_3SO_3^-$	Trifluoromethylsulfonate	[TfO]

Figure 3.1 Typical structures, names and abbreviated names of cations and anions involved in ILs usually applied in biocatalysis.

(for example organic solvent–IL or water–IL) for product recovery by extraction is the most usual procedure followed (see Table 3.1). However, many organic solvents used in extractions are known for their volatile, flammable and toxic properties, which clearly go against the ideal of greenness in a given process. Also, water in ionic liquids based on the [PF$_6$] anion results in the release of hydrogen fluoride at high temperatures (>100 C) [9]. An important parameter that characterizes ILs is their polarity, which has been studied by using several solvatochromic and fluorescent probes (for example Nile Red and Reichardt's betaine dye). For ILs based on 1-alkyl-3-methylimidazolium as cation, the polarity decreases slightly through the series [NO$_3$] > [BF$_4$] > [NTf$_2$] > [PF$_6$], although all ILs are

Table 3.1 Miscibility of different ILs with molecular solvents[a][7].

| Cation | Anion | Miscibility | | | | |
		H_2O	AcN	iPrOH	Hexane	Toluene
[EMim]	[BF$_4$]	+	+	−	+/−	−
[BMim]	[BF$_4$]	+	+	−	−	−
[HMim]	[BF$_4$]	+/−	+/−	+	−	−
[EMim]	[PF$_6$]	+	+	−	−	−
[BMim]	[PF$_6$]	−	+	−	−	−
[HMim]	[PF$_6$]	−	+	−	−	−
[BMim]	[NTf$_2$]	−	+	+	+/−	−
[HMim]	[NTf$_2$]	−	+	+	−	−
[OMim]	[NTf$_2$]	−	+	−	−	−
[O$_3$MeN]	[TfO]	−	+	+	−	+
[O$_3$MeN]	[NTf$_2$]	−	+	+	−	+/−
[O$_3$MeN]	[TFA]	+/−	+	+	+	+

[a] +, Totally miscible; +/−, partially miscible; −, immiscible. AcN, acetonitrile; iPrOH, 2-propanol. For IL abbreviations, see Figure 3.1.

in the same polarity region as 2-aminoethanol and lower than alcohols such as methanol, ethanol and butanol; for reviews on solvent properties of ILs, see [10]. Although ILs could be considered as polar solvents, they cannot be considered to be similar to those polar molecular solvents as regards biocatalysis. The effect of ILs on enzyme activity is not only determined by the overall solvent properties, because of the contribution of individual ions in their interactions with water molecules. Water is the key component of biocatalytic reaction media, because of the importance of enzyme-water interactions for the maintenance of the active conformation of the enzyme.

3.2.2
Enzymes in ILs

In spite of the many different kinds of enzymes, only hydrolases, dehydrogenases and oxidases (for example peroxidase, laccase and glucose oxidase) have been reported to be active in ILs [4–6]. As a function of the miscibility of water in ILs, two different types of reaction media have been studied: aqueous solutions of ILs and ILs in nearly anhydrous conditions.

Although aqueous solutions of ILs are outside the concept of ILs (liquids composed entirely of ions), the use of aqueous reaction media for enzyme-catalyzed reactions of industrial interest has mainly focused on enhancing the solubility of substrates and/or products with hydrophobic moieties (for example amino acid derivatives), and also on dehydrogenase- or oxidase-catalyzed redox reactions for hydrophobic substrates. For example, in the oxidation of anthracene catalyzed by laccase C and assisted by

1-hydroxybenzotriazole as mediator, the addition of 25% v/v [MBPy] instead of *tert*-butanol increased the yield of the oxidation product 15-fold [11].

In aqueous solution of ILs, the kind of IL and the reaction medium concentration are very important parameters, because of the enzyme deactivation that can easily be occur. For example, the formate dehydrogenase from *Candida boidinii* was observed to be totally inactive in aqueous buffer–[E_3NH][$MeSO_4$] mixtures at all the concentrations tried, whereas it maintained 55% residual activity when a 50% v/v aqueous buffer–[E_3MeN][$MeSO_4$] mixture was used, although it became inactive again at 75% v/v [12]. In the same context, the resolution of several *N*-acetylamino acid esters catalyzed by subtilisin was enhanced at 15% v/v [EPy][TFA] in water, whereas the enzyme activity was drastically reduced at higher IL concentrations [13] (Scheme 3.1). A similar behavior was observed for other enzymes, such as chloroperoxidase, hydroxynitrile lyase and β-galactosidase, which was attributed to the ability of ions to interact strongly with proteins, producing deactivation through the unfolding of the 3-D structure [6b].

The most interesting biotransformations in ILs were observed at low water content or in nearly anhydrous conditions because of the excellent performance of hydrolases in carrying out synthetic reactions, and also the possibility of designing two-phase reaction systems that easily permit product recovery. At low water content (<2% v/v), all the assayed water-immiscible ILs (for example [BMim][PF_6] and [BM_3N][NTf_2]) were shown to be suitable reaction media for biocatalytic reactions [4], whereas most of the ILs miscible with water clearly acted as enzyme deactivating agents (for example [BMim][Cl] and [BMim][lactate]); for examples of enzyme deactivation by water-miscible ILs, see [14]. The behavior of enzymes in non-aqueous media is strictly related to the degree of hydration of the protein and the ability of some water-miscible ILs to strip enzymes from essential water molecules, leading to deactivation by unfolding.

Several types of biocatalyst preparations (for example, aqueous solutions of enzymes, lyophilized free enzymes, crosslinked enzyme aggregates and immobilized enzymes) have been used for biotransformations in water-immiscible ILs at low water content (<2% v/v) (Figure 3.2). In these conditions, a large variety of biotransformations have been recorded for the case of lipases, involving, in the main, direct esterification, transesterification, enantioselective transesterification, enantioselective hydrolysis, enantioselective acylation of amines and sugars, ammoniolysis and perhydrolysis, reactions (Table 3.2) [4–6]. The catalytic activity shown by enzymes (for example lipases and proteases) is of the same order as or higher than that observed in conventional organic solvents. Furthermore, many examples can be found where both the stereoselectivity (for examples of enzyme deactivation by water-miscible ILs,

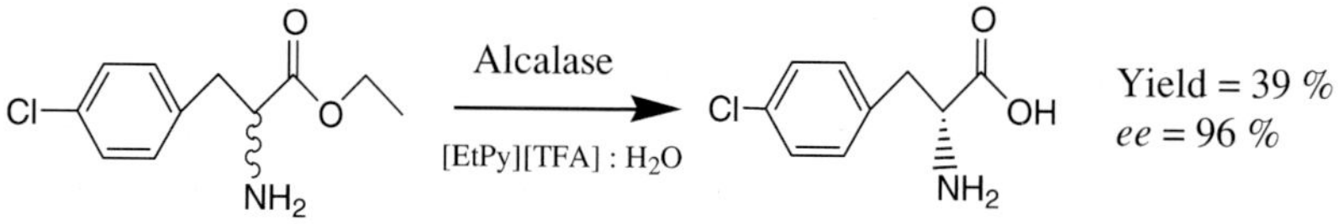

Scheme 3.1 Subtilisin-catalyzed resolution of 4-chlorophenylalanine ethyl ester using a 15% v/v [EtPy][TFA] solution in water as reaction medium [12].

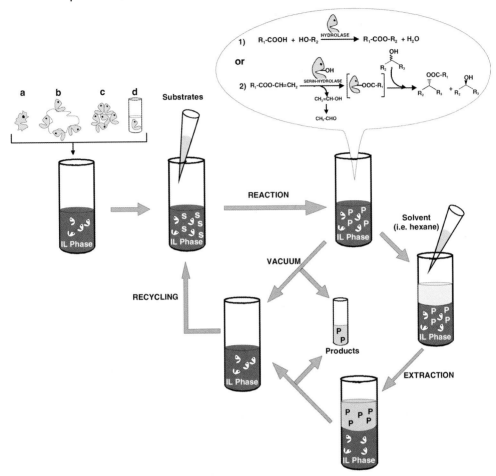

Figure 3.2 General representation of hydrolase-catalyzed direct esterification (1) or stereoselective transesterification (2) process, in monophasic ILs systems, including both the product separation and the recycling of the enzyme–IL system steps. (a) Lyophilized enzymes; (b) immobilized enzymes; (c) crosslinked enzyme aggregates (CLEAs); (d) aqueous solution of enzymes.

see [15]) and the stability towards reuse (for examples of enzyme stabilization by ILs, see [16]) displayed by enzymes were largely enhanced with respect to the same properties in organic solvents. In addition, the unique solvent properties of ILs, such as their low vapor pressure and their ability to form biphasic systems with molecular solvents, have been applied to recovering products by vacuum pressure [17] or liquid–liquid extraction [18], respectively. In this way, the resulting enzyme–IL system can be reused for a new operation cycle. At this point, it is necessary to point out that carrier-free enzyme molecules suspended in ILs behave as anchored or immobilized biocatalysts, because they cannot be separated by extraction with water or aqueous

Table 3.2 Some examples of enzyme-catalyzed reactions in nearly-dry ILs [4, 5].

IL	Enzyme	Reaction
[BMim][PF$_6$]	Thermolysin	Synthesis of (Z)-aspartame
[BMim][PF$_6$]	Lipase from	Synthesis of different alkyl esters by transesterification
[OMim][PF$_6$]	C. antarctica	Resolution of several sec-alcohols (e.g. 1-phenylethanol, 2-pentanol) with different acyl donors
[EMim][BF$_4$]	R. miehei	
[BMim][BF$_4$]	P. species	Resolution of chiral amines
[HMim][BF$_4$]	P. putida	Synthesis of geranyl acetate
[EMim][NTf$_2$]	C. rugosa	Synthesis of monoacylglycerides by glycerolysis from vegetable oils
[BMim][NTf$_2$]	P. cepacia	Synthesis of polyesters
[BuMe$_3$N][NTf$_2$]	T. lanuginosus	Synthesis of fatty acid esters of glucose and L-ascorbic acid
[HxMe$_3$N][NTf$_2$]		Synthesis of methyl glucosides and galactosides
[Oc$_3$MeN][NTf$_2$]		Synthesis of flavonoid esters
[BMMim][TfO]		Enantioselective acylation of methyl mandelate
[BMPy][BF$_4$]		Polymerization of ε-caprolactone
[BMim][PF$_6$]	Chymotrypsin	Transesterifications of L-tyrosine methyl ester
[EMim][NTf$_2$]		
[BMim][PF$_6$]	Penicillin G amidase	Acylation of S-phenylglycine
[OMim][PF$_6$]		
[BMim][BF$_4$]		
BMim][BF$_4$]	Epoxide hydrolases	Stereoselective hydrolysis of epoxides
[BMim][PF$_6$]		
[BMim][NTf$_2$]		
[BMim][PF$_6$]	Morphine dehydrogenase	Oxidation of codeine to codeinone
[BMMim][PF$_6$]		
[HMim][PF$_6$]		

buffers. ILs form a strong matrix that includes rather than dissolves enzyme molecules, meaning that they acts as a liquid immobilization support rather than as a solvent [19]. To eliminate enzyme molecules from the IL matrix, it is necessary to filter the mixture through membranes with cut-off points lower than the molecular weight of the protein [12]; for practical use of enzymes in ILs, see [20].

To illustrate the phase behavior of ILs, examples of biocatalytic processes in IL-water or IL-organic solvent biphasic systems are presented for the enantioselective reduction of 2-octanone by coupling two enzymatic reactions [21] and the synthesis of citronellyl esters [22]. For the first example (Figure 3.3), an NADP$^+$-dependent alcohol dehydrogenase, which resides in the aqueous phase of a buffer–[BMim][NTf$_2$] biphasic system, was able to catalyze the synthesis of (R)-2-octanol from 2-octanone (98% yield, 99% ee). This enzymatic reaction was coupled with the oxidation of 2-propanol to acetone for the continuous regeneration of the expensive cofactor NADPH + H$^+$/NADP$^+$ catalyzed by the same enzyme. In this example, the IL

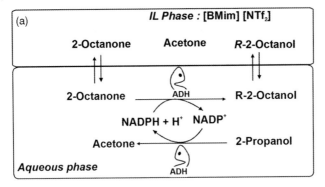

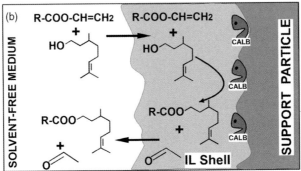

Figure 3.3 (a) Schematic diagram of the asymmetric reduction of 2-octanone catalyzed by an alcohol dehydrogenase (ADH), coupled to NADPH regeneration with 2-propanol, in IL–aqueous buffer biphasic systems [21]. (b) Scheme of immobilized *Candida antarctica* lipase B (CALB) particles coated with ionic liquids catalyzing the synthesis of citronellyl alkyl ester in solvent-free medium [22].

phase was used as a reservoir of the hydrophobic substrate (2-octanone) and the favorable partition coefficients of the products shift the biotransformation towards the synthesis of (R)-2-octanol. For the second example, several citronellyl esters (acetate, propionate, butyrate, caprate and laurate) were synthesized by immobilized *Candida antarctica* lipase B to provide high yields (>99%) using equimolar mixtures of citronellol and alkyl vinyl ester as substrates in solvent-free medium (Figure 3.3b). Coating the biocatalyst particles with alkylimidazolium-based ionic liquids (for example [OMim][PF$_6$]) gave rise to an organic: IL biphasic system and allowed the enzyme activity to increase up to twofold as a result of the favored partitioning of the substrate and product molecules between phases.

One of the most important features of biocatalysis in ILs concerns the excellent stability shown by enzymes in some ILs at low water content. Both the thermal stability and reusability of many different enzymes (for example lipases from *C. antarctica* and *C. rugosa*, esterase from *B. stearothermophilus*, α-chymotrypsin, papain), in both their free and immobilized forms, have been greatly improved in water-immiscible ILs (for example [BMim][PF$_6$] and [BMim][NTf$_2$]), and also in the water-miscible [BMim][BF$_4$] [16]. For the case of free enzyme, water-immiscible ILs were shown to be able to maintain the secondary structure and the native conformation of proteins, while the enzyme deactivation that occurs with many water-miscible ILs was related to protein unfolding, as demonstrated by several spectroscopic techniques, including fluorescence, circular dichroism and FT-IR; for examples of studies on protein structure in ILs, see [23]. To understand these structure–function relationships of enzymes in water-immiscible ILs, it is necessary to consider that the incorporation of other molecules (for example water) into the IL network involves changes to the physical properties of these materials and may lead to the formation of polar and non-polar regions [6]. Wet ILs are nanostructured materials and so neutral molecules are able to reside in less polar regions, whereas ionic or polar species undergo faster diffusion in the more polar or wet regions. In this context, enzymes in water-immiscible ILs should also be considered as filling hydrophilic gaps in the network, where the observed stabilization of enzymes could be attributed to the maintenance of this strong net around the protein [16a]. In this context, it was also observed that the presence of water-immiscible ILs greatly improves the stability of lipase in sol-gel immobilization processes [24].

3.3
Biocatalytic Processes in Supercritical Carbon Dioxide

Since 1985, SCFs have been applied as non-aqueous media for enzymatic reactions because of their unique solvent properties, which make them ideal for the development of clean chemical processes; for pioneering studies using SCFs in enzymatic catalysis, see [25]. SCFs are materials which operate at pressures and temperatures higher than their critical points (P_c and T_c) and have densities comparable to those of liquids, whereas their diffusivities and viscosities are similar to those of gases. These characteristics, which make them ideal solvents for use in extraction processes, also make them attractive as a medium for biocatalytic transformations, especially when reactions are limited by the rate of diffusion rather than by any intrinsic kinetics. However, also due to the unique catalytic properties of enzymes determined by their folded conformation, the use of SCFs in biocatalysis has been conditioned by the low stability displayed by enzymes. Carbon dioxide is the most widely used fluid in supercritical conditions for enzyme catalysis, other SCFs being less attractive because of their flammability (for example ethane and propane), high cost (CHF$_3$) or poor solvent power (SF$_6$), meaning that they have been used in only a few research studies; for reviews on enzyme catalysis in SCFs, see [26].

3.3.1
Basic Properties of scCO$_2$

Supercritical carbon dioxide (scCO$_2$) is the most popular SCF because its critical parameters are compatible with the conditions needed for enzyme catalysis

($P_c = 73.8$ bar, $T_c = 31.3\,°C$); it is non-toxic and non-flammable; furthermore, it is chemically inert under most conditions and has excellent solvent properties for non-polar solutes. Another important advantage related to downstream processing is that CO_2 is gaseous at atmospheric pressure, which means that simple depressurization is necessary to separate solutes from $scCO_2$, after which it can be pressurized for reuse. Although $scCO_2$ is a low-polarity solvent which dissolves hydrophobic compounds, its solvent power may be tuned with slight changes in pressure and temperature, for which reason it can be used to recover products from reagents and to modify the selectivity of reactions (Figure 3.4). Similarly, all of its density-dependent solvent properties (for example dielectric constant, relative permittivity and Hildebrand solubility parameter) may be substantially modified by small changes in pressure or temperature; for reviews on $scCO_2$ properties, see [27]. For example, for lipase-catalyzed ethyl oleate synthesis by esterification from oleic acid and ethanol in $scCO_2$, a recycling packed-bed enzyme reactor coupled with a series of four high-pressure separator vessels was applied. The system was operated in a pressure cascade by back-pressure valves, which allowed continuous recovery of the liquid product at the bottom of each separator and then recycling of non-reacted substrates [28].

Additionally, the solvent power of $scCO_2$ can be modified by adding a co-solvent as modifier (for example ethanol or 2-propanol) in order to change the polarity of the medium, which is potentially useful for controlling reactions by the precipitation of products. The most important limitation to applying $scCO_2$ on an industrial scale is

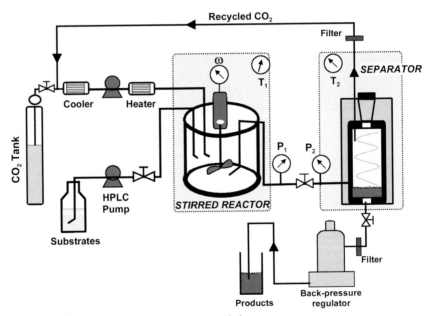

Figure 3.4 High-pressure stirred-tank reactor coupled to a separator tank for enzyme-catalyzed transformations, product recovery and reuse of $scCO_2$.

the fact that the high-pressure equipment needed is fairly costly. Moreover, as CO_2 exhibits a relatively high critical pressure, the energy budget for compression and purification of large volumes of CO_2 is also high [27].

3.3.2
Enzymes in scCO$_2$

Enzymes are not soluble in scCO$_2$ and several free or immobilized enzymes (lipases, trypsin, chymotrypsin, penicillin acylase, alcohol dehydrogenase, cholesterol oxidase, etc.) are suitable for catalyzing chemical transformations (for example esterification, hydrolysis, alcoholysis and asymmetric synthesis) in scCO$_2$ [25, 26]. Water content, pressure and temperature are the most important environmental factors affecting enzymatic catalysis in SCFs, especially activity, enantioselectivity and stability. As in the case with water-immiscible organic solvents, the role of water in enzyme-catalyzed reactions in scCO$_2$ needs to be taken into account. Although water is needed to maintain the active conformation of the enzyme, it can also act as a solvent of polar substances and/or be consumed or produced *in situ* in the system if, hydrolytic or esterification processes occur, respectively. As the solubility of water in scCO$_2$ is low (for example 0.31% v/v at 50 °C and 345 bar), if the water content increases during an enzymatic synthetic reaction (for example esterification), it is necessary to use dehydrating agents (for example molecular sieves) to prevent a decay in the synthetic activity [29]. The actual amount of water needed is specific to each solvent-substrate-enzyme system and must be maintained constant throughout the process [26a]. In this context, the support used to immobilize the enzymes and the hydrophobicity of the SCF also need to be considered. For example, an increase in activity has been observed for the lipase-catalyzed alcoholysis of methyl methacrylate on increasing the hydrophobicity of the SCF under the same conditions (45 °C and 110 bar): SF_6 > propane > ethane > ethylene > CHF_3 > CO_2 [30]. In the case of scCO$_2$, it has also been observed how its log *P* parameter changed from 0.9 to 2.0, which involved an increase in solvent hydrophobicity, as a consequence of increasing the pressure from 30 to 118 bar at 50 °C [27c]. However, one limitation of scCO$_2$ for enzyme catalysis is the fact that it preferentially dissolves hydrophobic compounds, although strategies involving complexation with phenylboronic acid or the addition of surfactants have been successfully applied to dissolve hydrophilic materials [31].

The use of lipases for the asymmetric synthesis of esters is one of the most important tools for organic synthesis. The unique properties of scCO$_2$ combined with the catalytic excellence of lipases have determined that the chiral resolution of a large number of racemates (for example 1-phenylethanol, glycidol and ibuprofen) has successfully been carried out. The high diffusivity of scCO$_2$ has led to it being successfully applied for other types of enzymatic reactions on polymeric substrates, such as lipase-catalyzed polymerizations (for example polyester synthesis) or depolymerization (for example production of ε-caprolactone from polycaprolactone) in scCO$_2$, and also hydrolytic reactions catalyzed by polysaccharide hydrolases (for example α-amylase-catalyzed corn starch hydrolysis) in H_2O–scCO$_2$ biphasic systems [27, 32]. Furthermore, the ability of scCO$_2$ to dissolve oils and fats has been widely applied to many lipase-catalyzed synthetic reactions, such as the esterification of fatty acids with alcohols (for example myristic acid + ethanol, palmitic acid + octanol and butyric acid + oleyl alcohol) or the modification of TAG by acidolysis or alcoholysis to produce MAG, DAG and TAG with a desired composition, and also the synthesis of sugar–fatty acid esters and fatty acid methyl esters (FAMEs) [26c].

The supercritical conditions, determined by both the pressure and temperature of the CO_2, directly affect the catalytic behavior of enzymes by changing the rate-limiting steps or modulating its selectivity [33]. For example, in the case of immobilized lipase-catalyzed enantioselective acetylation of *rac*-1-phenylethanol, it was reported how the increase in pressure, ranging from 8 to 19 MPa, resulted in a continuous decrease in the *E*-value from 50 to 10 [26b]. Another example in the same context shows how, for the esterification of *rac*-citronellol with oleic acid in $scCO_2$ catalyzed by lipase, an ester product with *ee* >99.9% can be obtained simply by manipulating the pressure and temperature in the vicinity of the critical point [34]. In the same way, in the case of immobilized lipase-catalyzed synthesis of butyl butyrate by transesterification, the synthetic activity increased exponentially as the $scCO_2$ density decreased, as a result of different combinations of pressure and temperature [35].

However, the poor stability exhibited by enzymes in SCFs is probably the main drawback of these solvents as regards the development of biocatalytic processes [27a]. In the case of $scCO_2$, several adverse effects on enzyme activity and stability have been described. These effects were attributed to local changes in the pH of the hydration layer, to conformational changes produced during the pressurization/depressurization steps and to the ability of CO_2 to form carbamates with free amine groups on the protein surface, resulting in changes in the secondary structure; for examples of enzyme deactivation in SCFs, see [36]. As a consequence, many attempts have been made to stabilize enzymes by modifying its microenvironment using, for example, immobilized enzymes, lipid-coated enzymes, crosslinked enzyme crystals (CLECs), crosslinked enzyme aggregates (CLEAs) and enzyme entrapment by sol–gel (Figure 3.5).

Enzyme attachment to solid supports is the most popular approach for using enzymes in SCFs, because supports usually stabilize enzymes under mechanical stress and facilitate their removal/recycling from the reaction media. Several procedures for immobilizing enzymes, and also commercial immobilized enzyme preparations, are available; for review on immobilized enzymes, see [37a]; for practical use of enzymes in SCFs, see [37b]. The coating of enzymes with lipids (up 200 ± 50 lipid molecules as a monolayer around the enzyme) is another approach used to improve catalytic efficiency because enzyme solubility is enhanced in supercritical conditions [38]. The use of carrier-free immobilized enzymes, such as crosslinked enzyme crystals (CLECs) and crosslinked enzyme aggregates (CLEAs), has also been reported as a way to obtain active and stable biocatalysts for supercritical reaction media [39]. Another strategy to stabilize enzymes is their immobilization by sol-gel entrapment in silica–aerogels, involving their inclusion in a rigid glass framework, which prevents enzyme deactivation by $scCO_2$ [24, 40]. However, despite all these strategies, the best results for enzyme-catalyzed reactions in $scCO_2$

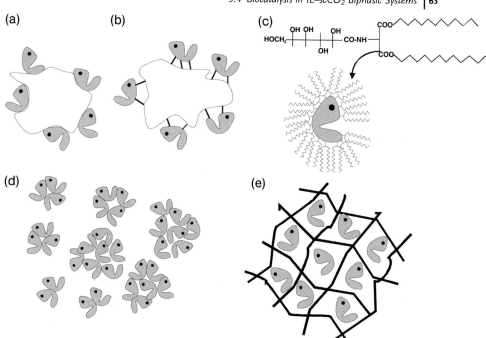

Figure 3.5 Scheme of several stabilized enzyme preparations for use in scCO₂: (a) adsorbed or (b) covalently-attached enzymes on solid supports [37]; (c) lipid-coated enzymes [38]; (d) crosslinked enzyme aggregates (CLEAs, [39]); (e) enzyme entrapped by sol–gel techniques [40].

have been obtained by using ILs as a protective/coating shell around the enzyme [4, 26c, 41].

3.4
Biocatalysis in IL–scCO₂ Biphasic Systems

As already stated, one of the goals of green chemistry is to develop clean chemical processes and it is the low volatility of ILs that makes them useful as green solvents. However, the recovery of solutes dissolved in ILs may represent a breakdown in the greenness of any such process if volatile organic solvents are used to extract them in liquid–liquid biphasic systems. The pioneering work of Brennecke's group in 1999 showed that ILs and scCO₂ form biphasic systems [42]. Additionally, although scCO₂ is highly soluble in the IL phase and is able to extract previously dissolved hydrophobic compounds, the same IL is not measurably soluble in the scCO₂ phase [42]. This discovery was crucial for further developments in non-aqueous green processes involving both biotransformation and extraction steps. The combination of both ILs and scCO₂ neoteric solvents as reaction media for biocatalysis was described in 2002 and represented the first operational strategy for the development of integral green processes in non-aqueous environments; for pioneering studies on enzymatic catalysis in IL–scCO$_2$ biphasic systems, see [43].

3.4.1
Phase Behavior of IL–scCO$_2$ Systems

The phase behavior of IL–scCO$_2$ systems has been studied for several ILs and in many supercritical conditions, and also in the presence of solutes dissolved in the IL phase [9]. Knowledge of this phase behavior is essential for developing any process because it determines the contact conditions between scCO$_2$ and solute and also reduces the viscosity of the IL phase, which enhances the mass-transfer rate of any the reaction system.

Earlier studies of [BMim][PF$_6$]–scCO$_2$ indicated that this system behaves as a biphasic system, in which no measurable amount of [BMim][PF$_6$] is soluble in the CO$_2$-rich phase, whereas a large amount of CO$_2$ is dissolved in the IL-rich phase [42]. Further studies showed how, as the pressure increases, the solubility of CO$_2$ in the IL-rich phase increases dramatically, reaching a solubility of up 0.32 mole fraction at 93 bar and 40 °C [44]. The same study clearly showed how the dependence of CO$_2$ solubility in [BMim][PF$_6$] on temperature was fairly low and there was only small decrease in solubility when the temperature was increased from 40 to 60 °C. Studies on the phase behavior of [BMim][PF$_6$]–scCO$_2$ under high-pressure conditions (up to 970 bar) found two distinct phases in all conditions studied [45]. Although CO$_2$ is fairly soluble at very high pressures, the mixtures never become a single phase. Under such conditions, the density of the CO$_2$ phase increases but, since the IL phase does not expand, the two phases will never become one phase. Additionally, it was shown how the water content of ILs has a significant effect on their phase behavior with CO$_2$, due to the increase in CO$_2$ solubility when ILs are previously dried. However, the ability of ILs to rehydrate by atmospheric humidity is very high. For example, the estimated water content of [BMim][PF$_6$] after drying is approximately 0.15% w/w, but it is able to absorb a few weight percent of water when exposed to the atmosphere [44]. Other IL–scCO$_2$ biphasic systems (for example [OMim][PF$_6$], [OMim][BF$_4$], [BMim][NO$_3$], [EMim][EtSO$_4$] and [BPy][BF$_4$]) showed a similar phase behavior to [BMim][PF$_6$], the solubility CO$_2$ in the IL-rich phase being highest for the ILs with fluorinated anions (i.e. [PF$_6$] and [BF$_4$]). Additionally, for the case of ILs based on the same anion (for example [NTf$_2$]), the solubility of CO$_2$ increased proportionally with the increase in the alkyl chain length of the cation [44, 45].

Although the solubility of ILs in scCO$_2$ is extremely low (for example, a 5×10^{-7} mole fraction of [BMim][PF$_6$] in CO$_2$ at 40 °C and 138 bar [44]), the scCO$_2$ phase may contain other components (for example, substrates and products) for industrial and catalytic applications, which may act as co-solvents able to enhance the solubility of ILs in the scCO$_2$ phase. Taking the [BMim][PF$_6$]–scCO$_2$ system as an example, the addition of ethanol or acetone dramatically increases the solubility of the IL in the supercritical phase as a result of the strong interaction of these co-solvents with the IL due to their strong polarity. The ability of co-solvents to increase the solubility of ILs (for example [BMim][PF$_6$] and [BMim][BF$_4$]) in scCO$_2$ follows the same order as the dipole moment of co-solvents (*i.e.* acetonitrile > acetone > methanol > ethanol > hexane) [46].

Because of the unique phase behavior of IL–CO_2 systems, organic compounds can be extracted from ILs by using $scCO_2$, which represents a green approach for the recovery of solvent-free solutes. The extraction of naphthalene from a [BMim][PF₆] solution (the solubility of naphthalene in [BMim][PF₆] is 0.3 mole fraction at 40 °C) by $scCO_2$ was the first example of solute recovery in green non-aqueous conditions, the process providing a product extract containing no detectable liquid solvent [42]. Further studies have demonstrated that a wide variety of aliphatic and aromatic solutes containing different substituent groups (for example halogen, alcohol, amide, ester, ketone) can be extracted from ILs by using $scCO_2$ with recoveries >95%. In this respect, it should be noted how compounds with a low solubility or even immiscibility with the IL phase (for example, benzene and chlorobenzene are immiscible with [BMim][PF₆]) required the least amount of CO_2 for recovery. The distribution coefficient (defined as the ratio of solute mole fractions in the $scCO_2$ and IL phases) is an important thermodynamic parameter for understanding the ternary phase behavior of IL–organic solute–CO_2 systems. Solutes with high polarities give low distribution coefficients because of their high affinity for IL and low affinity for CO_2, which makes extraction more difficult [47]. Another important event in the ternary phase behavior of IL–organic solute–CO_2 systems is the formation of three phases in supercritical conditions, as was demonstrated for the case of several organic compounds dissolved in different ILs (for example methanol dissolved in [BMim][PF₆]) in the presence of $scCO_2$ (Figure 3.6) [48]. By using the case of the methanol:[BMim][PF₆] system as an example, in which both compounds are completely miscible in all proportions at ambient conditions, it is possible to design an approach to separate ILs from organic compounds by using CO_2 in the case of dilute solutions of ILs. The pressurization of an organic compound–IL solution (L) with CO_2 results in the formation of a second liquid phase. The densest liquid is rich in IL (L₁), the next phase is rich in the organic compound (L₂) and the third vapor phase (V) is mostly CO_2 with some organic compounds. The pressure and

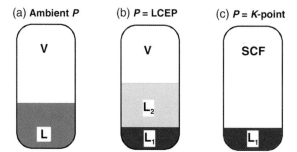

Figure 3.6 Schematic representation of the influence of pressure on the phase behavior of IL–organic solute–CO_2 ternary systems: (a) ambient pressure (*P*); (b) lower critical endpoint pressure (LCEP); (c) *K*-point pressure. (L, L₁ and L₂, liquid phases; V, gas phase; SCF, supercritical phase) [48].

temperature conditions under which the second liquid phase appears is called the lower critical end-point (LCEP). Under these conditions, the organic compound-rich phase expands significantly with increase in CO_2 pressure, whereas the IL-rich phase expands relatively little. Furthermore, the increase in CO_2 pressure may induce another critical point, called the *K*-point, at which the organic compound-rich phase (L_2) merges with the vapor phase (V), while the resulting $scCO_2$–organic compound phase contains no detectable IL. This interesting phase behavior has mechanistic and practical implications for both reaction and separation systems using ILs [48]. For example, it was demonstrated that a desired solute might be extracted from an IL using $scCO_2$ without any cross-contamination. Furthermore, the use of $scCO_2$ to separate ILs from their organic solvents, and the addition of CO_2 to separate hydrophobic and hydrophilic imidazolium-based ILs from aqueous solutions, are important applications of IL–$scCO_2$ systems [9]

3.4.2
Biocatalytic Processes in IL–scCO₂ Biphasic Systems

Despite the clear advantages of using ILs as non-aqueous reaction media for enzyme catalysis, a breakdown point in the greenness of any chemical process occurs if products are recovered by liquid–liquid extraction with organic solvents. The use of IL–$scCO_2$ biphasic systems as reaction media for enzyme catalysis has opened up new opportunities for the development of integral green processes in non-aqueous environments [41]. In this context, a new concept for continuous biphasic biocatalysis, whereby a homogeneous enzyme solution is immobilized in one liquid phase (catalytic phase), while substrates and products reside largely in a supercritical phase (extractive phase) directly providing products, was put forward as the first approach of integral green bioprocesses in non-aqueous media (Figure 3.7).

The system was tested for two different reactions catalyzed by *Candida antarctica* lipase B (CALB): the synthesis of butyl butyrate from vinyl butyrate and 1-butanol and the kinetic resolution of *rac*-1-phenylethanol at 150 bar over a range of temperatures (40–100 °C). Under these conditions, the enzyme showed an exceptional level of

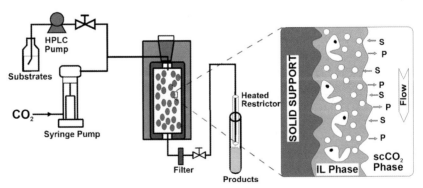

Figure 3.7 Setup of a continuous green enzyme reactor working in an IL–scCO₂ biphasic system [43a].

activity, enantioselectivity (*ee* >99.9) and operational stability after 11 cycles of 4 h of work [43a]. These results were corroborated under extreme conditions, such as 100 bar and 150 °C [49]. However, for these enzyme reactions in IL–scCO$_2$ systems, the transport of substrates from the supercritical to the enzyme–IL phase and then the release of products towards the supercritical phase are key parameters for controlling the efficiency of the reaction system. To analyze this phenomenon, two similar ILs based on the same ions, but with different degrees of hydrophobicity in the cation, [BuMe$_3$N][NTf$_2$] and [CN-PMe$_3$N][NTf$_2$], were successfully applied for the CALB-catalyzed continuous synthesis of six short-chain alkyl esters (butyl acetate, butyl propionate, butyl butyrate, hexyl propionate, hexyl butyrate and octyl propionate) in scCO$_2$ at 10 MPa and 50 °C. Using Hansen's solubility parameter (δ) as criterion to compare the hydrophobicity of the main alkyl chain of cations in ILs with respect to substrates and products, it was shown how the same values for this parameter in reagents and IL resulted in a clear improvement in productivity, as a consequence of favoring the mass transfer phenomena between IL and scCO$_2$ phases [50].

A further step towards green biocatalysis in IL–scCO$_2$ biphasic systems was taken by the appropriate selection of reagents, because of the selective separation of the synthetic product can be included as an integrated step in the full process [51]. By using vinyl laurate as acyl donor in the kinetic resolution of *rac*-1-phenylethanol catalyzed by immobilized CALB, the stereoselective synthetic product, (*R*)-1-phenyl-ethyl laurate, can be selectively separated from the non-reacted alcohol with scCO$_2$. This process takes advantage of the fact that the solubility of a compound in scCO$_2$ depends on both the polarity and vapor pressure. Thus, if the alkyl chain of an ester product is long enough, its low volatility should mean that it is less soluble in scCO$_2$ than the corresponding alcohol. By this strategy, the introduction of an additional separation chamber placed between the reactor and the back-pressure outlet of the system (as an example, see Figure 3.4) and the selection of an appropriate pressure and temperature provided the selective separation of the synthetic product from the resulting reaction mixture (66% yield, *ee* >99.9%).

Cutinase from *Fusarium solani pisi* immobilized on zeolite NaY was also tested in a [BMim][PF$_6$]–scCO$_2$ biphasic system for the kinetic resolution of 2-phenyl-1-propanol. The results pointed to a protective effect of the IL against the enzyme deactivation by scCO$_2$ and also higher activity than observed for the cutinase–IL system. This enhancement in activity was attributed to the CO$_2$ dissolved in the IL, which would decrease its viscosity and hence improve the mass transfer of substrates to the enzyme active site [52].

Two final examples of biocatalysis in IL–scCO$_2$ systems are worth mentioning because of the introduction of new improvements for the development of these systems for industrial purposes, such as multicatalytic processes and reaction systems with reduced amount of ILs. In the first case, continuous dynamic kinetic resolution processes were carried out in different IL–scCO$_2$ biphasic systems simultaneously using both immobilized CALB and silica modified with benzene-sulfonic acid as catalysts at 40 °C and 10 MPa (Figure 3.8a). Kinetic resolution with enzymes is the most widely used method for separating the two enantiomers of a racemic mixture. However, the main drawback of the enzymatic method is that the

(a)

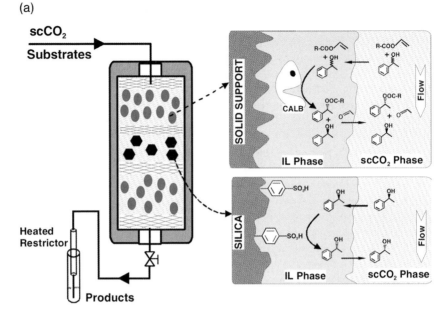

(b)

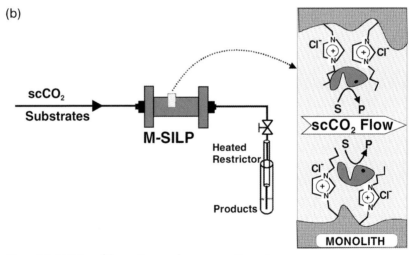

Figure 3.8 (a) Setup of the continuous chemoenzymatic reactor for dynamic kinetic resolution of *rac*-1-phenylethanol in IL–scCO$_2$ biphasic systems [53]. (b) Setup of the reactor with CALB immobilized on monolith-supported ionic liquid phase (M-SILP) for continuous operation under flow conditions in scCO$_2$ [55a].

chemical yield is limited to 50%, although this can be overcome by combining kinetic resolution with *in situ* racemization of the undesired enantiomer, using so-called dynamic kinetic resolution (DKR). The silica modified with benzenesulfonic acid was shown to act as an efficient heterogeneous chemical catalyst for the racemization of (*S*)-1-phenylethanol in different ionic liquid media ([EMim][NTf$_2$], [BMe$_3$N][NTf$_2$] and [BMim][PF$_6$]). Coating of both chemical and enzymatic catalysts with ILs greatly improved the efficiency of the process, providing a good yield (76%) of (*R*)-1-phenylethyl propionate product and good enantioselectivity (91–98% *ee*) in continuous operation [53].

Since some ILs have also been described as being not fully green solvents, because of their low biodegradability and high (eco)toxicological properties [54], reaction systems based on reduced amounts of ILs have been encouraged. In this context, a further step towards reducing the amount of ILs used in enzymatic processes in IL–scCO$_2$ biphasic systems resulted from the development of monolithic solid supports, to which the IL phase is covalently attached (Figure 3.8b). The adsorption of CALB on this linked IL phase gives excellent immobilized biocatalysts with enhanced activity and increased operational stability for the synthesis of citronellyl butyrate in scCO$_2$, compared with the original strategy based on enzymes coated with ILs [55]. In the same context of sustainability, the discovery that PEG and scCO$_2$ formed similar biphasic systems to that of ILs–scCO$_2$, since PEG has very low solubility in scCO$_2$ whereas CO$_2$ is highly soluble in PEG, has opened up a new alternative with respect to IL in this kind of enzymatic biphasic process [56]. Furthermore, PEG is less expensive than ILs and an accepted additive for food with a fully evaluated toxicity. By using the lipase-catalyzed kinetic resolution of *rac*-1-phenylethanol as a reaction model, preliminary studies have demonstrated the good suitability of this PEG–scCO$_2$ biphasic medium because enzyme activity and selectivity are maintained after 11 batches at 50 °C and 80 bar [57].

3.5
Future Trends

The technological application of enzymes can be enhanced greatly by using them in non-aqueous environments rather than their natural aqueous reaction media, because of the resulting expansion in the repertoire of enzyme-catalyzed transformations. Consequently, a number of potentially interesting applications of enzymes that are either impossible to use or of marginal benefit in water become commercially attractive in non-aqueous environments. Enzymes constitute the most important toolbox for green organic synthesis and the interest for application at an industrial scale is beyond doubt.

The unique properties of ILs, and the possibility that these properties can be tailored by appropriate selection of ions and their substituents, open the door to many more processing options than have been available using conventional organic solvents. Ionic liquids offer the opportunity to modify drastically all industrial chemical processes based on these conventional solvents in the near future.

Furthermore, the exceptional ability of some ILs to stabilize enzymes, which maintain their native and active conformation even under extremely harsh conditions, is an important added value of these neoteric solvents in the construction of a near green chemical industry. Furthermore, $scCO_2$ seems to be the perfect companion of ILs for the development of downstream steps in green synthetic processes, and also for cleaning and recovering ILs for reuse, because of the unique phase behavior of $IL-scCO_2$ systems. The combination of enzymes with $IL-scCO_2$ biphasic systems is the clearest strategy for developing integral green synthetic processes for industrial applications. The possibility of using more complex enzymes (for example oxidoreductases and lyases) is an almost unexplored field. Furthermore, multi-enzymatic and/or multi-chemoenzymatic processes in $IL-scCO_2$ for synthesizing pharmaceutical drugs are only at the beginning. The door for the green chemical industry of the near future is open.

Acknowledgments

Financial support was provided by grants from CICYT (Ref. CTQ2005-01571-PPQ), SENECA Foundation (Ref. 02910/PI/05) and BIOCARM (Ref. BIO-BMC 06/01-0002).

References

1 Anastas, P.T. and Warner, J.C. (1998) *Green Chemistry: Theory and Practice*, Oxford University Press, Oxford.

2 Clark, J.H. and Tavener, S.J. (2007) *Organic Process Research & Development*, **11**, 149–155.

3 Klibanov, A.M. (2001) *Nature*, **409**, 241–246.

4 Cantone, S., Hanefeld, U. and Basso, A. (2007) *Green Chemistry*, **9**, 954–971.

5 (a) Madeira Lau R., R., van Rantwijk, F., Seddon, K.R. and Sheldon, R.A. (2000) *Organic Letters*, **2**, 4189–4191; (b) Erbeldinger, M., Mesiano, A.J. and Russell, A.J. (2000) *Biotechnology Progress*, **16**, 1129–1131.

6 (a) Yang, Z. and Pan, W. (2005) *Enzyme and Microbial Technology*, **37**, 19–28; (b) van Rantwijk, F. and Sheldon, R.A. (2007) *Chemical Reviews*, **107**, 2757–2785; (c) Durand, J., Teuma, E. and Gómez, M. (2007) *Comptes Rendus Chimie*, **10**, 152–177.

7 (a) Dupont, J., de Souza, R.F. and Suarez, P.A. Z. (2002) *Chemical Reviews*, **102**, 3667–3691; (b) Wasserscheid, P., Welton, T. (eds), (2003) *Ionic Liquids in Synthesis*, Wiley-VCH Verlag GmbH, Weinheim.

8 (a) http://www.chemdat.info/ ionic_liquids/ionic_liquids.html; (b) http://ilthermo.boulder.nist.gov/ ILThermo/mainmenu.uix.

9 Keskin, S., Kayrak-Talay, D., Akman, U. and Hortaçsu, O. (2007) *Journal of Supercritical Fluids*, **43**, 150–180.

10 (a) Poole, C.F. (2004) *Journal of Chromatography. A*, **1037**, 49–82; (b) Reichardt, C. (2005) *Green Chemistry*, **7**, 339–351.

11 Hinckley, G., Mozhaev, V.V., Budde, C. and Khmelnitsky, Y.L. (2002) *Biotechnology Letters*, **24**, 2083–2087.

12 Kaftzik, N., Wasserscheid, P. and Kragl, U. (2002) *Organic Process Research & Development*, **6**, 553–557.

13 Zhao, H. and Malhotra, S.V. (2002) *Biotechnology Letters*, **24**, 1257–1260.

14 (a) Turner, M.B., Spear, S.K., Huddleston, J.G., Holbrey, J.D. and Rogers, R.D. (2003) *Green Chemistry*, **5**, 443–447; (b) Lau, R.M., Sorgedrager, M.J., Carrea, G., van Rantwijk, F., Secundo, F. and Sheldon, R.A. (2004) *Green Chemistry*, **6**, 483–487.

15 (a) Schofer, S.H., Kaftzik, N., Wasserscheid, P. and Kragl, U. (2001) *Chemical Communications*, 425–426; (b) Kim, K.W., Song, B., Choi, M.Y. and Kim, M.J. (2001) *Organic Letters*, **3**, 1507–1509; (c) Park, S. and Kazlauskas, R.J. (2001) *The Journal of Organic Chemistry*, **66**, 8395–8401; (d) Ulbert, O., Frater, T., Belafi-Bako, K. and Gubicza, L. (2004) *Journal of Molecular Catalysis B-Enzymatic*, **31**, 39–45.

16 (a) Lozano, P., De Diego, T., Carrié, D., Vaultier, M. and Iborra, J.L. (2001) *Biotechnology Letters*, **23**, 1529–1533; (b) Lozano, P., De Diego, T., Guegan, J.P., Vaultier, M. and Iborra, J.L. (2001) *Biotechnology and Bioengineering*, **75**, 563–569; (c) Persson, M. and Bornscheuer, U.T. (2003) *Journal of Molecular Catalysis B-Enzymatic*, **22**, 21–27; (d) Kaar, J.L., Jesionowski, A.M., Berberich, J.A., Moulton, R. and Russell, A.J. (2003) *Journal of the American Chemical Society*, **125**, 4125–4131; (e) Ulbert, O., Belafi-Bako, K., Tonova, K. and Gubicza, L. (2005) *Biocatalysis and Biotransformation*, **23**, 177–183.

17 (a) Park, S., Viklund, F., Hult, K. and Kazlauskas, R.J. (2003) *Green Chemistry*, **5**, 715–719; (b) Gubicza, L., Nemestothy, N., Frater, T. and Belafi-Bako, K. (2003) *Green Chemistry*, **5**, 236–239; (c) Lourenco, N.M.T., Barreiros, S. and Afonso, C.A.M. (2007) *Green Chemistry*, **9**, 734–736.

18 (a) Lozano, P., De Diego, T., Carrie, D., Vaultier, M. and Iborra, J.L. (2003) *Journal of Molecular Catalysis B-Enzymatic*, **21**, 9–13; (b) Itoh, T., Nishimura, Y., Ouchi, N. and Hayase, S. (2003) *Journal of Molecular Catalysis B-Enzymatic*, **26**, 41–45; (c) Kim, M.J., Kim,

H.M., Kim, D., Ahn, Y. and Park, J. (2004) *Green Chemistry*, **6**, 471–474.

19 (a) Itoh, T., Akasaki, E., Kudo, K. and Shikarami, S. (2001) *Chemistry Letters*, 262–263; (b) Lee, J.K. and Kim, M.J. (2002) *The Journal of Organic Chemistry*, **67**, 6845–6847

20 Lozano, P., de Diego, T. and Iborra, J.L. (2006) in J. M Guisán (ed.), *Immobilization of Enzymes and Cells. Methods in Biotechnology Series*, Vol. 22, Humana Press, Totowa, NJ, Chapter 22.

21 Eckstein, M., Villela, M., Liese, A. and Kragl, U. (2004) *Chemical Communications*, 1084–1085.

22 Lozano, P., Piamtongkam, R., Kohns, K., De Diego, T., Vaultier, M. and Iborra, J.L. (2007) *Green Chemistry*, **9**, 780–784.

23 (a) De Diego, T., Lozano, P., Gmouh, S., Vaultier, M. and Iborra, J.L. (2004) *Biotechnology and Bioengineering*, **88**, 916–924; (b) Baker, S.N., McCleskey, T.M., Pandey, S. and Baker, G.A. (2004) *Chemical Communications*, 940–941; (c) De Diego, T., Lozano, P., Gmouh, S., Vaultier, M. and Iborra, J.L. (2005) *Biomacromolecules*, **6**, 1457–1464; (d) van Rantwijk, F., Secundo, F. and Sheldon, R.A. (2006) *Green Chemistry*, **8**, 282–286; (e) Lou, W.Y., Zong, M.H., Smith, T.J., Wu, H. and Wang, J.F. (2006) *Green Chemistry*, **8**, 509–512.

24 Lee, S.H., Doan, T.T.N., Ha, S.H. and Koo, Y.M. (2007) *Journal of Molecular Catalysis B-Enzymatic*, **45**, 57–61.

25 (a) Hammond, D.A., Karel, M., Klibanov, A.M. and Krukonis, V.J. (1985) *Applied Biochemistry and Biotechnology*, **11**, 393–400; (b) Randolph, T.W., Blanch, H.W., Prausnitz, J.M. and Wilke, C.R. (1985) *Biotechnology Letters*, **7**, 325–328; (c) Nakamura, K., Chi, Y.M., Yamada, Y. and Yano, T. (1985) *Chemical Engineering Communications*, **45**, 207–212.

26 (a) Mesiano, A.J., Beckman, E.J. and Russel, A.J. (1999) *Chemical Reviews*, **99**, 623–633; (b) Matsuda, T., Harada, T. and Nakamura, K. (2005) *Current Organic Chemistry*, **9**, 299–315; (c) Hobbs, H.R.

and Thomas, N.R. (2007) *Chemical Reviews*, **107**, 2786–2820.

27 (a) Jessop, P.J. and Leitner, W. (eds), (1999) *Chemical Synthesis Using Supercritical Fluids*, Wiley-VCH Verlag GmbH, Weinheim. (b) Beckmann, E.J. (2004) *Journal of Supercritical Fluids*, **28**, 121–191; (c) Nakaya, H., Miyawaki, O. and Nakamura, K. (2001) *Enzyme and Microbial Technology*, **28**, 176–182.

28 Marty, A., Combes, D. and Condoret, J.S. (1994) *Biotechnology and Bioengineering*, **43**, 497–504.

29 Hakoda, M., Shiragami, N., Enomoto, A. and Nakamura, K. (2002) *Bioprocess and Biosystems Engineering*, **24**, 355–361.

30 Kamat, S., Barrera, J., Beckman, E.J. and Russell, A.J. (1992) *Biotechnology and Bioengineering*, **40**, 158–166.

31 (a) Castillo, E., Marty, A., Combes, D. and Condoret, J.S. (1994) *Biotechnology Letters*, **16**, 169–174; (b) Knez, Z. and Habulin, M. (2002) *Journal of Supercritical Fluids*, **23**, 29–42.

32 Ghanem, A. (2007) *Tetrahedron*, **63**, 1721–1754.

33 Rezaei, K., Temelli, F. and Jenab, E. (2007) *Biotechnology Advances*, **25**, 272–280.

34 Ikushima, Y., Saito, N., Hatakeda, K. and Sato, O. (1996) *Chemical Engineering Science*, **51**, 2817–2822

35 Lozano, P., Villora, G., Gómez, D., Gayo, A.B., Sánchez-Conesa, J.A., Rubio, M. and Iborra, J.L. (2004) *Journal of Supercritical Fluids*, **29**, 121–128.

36 (a) Kamat, S., Critchley, G., Beckman, E.J. and Russell, A.J. (1995) *Biotechnology and Bioengineering*, **46**, 610–620; (b) Lozano, P., Avellaneda, A., Pascual, R. and Iborra, J.L. (1996) *Biotechnology Letters*, **18**, 1345–1350; (c) Giessauf, A., Magor, W., Steinberger, D.J. and Marr, R. (1999) *Enzyme and Microbial Technology*, **24**, 577–583; (d) Habulin, M. and Knez, Z. (2001) *Journal of Chemical Technology and Biotechnology*, **76**, 1260–1266; (e) Striolo, A., Favaro, A., Elvassore, N., Bertucco, A. and Di Notto, V. (2003) *Journal of Supercritical Fluids*, **27**, 283–295.

37 (a) Mateo, C., Palomo, J.M., Fernandez-Lorente, G., Guisan, J.M. and Fernandez-Lafuente, R. (2007) *Enzyme and Microbial Technology*, **40**, 1451–1463; (b) Lozano, P., de Diego, T. and Iborra, J.L. (2006) in *Immobilization of Enzymes and Cells - Methods in Biotechnology Series*, Vol. 22, (Ed. J. M Guisán), Humana Press Inc., Totowa, Chapter 24.

38 (a) Mori, T. and Okahata, Y. (1998) *Chemical Communications*, 2215–2216; (b) Mori, T., Funasaki, M., Kobayashi, A. and Okahata, Y. (2001) *Chemical Communications*, 1832–1833.

39 Sheldon, R.A. (2007) *Advanced Synthesis and Catalysis*, **349**, 1289–1307.

40 Novak, Z., Habulin, M., Kremelj, V. and Knez, Z. (2003) *Journal of Supercritical Fluids*, **27**, 169–178.

41 Dzyuba, S.V. and Bartsch, R.A. (2003) *Angewandte Chemie-International Edition*, **42**, 148–150.

42 Blanchard, L.A., Hancu, D., Beckman, E.J. and Brennecke, J.F. (1999) *Nature*, **399**, 28–29.

43 (a) Lozano, P., De Diego, T., Carrié, D., Vaultier, M. and Iborra, J.L. (2002) *Chemical Communications*, 692–693; (b) Reetz, M.T., Wiesenhofer, W., Francio, G. and Leitner, W. (2002) *Chemical Communications*, 992–993.

44 Blanchard, L.A., Gu, Z.Y. and Brennecke, J.F. (2001) *The Journal of Physical Chemistry. B*, **105**, 2437–2444.

45 Aki, S.N.V.K., Mellein, B.R., Saurer, E.M. and Brennecke, J.F. (2004) *The Journal of Physical Chemistry. B*, 20355–20365.

46 (a) Wu, W.Z., Zhang, J.M., Han, X.B., Chen, J.W., Liu, Z.M., Jiang, T., He, J. and Li, W.J. (2003) *Chemical Communications*, 1412–1413; (b) Wu, W.Z., Li, W.J., Han, B.X., Jiang, T., Shen, D., Zhang, Z.F., Sun, D.H. and Wang, B. (2004) *Journal of Chemical and Engineering Data*, **49**, 1597–1601.

47 Blanchard, L.A. and Brennecke, J.F. (2001) *Industrial & Engineering Chemistry Research*, **40**, 287–292.

48 (a) Scurto, A.M., Aki, S.N.V.K. and
Brennecke, J.F. (2002) *Journal of the
American Chemical Society*, **124**,
10276–10277; (b) Scurto, A.M., Aki, S.N.
V.K. and Brennecke, J.F. (2003) *Chemical
Communications*, 572–573; (c) Aki,
S.N.V.K., Scurto, A.M. and Brennecke, J.F.
(2006) *Industrial & Engineering Chemistry
Research*, **45**, 5574–5585.

49 Lozano, P., De Diego, T., Carrié, D.,
Vaultier, M. and Iborra, J.L. (2003)
Biotechnology Progress, **19**, 380–382.

50 Lozano, P., De Diego, T., Gmouh, S.,
Vaultier, M. and Iborra, J.L. (2004)
Biotechnology Progress, **20**, 661–669.

51 Reetz, M.T., Wiesenhofer, W., Francio, G.
and Leitner, W. (2003) *Advanced Synthesis
and Catalysis*, **345**, 1221–1228.

52 Garcia, S., Lourenco, N.M. T., Lousa, D.,
Sequeira, A.F., Mimoso, P., Cabral, J.M. S.,
Afonso, C.A. M. and Barreiros, S. (2004)
Green Chemistry, **6**, 466–470.

53 Lozano, P., De Diego, T., Larnicol, M.,
Vaultier, M. and Iborra, J.L. (2006)
Biotechnology Letters, **28**, 1559–1565.

54 Cho, C.W., Pham, T.P. T., Jeon, Y.C.,
Vijayaraghavan, K., Choe, W.S. and Yun,
Y.S. (2007) *Chemosphere*, **69**, 1003–1007.

55 (a) Lozano, P., Garcia-Verdugo, E.,
Piamtongkam, R., Karbass, N., De Diego, T.,
Burguete, M.I., Luis, S.V. and Iborra, J.L.
(2007) *Advanced Synthesis and Catalysis*, **349**,
1077–1084; (b) Riisager, A., Fehrmann, R.,
Haumann, M. and Wasserscheid, P. (2006)
Topics in Catalysis, **40**, 91–102.

56 Heldebrant, D.J. and Jessop, P.G. (2003)
Journal of the American Chemical Society,
125, 5600–5601.

57 Reetz, M.T. and Wiesenhofer, W. (2004)
Chemical Communications, 2750–2751.

4
Thiamine-Based Enzymes for Biotransformations

Martina Pohl, Dörte Gocke, and Michael Müller

4.1
Introduction

Chiral and prochiral α-oxyfunctionalized compounds, in particular carboxylic acids, aldehydes, ketones, alcohols are indispensable building blocks for asymmetric synthesis. 2-Hydroxy ketones encompassing a chiral and prochiral center are therefore versatile building blocks for organic synthesis (Scheme 4.1), as they can be easily converted into other functionalities, such as diols, epoxides, amino alcohols and diamines. Such compounds can be accessed chemically (non-enzymatically) by classical benzoin condensation, also employing biomimetic catalysts such as chiral thiazolium and triazolium salts [1–7].

This reaction is catalyzed under extremely mild conditions by several biocatalysts such as acetohydroxy acid synthase (AHAS), benzaldehyde lyase (BAL), benzoyl-formate decarboxylase (BFD), pyruvate decarboxylase (PDC) and phenylpyruvate decarboxylase (PhPDC).

This chapter gives an overview of the state of the art concerning characterization of these enzymes, with an emphasis on structure–function studies to design these catalysts for special tasks in chemoenzymatic synthesis.

Thiamine diphosphate (ThDP)-dependent enzymes are widespread among several central points of anabolism and catabolism (e.g. pentose phosphate pathway, tricarboxylic acid pathway), where ThDP mediates the formation and cleavage of C–C, C–S, C–N, C–O and C–P bonds (Table 4.1) (for details, see reviews [8, 9]). The evolutionary relationship of ThDP-enzymes has recently been investigated [10]. This group of enzymes can be roughly categorized into those with decarboxylase activity, which can further be divided into non-oxidative and oxidative decarboxylation, and enzymes with transferase-type activity. Both reaction types rely on a similar mechanism with ThDP as the reactive agent. An overview about the actually known ThDP-enzymes, which are not part of multienzyme complexes, is given in Table 4.1.

Handbook of Green Chemistry, Volume 3: Biocatalysis. Edited by Robert H. Crabtree
Copyright © 2009 WILEY-VCH Verlag GmbH & Co. KGaA, Weinheim
ISBN: 978-3-527-32498-9

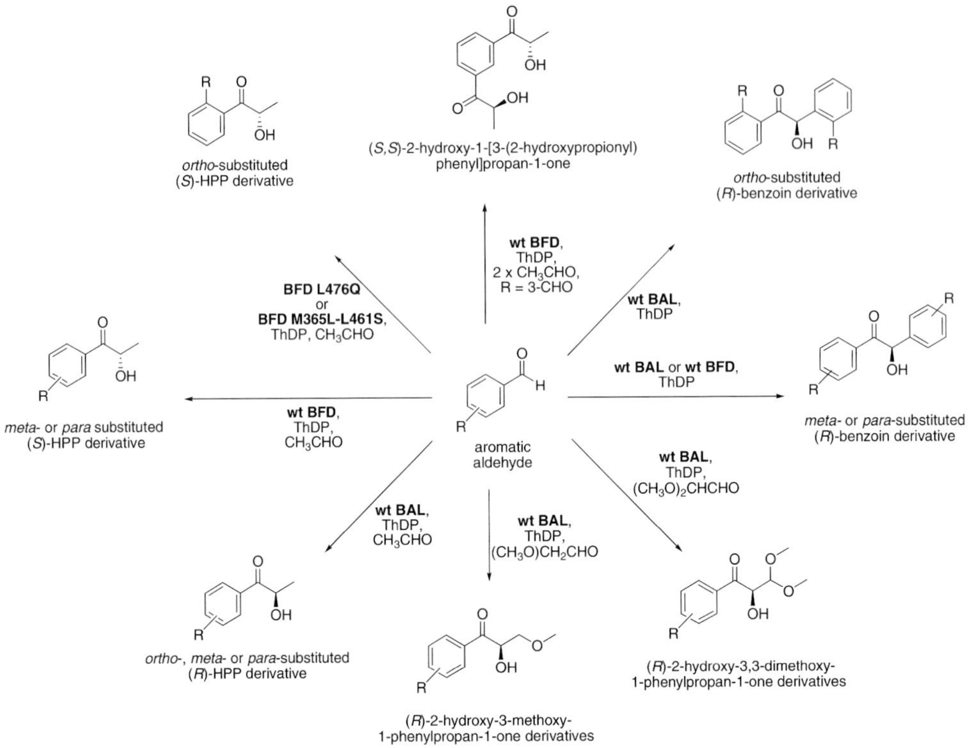

Scheme 4.1 A survey of selected carboligation reactions catalyzed by ThDP-dependent enzymes.

4.1.1
Thiamine Diphosphate

Thiamine diphosphate (ThDP) (Scheme 4.2), the biologically active form of vitamin B_1 (thiamine), is essential for catalytic activity of ThDP-dependent enzymes. It is non-covalently bound to the active site via a divalent metal ion, usually Mg^{2+}.

In contrast to yeasts, plants and bacteria, mammals are not able to synthesize this vitamin and thus have to feed thiamine by a balanced diet to produce the vitally important ThDP. Vitamin B_1 is essential for the metabolism of carbohydrates (to produce energy) and for nerve and heart function. A consequence of thiamine deficiency in humans is beriberi.

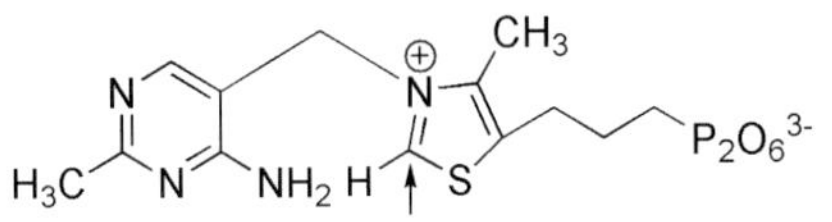

Scheme 4.2 Structure of ThDP. The catalytically important C2 atom of the thiazolium ring is indicated by an arrow.

Table 4.1 Selected information on ThDP-enzymes with respect to their origin, 3D-structure and their application in biotransformations[a].

Enzyme EC	Organism/organelle	Pathway	Carboligation studies [refs 2004–2007]	pdb code
PDC 4.1.1.1	Bacteria: Zymomonas mobilis, Zymobacter palmae, Acetobacter pasteurianus Sarcina ventriculi	Glycolysis Oxidative lactic acid metabolism	ZmPDC [53] ApPDC [13]	1 pdc[b] 1 vbi
	Diverse fungi and yeast strains, especially S. cerevisiae		ScPDC [105, 178, 179] Various filamentous fungi [102] Various yeasts [100, 101, 114, 180, 181]	1 pvd, 1 pyd[b]
	Various plants		n.a.	
KdcA 4.1.1.72	Lactococcus lactis strains Bacillus subtilis	Branched-chain keto acid metabolism	LlKdcA [50]	2vbg, 2vb f[b]
BFD 4.1.1.7	Pseudomonas putida Pseudomonas stutzeri Diverse bacteria	Mandelate catabolism	PpBFD [53]	1 bfd, 1 mcz[b]
PhPDC 4.1.1.43	S. cerevisiae, Azospirillum brasilense Achromobacter eurydice, Acinetobacter calcoaceticus, Thauera aromatica	Ehrlich pathway	n.a.	1nxw[b]
InPDC 4.1.1.74	Enterobacter cloacae Diverse. bacteria	Tryptophan catabolism	n.a.	1ovm
AHAS 4.1.3.18	E. coli, S. cerevisiae, other bacteria, fungi, diverse plants, mammalians (chloro-plasts, mitochondria)	Branched-chain amino acid biosynthesis	EcAHAS I–III [62, 154, 159, 162–164]	1jsc[b,c]
ALS 2.2.1.6	Klebsiella pneumoniae	2,3-Butanediol pathway	n.a.	1ozf[b]

(Continued)

Table 4.1 (*Continued*)

Enzyme EC	Organism/organelle	Pathway	Carboligation studies [refs 2004–2007]	pdb code
BAL 4.1.2.38	*Pseudomonas fluorescens*	n.d.	*Pf*BAL [48, 49, 51, 52, 54–57, 59–61, 182]	2ag0, 2uz1[b]
TK 2.2.1.1	*E. coli*, *S. cerevisae*, other bacteria, yeasts, plants, mammalians	Pentose phosphate pathway	*Ec*TK [183]	1trk[b], 1qgd[b]
CEAS	*Streptomyces clavuligerus*	Clavulanic acid biosynthesis	n.a.	1upa[b]
GXC 4.1.1.47	*E. coli*	Glyoxylate catabolism	n.a.	2pan
DXS 4.1.3.37	Various bacteria, plants	Isoprenoid biosynthesis	n.a.	2o1s

[a]Abbreviations, ALS, acetolactate synthase; AHAS, acetohydroxy acid synthase; BAL, benzaldehyde lyase; BFD, benzoylformate decarboxylase; CEAS, N^2-(2-carboxyethyl) arginine synthase; DXS, 1-deoxy-D-xylulose-5-phosphate synthase; GXC, glyoxylate carboligase = tartronate semialdehyde synthase; InPDC, indole 3-pyruvate decarboxylase; KdcA, branched-chain keto acid decarboxylase; PDC, pyruvate decarboxylase; PhPDC, phenylpyruvate decarboxylase; TK, transketolase; n.a., not applicable.

[b]Further 3D structures of variants and/or with substrates, effectors or ThDP analogues are available from the pdb databank.

[c]Structures of enzymes from other species are available from the pdb databank.

4.1.2
Enzyme Structures

Although sequence similarities among ThDP-dependent enzymes are usually less than 20%, many of the tertiary structures are remarkably similar. Most likely the common binding fold emerged by divergent evolution with a selective pressure on the geometric position of the small number of residues binding ThDP, but tolerating other substitutions elsewhere [8].

Crystal structures of many different classes of ThDP-enzymes have been solved; among these are several 2-keto acid decarboxylases, and also the benzaldehyde lyase from *Pseudomonas fluorescens* (*Pf* BAL) (Table 4.1). All of them show a general fold consisting of three domains of α/β-topology [10, 11].

Two monomers form a dimer with two ThDPs bound at the dimer's interface. The pyrimidine ring of the cofactor interacts with the pyrimidine ring-binding domain of one subunit, whereas the residual parts of ThDP interact with the diphosphate-binding domain of the neighboring subunit [12–16]. ThDP is bound in a conserved V-shape (Scheme 4.2), forced by, among others, a large hydrophobic residue next to the thiazolium ring [11, 17] (see also Table 4.7). Deviations from this generally conserved enzyme-bound ThDP conformation have been observed in the 3D structure of acetolactate synthase (ALS) from *Klebsiella pseumoniae* [18].

ThDP-dependent enzymes are mostly active either as dimers, for example trans-ketolase [19] and the branched-chain keto acid decarboxylase [20], or as tetramers, for example most 2-keto acid decarboxylases [12–14, 16, 21–23], acetohydroxy acid synthases [18, 24] and benzaldehyde lyase [15, 25]. The tetramer can be best described as a dimer of dimers [26]. In addition to the tetrameric structure, some PDCs have a tendency to dissociate into dimers [27] or to form higher association states such as octamers [28, 29]; in some plants oligomers up to 300–500 kDa have been observed [30]. Further, ThDP-enzymes are part of larger complexes, such as 2-oxo acid multienzyme complexes [31]. There is much evidence that the different catalytic sites of one enzyme do not operate independently but show a certain degree of interaction [26, 31–35].

4.1.3
Reaction Mechanism

4.1.3.1 Lyase and Carboligase Activity Occur at the Same Active Site

ThDP-dependent enzyme reactions start with a chemically challenging breaking of a C–C or C–H bond adjacent to a carboxyl group in the substrate (Scheme 4.3b and c). To mediate this activity, the C2 carbon of the thiazolium ring ThDP first needs to be activated by deprotonation to form a potent nucleophilic ylide. Pioneering work in this field has been published by, among others, Breslow (1957) [36] and Kern et al. (1997) [37]. A very good survey of known mechanisms of ThDP-enzymes is given in a recent review [8].

This deprotonation is supported by a conserved glutamate residue present in all ThDP-dependent enzymes, except glyoxylate carboligase [38], which induces the

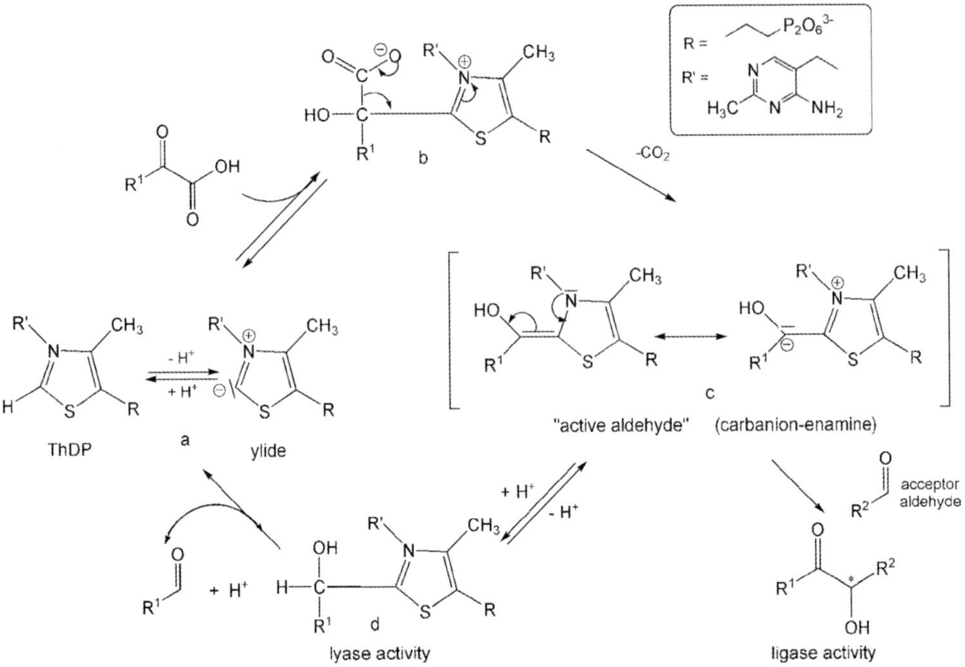

Scheme 4.3 Reaction mechanism of the lyase and ligase activity of 2-keto acid decarboxylases.

formation of a $1',4'$-imino tautomer in the pyrimidine ring. The constrained V-conformation of the cofactor is essential for this step since the C2 proton is positioned at a reactive distance to the $4'$-imino group of the neighboring pyrimidine ring (Scheme 4.2).

In the first step of a **decarboxylase reaction** (Scheme 4.3, left cycle) the carbonyl group of the 2-keto acid substrate reacts with the ylide-form of ThDP (b) yielding the corresponding 2-hydroxy acid adduct. After CO_2 is released, a carbanion-enamine, the so-called *active aldehyde*, is formed as a reactive intermediate (c). Subsequently, the enamine intermediate is protonated (d) and the corresponding aldehyde is released, thereby reconstituting the ylide.

Carboligation reactions result from the binding of a carbonyl compound such as an aldehyde as the second substrate (acceptor aldehyde), leading to the formation of 2-hydroxy ketones (e). In this case again a proton donor is required to neutralize the resulting negative charge. Therefore, all ThDP-dependent enzymes need a proton acceptor in the first half of the reaction cycle and a proton donor in the second half. While the proton acceptor for the first step is the $1',4'$-imino tautomer of ThDP in combination with an almost invariant glutamate residue (see above), the proton relay system for the second step is different in these enzymes. As was shown for the PDCs [8], *Pp*BFD [14, 39], *Ab*PhPDC [16, 40] and *Pf* BAL [15, 41], one or two histidine residues operate as proton relays. The thiazolium-proximal histidines are positionally conserved whereas other ThDP-dependent enzymes contain a polar residue (Arg or Gln) in these positions [8, 40] (for details, see Section 4.5.3). Since binding of an aldehyde as the donor substrate without a preliminary decarboxylation step is also possible, the decarboxylation of a 2-keto acid is not mandatory for carboligation.

Conclusively, this reaction mechanism explains the two possible products which can occur: if the acceptor is a proton a simple decarboxylase reaction takes place and an aldehyde is released; if the acceptor is an aldehyde a carboligation reaction can occur forming a 2-hydroxy ketone. As both reactions take place at the same active site, the steric and electronic properties of the active center influence both reactions similarly. Therefore, the investigation of the substrate range of decarboxylation is of interest in order to deduce information about the acyl donor (first substrate) spectrum for the carboligase activity of ThDP-enzymes. If a 2-keto acid is a substrate for the decarboxylation, the binding of the corresponding aldehyde to the C2 atom of ThDP located in the active center is reasonable, meaning that this aldehyde may be also a possible donor aldehyde in enzyme-catalyzed carboligation reactions. Factors influencing chemo- and stereoselectivity of the carboligation reaction are discussed in Section 4.5.

4.2
Carboligation: Chemo- and Stereoselectivity

4.2.1
Carboligations with Two Different Aldehydes

The mixed carboligation of benzaldehyde and acetaldehyde has been thoroughly studied for a broad range of ThDP-enzymes and is used as a preliminary test reaction to obtain valuable information about their chemo- and enantioselectivity [42–53]. The possible product range accessible from such carboligations is shown in Scheme 4.4. If both substrates are identical, only one (regioisomeric) product is gained, which can occur in the R- or S-configuration or a mixture of both.

The situation is more complex in the case of mixed carboligations of two different substrates, which is explained below.

4.2.1.1 Chemoselectivity
If a mixed product appears in biotransformations containing benzaldehyde and acetaldehyde as substrates, it can be either phenylacetylcarbinol (PAC), obtained from acetaldehyde as the donor and benzaldehyde as the acceptor, or 2-hydroxypropiophenone (2-HPP), resulting from benzaldehyde as the donor and acetaldehyde as the acceptor (Figure 4.1).

Whether 2-HPP or PAC is obtained by the respective enzyme depends on the chemoselectivity of the biocatalyst and may depend on the choice of substrates (e.g. benzoylformate versus benzaldehyde, pyruvate versus acetaldehyde) and their respective concentrations. Due to the variation of the shapes and amino acid residues

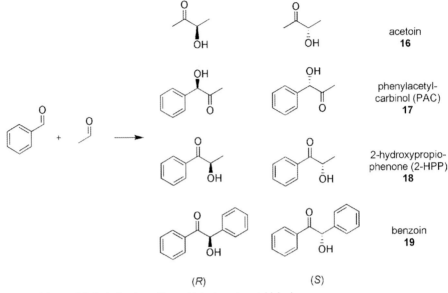

Scheme 4.4 Carboligation of benzaldehyde and acetaldehyde.
Four different products can occur, each in the R- or
S-configuration.

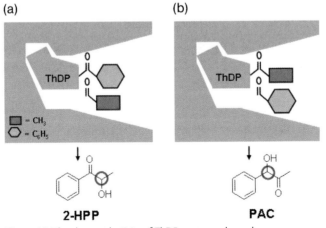

Figure 4.1 The chemoselectivity of ThDP-enzymes depends on
the binding order of the substrates in the active site. If the
benzaldehyde is bound to the C2 atom of ThDP, it functions as the
donor aldehyde and the aliphatic acetaldehyde as the acceptor
aldehyde yielding 2-HPP (a). If acetaldehyde is the donor and
benzaldehyde the acceptor PAC is formed (b).

forming the active sites of ThDP-dependent 2-keto acid decarboxylases the chemo-selectivity of these biocatalysts can differ significantly, which is discussed in detail in Section 4.5.

4.2.1.2 Stereoselectivity

In recent years, the ability of ThDP-enzymes to catalyze the stereoselective carboliga-tion of aldehydes yielding diverse chiral 2-hydroxy ketones has been intensively stu-died with different wild-type enzymes and variants thereof [43, 44, 48, 49, 52, 54–65]. From these studies, it became apparent that the stereocontrol is strict only if the carboligation reaction encompasses at least one aromatic aldehyde, whereas with two aliphatic aldehydes only moderate enantiomeric excesses could be obtained [52, 55]. It should be emphasized that in most mixed carboligations employing either solely aromatic or aromatic and aliphatic aldehydes the R-products are formed exclusively. There is currently only one exception from this rule: BFD from *Pseudomonas putida* (*Pp*BFD) catalyzes the formation of (*S*)-2-HPP (*ee* 92%) from benzaldehyde and acetaldehyde (Figures 4.1 and 4.2), but yields (*R*)-benzoin (*ee* 99%) in self-ligation reactions with benzaldehyde [46, 66] (see Table 4.5A). The structural basis for the differences in stereo control have recently been investigated by comparing the active

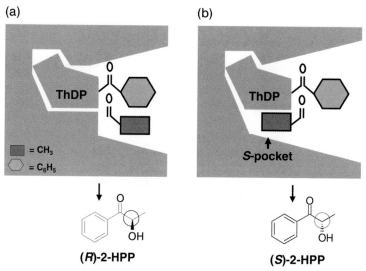

(a)   (b)

(*R*)-2-HPP   (*S*)-2-HPP

Figure 4.2 Molecular models showing the assumption for the orientation of the acceptor aldehyde in *Pf*BAL (a) and *Pp*BFD (b). In both cases benzaldehyde is assumed to adapt a coplanar conformation relative to the thiazolium ring, due to mesomeric effects. The orientation of the oxygen of the donor aldehydes (benzaldehyde, bright gray) and the acceptor aldehydes (acetaldehyde, dark gray) is directed upwards towards H29 in *Pf*BAL (mediated by a water molecule) and H70 in *Pp*BFD, acting as proton relays [67]. As there is no other possibility, in *Pf*BAL the acceptor aldehyde must be always arranged parallel to the donor aldehyde, as demonstrated for acetaldehyde leading to the formation of (*R*)-HPP. In the case of *Pp*BFD, the *S*-pocket allows the antiparallel orientation of the acceptor acetaldehyde relative to the donor benzaldehyde, resulting in the formation of (*S*)-HPP (according to [67]).

sites of *Pp*BFD with benzaldehyde lyase from *Pseudomonas fluorescens* (*Pf*BAL), which is strictly *R*-specific [67]. Modeling studies revealed a so-called *S*-pocket in *Pp*BFD, which could bind small acceptor aldehydes such as acetaldehyde, whereas no such pocket exists in *Pf*BAL (Figure 4.2). According to the modeling studies, the *S*-pocket in *Pp*BFD is just large enough to adopt the side-chain of acetaldehyde, whereas larger aldehydes cannot bind to this pocket. Consequently, larger aldehydes are arranged parallel to the ThDP-bound donor aldehyde, resulting in the formation of (*R*)-2-hydroxy ketones. Hence *R*-selectivity seems to be an intrinsic property of all 2-keto acid decarboxylases and *Pf*BAL. Using site-directed mutagenesis, the *S*-pocket in *Pp*BFD was recently enlarged, yielding BFD variants able to catalyze the enantioselective carboligation of larger aliphatic aldehydes [68].

4.3
Selected Enzymes

This section gives a review of the literature on different 2-keto acid decarboxylases, benzaldehyde lyase and acetohydroxy acid synthase.

4.3.1
2-Keto Acid Decarboxylases

Currently five different classes of ThDP-dependent 2-keto acid decarboxylases have been studied concerning their carboligation potential. A survey of the recent literature and available 3D structures is given in Table 4.1.

4.3.1.1 Pyruvate Decarboxylases
Pyruvate decarboxylases (PDCs; EC 4.1.1.1) catalyze the non-oxidative decarboxylation of pyruvate to acetaldehyde. They are the key enzymes in fermentative ethanol pathways and commonly found in

1. plants (especially in the seeds); for example *Oryza sativa* (rice) [69–71], *Zea mays* (maize) [72–74] and *Pisum sativum* (pea) [75, 76];
2. yeasts; for example brewer's yeast [77], *Saccharomyces cerevisiae* (*Sc*PDC) [78], *Saccharomyces kluyveri* [79], *Kluyveromyces marxianus* [80], *Kluyveromyces lacis* [27, 81] and *Zygosaccharomyces bisporus* [82, 83];
3. fungi; for example *Neurospora crassa* [84], *Aspergillus nidulans* [85] and *Rhizopus oryzae* [86];
4. bacteria; Gram-negative bacteria, *Zymomonas mobilis* (*Zm*PDC) [87], *Zymobacter palmae* (*Zp*PDC) [88] and *Acetobacter pasteurianus* (*Ap*PDC) [89]; Gram-positive bacterium, *Sarcina ventriculi* (*Sv*PDC) [90].

By contrast, no PDCs have been found in mammals.

Pyruvate decarboxylases are key enzymes in the production of biofuels from bulk plant materials such as sugars and cellulose. Currently, bacterial strains with engineered ethanol pathways, for example *Escherichia coli* strains with an inserted

ZmPDC, are being investigated for industrial applications [91]. Moreover, PDCs are able to catalyze a benzoin-condensation such as carboligation of aldehydes ('acyloin condensation').

A comparison of kinetic and stability data for various 2-keto acid decarboxylases is presented in Table 4.2 and a comparative survey of the substrate spectra and kinetic parameters for the decarboxylation reaction is given in Tables 4.3 and 4.4.

PDC from *Saccharomyces cerevisiae* (ScPDC)   ScPDC has been intensively studied concerning the mechanism of the decarboxylase reaction. Especially the substrate activation observed in brewer's yeast and ScPDC have been thoroughly studied [30, 92]. The sigmoidal v–[S] plot for yeast PDCs was described 30 years ago [93, 94]. Similar results have been obtained with plant PDCs [95] and with the prokaryotic PDC from *Sarcina ventriculi* (SvPDC) [96]. Although the binding site for the activator is still a matter of discussion, it is evident that its binding leads to structural changes in the tetrameric arrangement, resulting in a more compact active structure [97]. The substrate range of the decarboxylase reaction of ScPDC has recently been studied relative to bacterial PDC from *Zymomonas mobilis*, *Zymobacter palmae* and *Acetobacter pasteurianus*, demonstrating a broad substrate range comparable to ApPDC (Table 4.3), with the lowest specific activity towards pyruvate relative to the bacterial PDCs (Table 4.4) [13].

The carboligase potential of the PDC from yeast, in particular brewer's yeast, has been known for more than 100 years [98] and used for the production of (*R*)-PAC (**17**) (Scheme 4.4), a pre-step in the synthesis of ephedrine by whole cell biotransformation [99]. The investigation and optimization of the PAC synthesis are still a matter of research [100–102]. Both brewer's yeast and ScPDC show high enantioselectivity towards the *R*-product and high chemoselectivity. All pyruvate decarboxylases prefer small aliphatic donor aldehydes (in the form of the respective 2-keto acid) and aliphatic or aromatic acceptor aldehydes. Thus, acetoin and PAC (Scheme 4.4) are typical ligation products of PDCs [103–105] (Table 4.5).

PDC from *Zymomonas mobilis* (ZmPDC)   In addition to the yeast PDCs, the enzyme from *Zymomonas mobilis* has been intensively investigated. *Zymomonas mobilis* is a model microorganism for the Entner–Doudoroff pathway in which glucose is converted anaerobically to ethanol and CO_2. Analysis of purified ZmPDC concerning kinetics, pH stability, salt effects and thermostability revealed Michaelis–Menten-like kinetics for the decarboxylation of pyruvate (Table 4.4) [106]. In contrast to ScPDC, no substrate activation was observed. First mutagenesis studies were performed based on a structural model derived from the 3D structure of ScPDC [21] to investigate the reaction mechanism and to identify hot spots in the active site [53, 107–112]. The crystal structure of ZmPDC was finally published in 1998 [12], demonstrating a compact tetrameric structure, which resembles the quaternary structure of ScPDC upon substrate activation [30].

The carboligase activity of ZmPDC was first described by Bringer-Meyer and Sahm in 1988 [113], investigating the formation of PAC and acetoin with partially purified ZmPDC.

Table 4.2 Optima and stabilities of various 2-ketoacid decarboxylases[a].

	LlKdcA [50]	ApPDC [13]	ZpPDC [13]	ScPDC [13]	ZmPDC [106]	PpBFD [46]
Optimum pH (initial rate conditions)	6–7	3.5–6.5	4.5–8 Maximum at 7	5–7	5.5–8.0 Maximum at 6–6.5	5.5–7.0 Maximum at 6.2
pH-dependent stability half-life time	pH 5–7: no activity loss within 60 h pH 8: 40 h pH 4: < 2 h	pH 5–7: no activity loss within 60 h pH 4: 2.3 h	pH 6.5/8: stable for several days pH 4/9: complete activity loss within 2 h	pH 5: 80 h pH 7: 53 h pH 8: 13 h	n.d. in potassium phosphate buffer	pH 5.5: deactivation 0.3% min^{-1} pH 10: deactivation 0.1% min^{-1}
Optimum temperature (°C) (initial rate conditions)	50	65	55	43	60	68
Temperature dependent stability half-life time	40 °C: 80 h 50 °C: 9 h 55 °C: 4 h	20 °C: 193 h 30 °C: 144 h 40 °C: 34 h 50 °C: 12 h 60 °C: 2 h 70 °C: 0.4 h	30 °C: 150 h 40 °C: 40 h 50 °C: 10 h 60 °C: 0.4 h	20 °C: 235 h 30 °C: 78 h 35 °C: 62 h	50 °C: 24 h	60 °C: 2 h 80 °C: 15–20 min
Activation energy for decarboxylation (kJ mol^{-1}) Substrate	8.5 3-Methyl-2-oxobutanoic acid	27.1 Pyruvate	41 Pyruvate	n.d.	43 Pyruvate	38 Benzoylformate
Stability in organic solvents (half-life times measured at 30 °C)	Stable in 20% v/v DMSO (half-life 150 h); Unstable in 15% v/v PEG 400 (half-life 6 h)	Stability enhanced in 30% v/v DMSO (half-life 430 h)	n.d.	Stable in 20–30% v/v DMSO	n.d.	n.d.

[a]Optima were determined under initial rate conditions within 90 s using a direct decarboxylase assay. For stability investigations, the enzymes were incubated under the respective conditions. Samples, withdrawn at appropriate time intervals, were subjected to residual activity determination using the coupled decarboxylase assay. n.d., not determined.

Table 4.3 Substrate range of the decarboxylation reaction of various 2-keto acid decarboxylases[a].

2-Keto acid	Specific activity [U mg^{-1}]					
	L/KdcA [50]	ApPDC [13]	ZmPDC [53]	ScPDC [13]	ZpPDC [13]	PpBFD [68]
1	1.9	93	120	43.4	147	0.4
2	14.1	60.5	79	16.9	85.1	21.4
3	22.7	12.2	13	18.8	20.7	12.5
4	20.1	4.1	0.2	5.3	6.2	9.2
5	0.0	0.3	n.d.	0.0	0.6	7.2
6	152	n.d.	0.0	6.9	10.4	0.0
7	40	0.8	0.3	0.3	2.8	3.3
8	57.9	0.9	0.0	0.0	2.8	3.6
9	28.9	1.2	0.0	0.0	0.4	6.9
10	0.0	0.8	n.d.	0.0	0.6	0.0
11	12.8	3.4	0.0	0.0	0.3	420.1
12	13.1	0.4	0.0	0.0	1.8	0.0
13	2.4	1.0	0.0	1.7	0.3	4.4

(*Continued*)

Table 4.3 (*Continued*)

2-Keto acid	Specific activity [U mg^{-1}]					
	*Ll*KdcA [50]	*Ap*PDC [13]	*Zm*PDC [53]	*Sc*PDC [13]	*Zp*PDC [13]	*Pp*BFD [68]
14	1.7	0.0	0.0	0.2	0.2	0.0
15	1.3	0.0	n.d.	0.0	0.5	1.5

[a]Activity was measured using the coupled decarboxylase assay at 30 °C. Substrate concentration: 30 mM (**1–14**), 1 mM (**15**). Buffer conditions are given in Table 4.2, bottom row. The natural substrates (bold data) resulted in highest specific activity and were used for further kinetic studies (Table 4.4). n.d., not determined.

Table 4.4 Kinetic parameters for the decarboxylation of 2-keto acids by various 2-keto acid decarboxylases[a].

Enzyme	Substrate	V_{max} [U mg^{-1}]	K_M [mM]	Shape of the v–[S] curve
*Ap*PDC[b] [13]	**1**	110 ± 1.9	2.8 ± 0.2	Michaelis–Menten kinetics
*Zp*PDC[b][13]	**1**	116 ± 2.0	2.5 ± 0.2	Michaelis–Menten kinetics
*Zm*PDC[c][53]	**1**	120–150	1.1 ± 0.1	Michaelis–Menten kinetics
*Sc*PDC[b][13]	**1**	59.9 ± 6.2	12.1 ± 1.5	Sigmoidal ($h = 1.9 ± 0.2$) and substrate surplus inhibition ($K_s = 87.5 ± 10.7$)
*Ll*KdcA[b][50]	**6**	181.6 ± 1.7	5.02 ± 0.2	Michaelis–Menten kinetics
	1	3.7 ± 0.3	29.8 ± 6.4	Michaelis–Menten kinetics
	12	15.7 ± 0.2	0.127 ± 0.007	Michaelis–Menten kinetics

Table 4.4 (*Continued*)

Enzyme	Substrate	V_{max} [U mg^{-1}]	K_M [mM]	Shape of the v–[S] curve
	15	1.55 ± 0.6	0.234 ± 0.024	Michaelis–Menten kinetics
AbPhPDC^d[145]	**12**	5.56 ± 0.13	1.08 ± 0.09	Michaelis–Menten kinetics (h = 1)
	15	0.07 ± 0.002	0.13 ± 0.01	Sigmoidal (h = 1.85)
PpBFD^e[46, 68]	**11**	400 ± 7	0.37 ± 0.03	Michaelis–Menten kinetics

^{a}All data were deduced using the coupled decarboxylase assay. Kinetic parameters were calculated by non-linear regression using the Michaelis–Menten equation in Origin 7G SR4 (OriginLab Corp., Northampton, MA, USA), except for ScPDC and AbPhPDC, where the Hill equation was used (for further information see [13]); h = Hill coefficient, K_s= inhibition constant.
^{b}Reaction conditions as described in Table 4.2 (bottom).
^{c}50 mM MES–KOH-buffer pH 6.5, 2.5 mM MgSO$_4$, 0.1 mM ThDP [53].
^{d}10 mM MES buffer pH 6.5, 2.5 mM MgSO$_4$, 0.1 mM ThDP.
^{e}50 mM potassium phosphate buffer pH 6.5, 2.5 mM MgSO$_4$, 0.1 mM ThDP.

Site-directed mutagenesis was used to improve carboligase activity towards PAC and to identify residues which are important for the enantioselectivity of the carboligation [53, 111, 112]. From these studies, isoleucine 472 was determined to be a hot spot influencing enantio- and chemoselectivity [53, 112]. Tryptophan 392 was identified as a key residue limiting the access to the active center for sterically demanding aromatic substrates [107, 111]. The highly potent variant ZmPDCW392M was tested in a continuous enzyme–membrane reactor, giving space–time-yields of 81 g L^{-1} d^{-1} [114–116].

PDC from *Acetobacter pasteurianus* (ApPDC)   *Acetobacter pasteurianus* (ATCC 12874) is an obligate oxidative bacterium. In physiological studies, ApPDC was found to be a central enzyme for the oxidative metabolism, which is in contrast to the fermentative ethanol pathways in plants, yeasts and other bacterial PDCs described above [89]. *Acetobacter pasteurianus*, a common spoilage organism of fermented juices, derives energy and carbon from the oxidation of D- and L-lactic acid. During growth, ApPDC cleaves the central intermediate pyruvate to acetaldehyde and CO$_2$.

Recombinant, purified ApPDC was described with respect to its pH and temperature optima for the decarboxylase reaction [29]. With pyruvate, Michaelis–Menten kinetics were detected similarly to ZmPDC (Table 4.4).

Recently, for ApPDC the catalytic properties concerning decarboxylation and carboligation and the 3D structure have been studied [13]. Compared with ZmPDC, the active site is larger, which results in considerable activity towards the decarboxylation of aromatic 2-keto acids, such as benzoylformate (Table 4.3), and in less strict stereocontrol during carboligation, for example for the formation of (R)-PAC and derivatives thereof (see Table 4.5). Structure–function studies concerning the size of the active site and stereoselectivity are discussed in Section 4.5.

PDC from *Zymobacter palmae* (ZpPDC)   The ethanologenic bacterium *Zymobacter palmae* (ATCC 51623) was originally isolated from palm sap [88], producing ethanol from hexose sugars and saccharides like most other PDCs [117]. The gene was cloned and ZpPDC was characterized with respect to its putative application as a biocatalyst for fuel ethanol production from renewable biomass [29]. As has been observed for ZmPDC and ApPDC, ZpPDC shows Michaelis–Menten-like kinetic for pyruvate (Table 4.4). Concerning the substrate range for the decarboxylation reaction, the enzyme resembles ApPDC (Table 4.3). The carboligase activity of this enzyme has been tested in whole cell biotransformations [118] and is currently under investigation by the present authors.

4.3.1.2   Branched-Chain Keto Acid Decarboxylases

Branched-chain keto acid decarboxylases (EC 4.1.1.72) have been described from *Bacillus subtilis* [119], *Lactococcus lactis* IFPL730 [120] and *Lactococcus lactis* subsp. *cremoris* (LlKdcA) [121]; the last is suggested to be the key enzyme for the metabolism from leucine to 3-methylbutanal, which is crucial for flavor formation during cheese ripening [122]. The 3D structure of the LlKdcA has been solved recently, demonstrating that this enzyme is active as a dimer, which is in contrast to other known ThDP-dependent decarboxylases [20]. LlKdcA has an exceptionally broad substrate range encompassing linear and branched-chain aliphatic and aromatic 2-keto acids [50, 121, 123], showing the highest maximum velocity with 3-methyl-2-oxobutanoic acid (**6**) (Table 4.3). Although the maximum activity in the presence of phenylpyruvate (**12**) is only about 9% of the velocity towards the natural substrate **6**, the K_M value is 40 times lower. This high affinity for phenylpyruvate and also for indole-3-pyruvate is of special interest for carboligase reactions with C–H-acidic aldehydes (see Table 4.6), as discussed in Section 4.4.

4.3.1.3   Benzoylformate Decarboxylases

Benzoylformate decarboxylase (BFD; EC 4.1.1.7) activity was first described in the bacterium *Pseudomonas putida* (ATCC 12633) by Hegeman [124]. It is the third enzyme of the mandelic acid catabolism, where mandelic acid is converted to benzoic acid in order finally to be metabolized in the β-ketoadipate pathway and the citric acid cycle. This enables the organism to grow on (R)-mandelic acid as a sole carbon source. The physiological function of BFD is the non-oxidative conversion of benzoylformate to benzaldehyde and CO_2. The coding gene from *Pseudomonas putida* was cloned in 1990 [125]. Among the several putative genes encoding potential BFDs, besides PpBFD, only four were characterized concerning their activity, BFD from *Bradyrhizobium japonicum* [126], *Pseudomonas aeruginosa* [127], *Pseudomonas stutzeri* [128] and *Acinetobacter calcoaceticus* [127]. Henning *et al.* [129] isolated three further enzymes with BFD activity, two of them originating from a chromosomal library of *Pseudomonas putida* ATCC 12633 and the third from an environmental DNA library of a metagenome screening. Their physiological activity is still unknown.

The enzyme from *Pseudomonas putida* (*Pp*BFD) is by far the best characterized BFD. The substrate range for its physiological decarboxylase activity has been thoroughly studied, showing benzoylformate as the preferred substrate [46, 53] (Tables 4.3 and 4.4).

In 1992, Wilcocks and co-workers first described the ability of whole *Pseudomonas putida* cells to catalyze C–C bond formation of benzaldehyde, obtained after decarboxylation of benzoylformate and acetaldehyde yielding (*S*)-2-HPP (**18**) (*ee* 91–92%) [66, 130] (Table 4.5A). Detailed studies of the substrate range for the carboligation reaction using purified wild-type enzyme followed [42, 46, 131]. *Pp*BFD is able to catalyze the ligation of a broad range of different aromatic, heteroaromatic, cyclic aliphatic and olefinic aldehydes with acetaldehyde yielding exclusively the HPP-analogue (Scheme 4.4, Figure 4.1). Hence the selectivity for aromatic aldehydes as donor aldehydes is very high [42], which is also reflected in the relative activities of *Pp*BFD concerning the formation of (*S*)-2-HPP (7 U mg^{-1}; *ee* 92%), (*R*)-benzoin (0.25 U mg^{-1}; *ee* 99%) and (*R*)-acetoin (0.01 U mg^{-1}, *ee* 35%) [46, 52, 53] (Table 4.5). Depending on the substitution pattern of the aromatic ring, diverse HPP analogues are accessible in high yields and with good to high optical purity. Selectivity, activity and stability of *Pp*BFD have been optimized using reaction engineering. Best results have been obtained by adjusting very low benzaldehyde concentrations in a continuous reactor [46]. Directed evolution yielded variants with a mutation in position of leucine 476 with enhanced ligase activity and stereoselectivity for the formation of (*S*)-HPP, improved stability towards organic solvents and an enlarged donor substrate range [132, 133]. Based on the X-ray structure of the enzyme [14, 39], site-directed mutagenesis studies yielded variants with increased carboligase activity towards benzoin derivatives (*Pp*BFDH281A) [42] and also with an altered substrate range for the decarboxylation reaction (*Pp*BFDA460I), which additionally showed improved stereoselectivity during carboligation of benzaldehyde and acetaldehyde [53]. Recently, a further hot spot in *Pp*BFD was identified with leucine 461, a residue defining an important region in the acceptor binding site during carboligation [67]. Site-directed mutagenesis of this residue towards alanine and glycine yielded variants with an increased substrate range for the mixed (*S*)-2-hydroxy ketones from aromatic donor aldehydes and aliphatic acceptor aldehydes [68]. For more details, see Sections 4.4 and 4.5.

4.3.1.4 Phenylpyruvate Decarboxylases/Indole-3-pyruvate Decarboxylases

Indole-3-pyruvate decarboxylases (InPDC; EC 4.1.1.74) are involved in the biosynthesis of auxins, growth-promoting phytohormones in plants. Several biosynthetic pathways for the production of indole-3-acetic acid are discussed [134]. The best known is the indole-3-pyruvate pathway, where tryptophan is assumed to be first transaminated to indole-3-pyruvate, which is subsequently decarboxylated to indole-3-acetaldehyde by InPDC and finally oxidized to the auxin indole-3-acetic acid [135, 136]. In 1991, Koga *et al.* described the cloning and characterization of InPDC from *Enterobacter cloacae* (*Ec*InPDC) [137]. The highest decarboxylase activity was detected with indole-3-pyruvate whereas the activity with phenylpyruvate was negligible. Kinetic studies showed no indication of substrate activation [138]. The structure shows the typical homo-tetrameric fold of ThDP-dependent enzymes [23].

A similar gene was found in *Azospirillum brasilense* (*Ab*In3PDC) [134], a Gram-negative nitrogen-fixing rhizobacterium living in association with the roots of grasses and cereals due to its growth-promoting effect on root hairs [139]. *Ab*InPDC is also involved in the biosynthesis of phenylacetic acid from phenylalanine, an auxin with antimicrobial activity [140], demonstrating that the promiscuity of this enzyme may have an impact in two different pathways.

Phenyl pyruvate decarboxylases (PhPDC; EC 4.1.1.43) are involved in the Ehrlich pathway using amino acids as sole amino sources. The catabolism of phenylalanine comprises similar steps to the indole-3-pyruvate pathway: an initial transamination to phenylpyruvic acid is followed by a PhPDC-catalyzed decarboxylation to phenylacetaldehyde. However, the final step includes a reduction to phenylethanol with a rose-like aroma, which is interesting for the beverage and food industry [141]. PhPDCs have been described in bacteria species of *Acromobacter*, *Acinetobacter* and *Thauera* [142] and in the yeast *Saccharomyces cerevisiae* [141]. Their expression is inducible by growing the organism on phenylalanine, tryptophan or mandelate. Decarboxylase activity could be determined with phenyl pyruvate, indole 3-pyruvate and further 2-keto acids with more than six carbon atoms in a straight chain [142]. Asymmetric carboligase activity of PhPDC was investigated by Guo *et al.* using phenylacetaldehyde (derived by *in situ* decarboxylation of phenylpyruvate) and acetaldehyde as substrates, yielding the PAC analogue 3-hydroxy-1-phenyl-2-butanone by whole-cell catalysis of *Acromobacter eurydice*, *Pseudomonas aromatica* and *Pseudomonas putida* [143]. With purified *Acromobacter eurydice* PhPDC the ligation of phenylacetaldehyde (phenylpyruvate) with various aldehydes yielded also (*R*)-PAC derivatives with *ee*s of 87–98% [144].

A new enzyme of this class has recently been described in *Azospirillum brasilense* (*Ab*PhPDC), which was first classified as an indole-3-pyruvate decarboxylase due to its significantly lower specific activity towards indole 3-pyruvate (**15**) relative to phenyl pyruvate (**12**) [145]. Remarkably, *v*–[S] plots of *Ab*PhPDC for different 2-keto acids differ remarkably (Table 4.3), showing allosteric kinetic effects depending on the substrate [145]. Recently, Versées and co-workers demonstrated the mechanism of the substrate-induced activation mechanism by co-crystallization of *Ab*PhPDC with the substrate and a covalent reaction intermediate analogue of ThDP [40]. The carboligase properties of this enzyme are currently under investigation by the present authors.

4.3.2
Benzaldehyde Lyases

Up to now, the only biochemically characterized benzaldehyde lyase (BAL; EC 4.1.2.38) was found in *Pseudomonas fluorescens* Biovar I (*Pf*BAL) [146, 147]. The

Gram-negative bacterium was isolated from wood splits of a cellulose factory, showing the ability to grow on lignin-derived compounds such as anisoin (4,4′-dimethoxybenzoin) and benzoin as the sole carbon and energy sources. Due to the lyase activity of PfBAL, the organism can cleave these 2-hydroxy ketones yielding aldehydes, which are probably metabolized in the β-ketoadipate pathway.

Detailed analysis of the stability and lyase activity of recombinant, purified PfBAL was published by Janzen et al., revealing maximum specific activity for the cleavage of rac-benzoin of 74 U mg^{-1} ($K_M = 0.05$ mM) [48]. Due to its high R-selectivity PfBAL can be used for kinetic resolution, yielding enantiopure (S)-benzoin, which is otherwise difficult to access [148]. In addition to (R)-benzoin, the enzyme is able to cleave also (R)-2-HPP ($V_{max} = 3.6$ mM; $K_M = 0.3$) [48]. In contrast to the other ThDP-dependent enzymes described above, the decarboxylase activity of PfBAL is negligible (0.02 U mg^{-1} for benzoyl formate); however, the decarboxylase activity could be significantly enhanced by site-directed mutagenesis, yielding a variant PfBALA26S with a significant decarboxylase activity (2.8 U mg^{-1}) [41, 48].

Demir et al. later confirmed that the lyase reaction is reversible, enabling PfBAL also to form C–C bonds. Especially the self-ligation of benzaldehyde towards benzoin is catalyzed with high activity (336 U mg^{-1}) and very high enantioselectivity (ee >99% R), making this enzyme very interesting for industrial processes [148]. For benzoin formation, the substrate range of aldehydes substituted in ortho, meta and para positions is rather broad [65]. Mixed carboligations of several aromatic and aliphatic aldehydes result in the formation of (R)-2-HPP analogues in addition to the benzoin derivative (Scheme 4.4) [65, 148]. Also phenylacetaldehyde is accepted as a donor and/or acceptor aldehyde [63]. In addition to acetaldehyde, mono- and dimethoxy-acetaldehyde and formaldehyde have been successfully applied as aliphatic acceptor aldehydes [7, 49, 65, 148, 149].

The transformation of aromatic aldehydes requires the use of cosolvents to improve the solubility of the aromatic substrates and products in certain cases. Dimethyl sulfoxide (DMSO) was shown to be extremely useful, not only for the solubility of the aromatic compounds but also for the stability of PfBAL during biotransformations [47]. Alternatively, methyl tert-butyl ether can be used as a beneficial alternative, permitting a simplified work-up procedure [150]. Recently, the carboligation of different mixed 2-hydroxy ketones was studied in continuously operated membrane reactors using aqueous buffer systems with 30% v/v DMSO [56, 57, 151].

Further studies concerned the application of immobilized PfBAL in cryogel beads which can be used in organic solvents. This two-phase system was successfully applied in both batch and continuous reactions for the synthesis of (R)-3,3′-furoin [60, 61]. The crystal structure of this highly versatile biocatalyst was solved in 2005 [15]. It opened the way to rationalize effects observed with mutations in the active site which have been produced based on structural models [41, 48] and a structure-guided explanation for the differences in stereoselectivity during carboligation observed between PfBAL and PpBFD [67].

4.3.3
Acetohydroxy Acid Synthases

Acetohydroxy acid synthases (AHAS; EC 4.1.3.18) are found in plants, fungi and bacteria [152, 153]. They catalyze the first common step in branched-chain amino acid biosynthesis. Thereby acetaldehyde, formed by initial decarboxylation of pyruvate, is ligated with either 2-ketobutyrate or pyruvate as an acceptor [34, 154]. Due to its presence in plants, AHAS is a target for sulfonylurea and imidazolinone herbicides [155]. AHAS are composed of two types of subunits and a similar arrangement has been found recently for the yeast and plant enzymes. One type of subunit contains the catalytic machinery, whereas the other has a regulatory function [156, 157]. In bacteria, the enzyme has a large subunit containing the catalytic machinery and a non-catalytic FAD [158] in addition to a small subunit with a regulatory role. Two isoenzymes of *E. coli* have been intensively studied concerning their potential to catalyze the formation of chiral 2-hydroxy ketones, such as acetoin (Scheme 4.4, **16**) and PAC (**17**) [62, 159–164]. As was observed with most other ThDP-enzymes, both isoenzymes are strictly *R*-specific concerning the formation of PAC. *Ec*AHAS I proved to be especially useful for 2-hydroxy ketone formation and the substrate range concerning the formation of chiral mixed 2-hydroxy ketones from aromatic or heteroaromatic donor aldehydes and acetaldehyde (generated by decarboxylation of pyruvate) was studied in detail [62].

4.4
Enzymes for Special Products

4.4.1
Mixed Carboligation of Benzaldehyde and Acetaldehyde

For all PDCs investigated, phenylacetylcarbinol (PAC) (Scheme 4.4, Table 4.5, **17**) was identified as the main product, with small amounts of acetoin (**16**). This result is in accord with those obtained by whole cell transformation experiments with *Zp*PDC [118]. The same holds for studies with AHAS from *E. coli* [160]. Consequently, all of them exclusively bind acetaldehyde as the donor aldehyde and benzaldehyde as the acceptor, if the acetoin formation is suppressed by an appropriate choice of the reaction conditions (e.g. excess of benzaldehyde). Concerning the stereoselectivity of the carboligation, (*R*)-PAC is formed enantioselectively by *Zm*PDC (*ee* 98%) [53], whereas *Sc*PDC (*ee* 90%) [165] and *Ap*PDC (*ee* 91%) [13] show lower stereoselectivity (Table 4.5A). A possible structural explanation for this behavior is part of a recent publication [13] and is discussed in Section 4.5.

   *Pp*BFD, however, accepts benzaldehyde as a donor and acetaldehyde as an acceptor, yielding predominantly 2-hydroxypropiophenone (HPP, **18**) with excess of the *S*-enantiomer (Table 4.5A).

   *Ll*KdcA shows almost no chemoselectivity in carboligase reaction with acetaldehyde and benzaldehyde. As demonstrated in Table 4.5, both aldehydes may act as

Table 4.5 Selected chiral 2-hydroxy ketones accessible by carboligation of different aliphatic aldehydes with benzaldehyde and 3,5-dichlorobenzaldehyde catalyzed by different ThDP-dependent enzymes[a].

A

Enzyme	Donor	Acceptor	Acetoin ee **16**	PAC product ee **17**	HPP product ee **18**	Benzoin ee **19**
ZmPDC [53]			n.d.	98% R	—	—
ScPDC [165]			—	90% R	—	—
ApPDC [13]			n.d.	91% R	—	n.d.[b]
LlKdcA [50]			n.d.	92% R	93% R	90% R[b]
PpBFD [46, 53]			35% R	—	92% S	99% R
PpBFDL461A [68]			—	—	98% S	—

B

Enzyme	Donor	Acceptor	Acetoin ee **20**	PAC product ee **21**	HPP product ee **22**	Benzoin ee **19**
ApPDC [13]			n.d.	82% R	n.d.	—
LlKdcA [50]			n.d.	> 98% R	—	—
PpBFD [68]			n.d.	98% R	21% R	99% R
PpBFDL461A [68]			n.d.	—	93% S	n.d.
PpBFDL461G [68]			n.d.	—	97% S	n.d.

(Continued)

Table 4.5 (Continued)

C

Products: **23**, **24**, **25**, **19**

	23	24	25	19
ApPDC [13]	—	81% R	—	—
LlKdcA [50]	—	98% R	—	—

D

Products: **16**, **26**, **27**, **28**

	16	26	27	28
LlKdcA [50]	—	—	96.5% R	—

E

Product: **16**

ZmPDC [53, 169]	50–60% S
ApPDC [13]	31% S
ScPDC [169]	20% R
LlKdcA [50]	47% R
PpBFD [46, 52, 53]	35% R
PfBAL [52]	40% R

F

Product: **29**

PpBFD [52]	85–86% R
PfBAL [52]	86–89% R
LlKdcA [50]	30–47% S

[a] n.d., product was formed, but ee was not determined.
[b] Traces of benzoin have been observed only, also in absence of the aliphatic aldehydes.

either a donor or acceptor, yielding almost equimolar amounts of **17** and **18**, which is probably a consequence of the steric properties in the active site, as discussed in Section 4.5.

4.4.2
Mixed Carboligation of Larger Aliphatic and Substituted Aromatic Aldehydes

Mixed carboligase reactions with substituted benzaldehydes and different aliphatic aldehydes in order to gain versatile PAC analogue products have been investigated using *Ll*KdcA and *Ap*PDC as biocatalysts (Table 4.5). The (*R*)-PAC analogue of benzaldehyde and propanal (**21**) as well as benzaldehyde and cyclopropylcarbaldehyde (**24**) are selectively formed by *Ll*KdcA catalysis (Table 4.5B,C).

Chemo- and stereoselectivity are strongly influenced by the size of the aromatic aldehyde, as demonstrated in mixed carboligations of 3,5-dichlorobenzaldehyde and acetaldehyde (Table 4.5D), by the electronic properties of the substrates [42] and, of course, by the enzyme–substrate combination. Although *Ll*KdcA catalyzes the carboligation of 3,5-dichlorobenzaldehyde and acetaldehyde, the resulting product is not the expected PAC analogue but the HPP derivative **27** (Table 4.5D) [50]. The capability to influence the chemoselectivity of the carboligation reaction by an appropriate combination of the donor and acceptor aldehyde using *Ll*KdcA is pronounced among the currently known ThDP-dependent enzymes with carboligase activity and shows the great advantage of this enzyme to access both HPP and PAC analogous 2-hydroxy ketones. These observations have recently been explained based on the 3D structure of *Ll*KdcA [20].

4.4.3
Self-Ligation of Aromatic Aldehydes

The highest benzoin ligase activity was observed with *Pf*BAL (336 U mg^{-1}) [48]. *Pp*BFD catalyzes (*R*)-benzoin formation if benzaldehyde or benzaldehyde derivatives are used as the sole substrate, but with a distinctly lower rate compared with the (*S*)-2-HPP formation in the presence of acetaldehyde [131]. However, benzoin forming activity of *Pp*BFD was improved by site-directed mutagenesis [42]. Further, the PDCs and *Ll*KdcA catalyze benzoin formation from benzaldehyde in almost negligible amounts. With *Ap*PDC, benzoin formation was observed, demonstrating that benzaldehyde can bind to the C2 atom of ThDP as a donor aldehyde, although the activity is very low.

4.4.4
Self-Ligation of Aliphatic Aldehydes

The formation of acetoin starting from either acetaldehyde or pyruvate has been described for *Sc*PDC [166], some further yeast strains [83, 167] and *Zm*PDC [113, 168]. Although *Zm*PDC is strictly *R*-specific concerning the formation of PAC and PAC analogues, it catalyzes the formation of predominantly (*S*)-acetoin (*ee* 50–60%)

[53, 169]. The latter is in contrast to *Sc*PDC catalyzing the formation of (*R*)-acetoin in slight excess (*ee* 20%) [169] (Table 4.5E).

The formation of further aliphatic 2-hydroxy ketones has recently been described using *Pf*BAL, *Pp*BFD [52] and *Ll*KdcA [50] as biocatalysts. Highest stereoselectivities have been observed with branched-chain aldehydes, for example isovaleraldehyde, yielding enantiocomplementary products with *Pf*BAL, *Pp*BFD [52] and *Ll*KdcA [50] (Table 4.5F).

Nevertheless, the enantioselectivity of this reaction is in all cases only low to moderate and most probably a consequence of less stabilization of small aldehydes in the active site, which is further discussed in Section 4.5.

4.4.5
Carboligation of Unstable Aldehydes

CH-acidic aldehydes such as phenylacetaldehyde and indole-3-acetaldehyde are prone to enolization and therefore are difficult substrates for both enzymatic and chemical transformations [170]. Moreover, indole-3-acetaldehyde is very unstable and decomposes rapidly, which makes its direct application in biotransformations very difficult [171]. Phenylacetaldehyde is commercially available; however, in aqueous buffer aldol reaction occurs spontaneously. However, these problems can easily be overcome by *in situ* production of these aldehydes through enzymatic decarboxylation of the corresponding 2-keto acids. Referring to the reaction mechanism (Section 4.1.3, Scheme 4.3), the decarboxylation of a 2-keto acid results in a reaction intermediate (activated aldehyde) with the corresponding aldehyde bound to the cofactor ThDP, which can further react with a suitable electrophile/acceptor aldehyde.

4.4.5.1   *Ll*KdcA Catalyzes the Mixed Carboligation of CH-Acidic Aldehydes and Acetaldehyde

*Ll*KdcA and both phenylpyruvate decarboxylases from *Saccharomyces cerevisiae* and *Azospirillum brasilense* are able to decarboxylate phenylpyruvate. *Ll*KdcA and *Ab*PhPDC were also shown to decarboxylate indole-3-pyruvate with low activity (Table 4.3) [50, 145, 172]. Carboligase data presented in Table 4.6 demonstrate that *Ll*KdcA is able to catalyze the formation of the HPP-analogue **31** from acetaldehyde and indole-3-pyruvate as substrates with high chemoselectivity [50]. *Ll*KdcA further catalyzes the formation of a mixture of the HPP (**33**) and PAC analogous (**34**) products in mixed carboligations of phenylpyruvate and acetaldehyde (Table 4.6).

4.4.6
Accessing (*S*)-2-Hydroxy Ketones

As has been stated before, most of the known ThDP-enzymes are *R*-specific in mixed carboligation reactions with aromatic and aliphatic aldehydes, with *Pp*BFD being the only exception, coupling benzaldehyde and acetaldehyde to (*S*)-2-HPP with an *ee* of 92% (Table 4.5A) [46]. However, *S*-selective carboligation only works with acetaldehyde as the acceptor and fails with larger acceptor aldehydes such as propanal

Table 4.6 C-C coupling of CH-acidic aldehydes with LlKdcA [50].

Substrates	Possible products			
A	16	30	31	32
LlKdcA	Preparative scale, product yield: 23%	—	100%	—
B	16	33	34	35
LlKdcA	Preparative scale, product yield: 61%	—	20%	80%	—

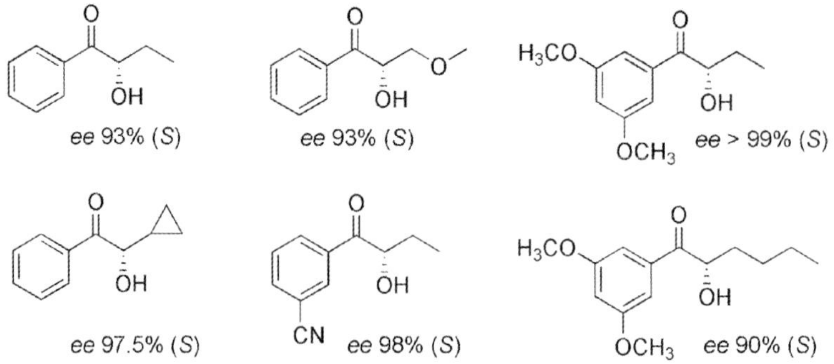

Scheme 4.5 Biocatalytically accessible 2-hydroxy ketones with high excess of the S-enantiomer catalyzed by *Pp*BFDL461A [68].

(Table 4.5B). In order to increase the range of accessible *S*-products, the molecular reasons for *R*-selectivity of ThDP-enzymes were investigated in more detail in order to elucidate the underlying mechanistic principle, which allows the introduction of *S*-selectivity by site-directed mutagenesis. Leucine 461 was identified as the main factor determining the size of the *S*-pocket in *Pp*BFD by molecular modeling studies [67]. The hypothesis was confirmed by previous studies concerning the expansion of the substrate range of *Pp*BFD to accept *ortho*-substituted benzaldehydes as donors, where a variant *Pp*BFDM365L/L461S was identified catalyzing the formation of various *ortho*-substituted (*S*)-2-HPP-analogues [133].

Shaping the *S*-pocket by site-directed mutagenesis of leucine 461 towards alanine and glycine in *Pp*BFD resulted in variants which open access to various (*S*)-hydroxy ketones with high enantioselectivity [68] (Table 4.5B, Scheme 4.5).

The biochemical data suggest that there is an optimal size for the acceptor aldehyde yielding high enantio- and chemoselectivity. If the acceptor is too large for the *S*-pocket, it is not optimally stabilized and a parallel arrangement relative to the donor aldehyde is an alternative (Figure 4.2). Compared with benzaldehyde, the stereo control in mixed carboligations with propanal is even better with 3,5-dimethoxy-benzaldehyde and 3-cyanobenzaldehyde (Scheme 4.5). This shows that not only the perfect stabilization of the acceptor aldehyde but also the interplay of donor and acceptor aldehydes fitting into the active site influence the chemo- and enantio-selectivity of the biocatalysts [68].

4.5
Investigation of Structure–Function Relationships

4.5.1
Deducing General Principles for Chemo- and Enantioselectivity

Based on the now available broad structural and biochemical database, structure–function studies are possible. The newly obtained 3D structures of *Ap*PDC and

*Ll*KdcA were superimposed with the structures of *Sc*PDC, *Zm*PDC, *Pp*BFD and *Pf*BAL and the active sites and the channels connecting the active sites with the protein surface have been compared. An overview of the assignment of amino acids lining the active sites at similar positions in the various enzymes is given in Table 4.7.

As a result of the information gained by the superimposition, schematic models of the different steric properties in the substrate channel and the active sites of the enzymes were deduced, which allow visualization of hot spots regarding catalytic activity and selectivity. As demonstrated in Figure 4.3, differences are visible in the size of the donor binding site and the substrate channel and also in the presence and form of an *S*-pocket.

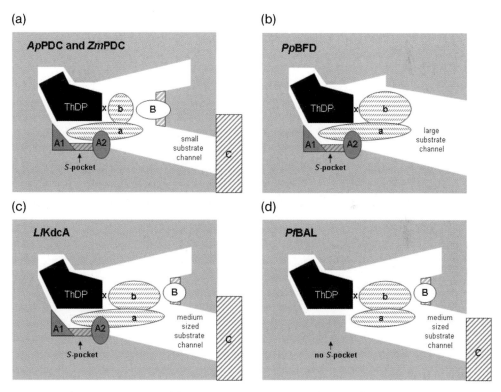

Figure 4.3 Schematic presentation of the substrate channel and the active site of *Ap*PDC, *Zm*PDC (a), *Pp*BFD (b), *Ll*KdcA (c) and *Pf*BAL (d). ThDP is bound in a V-conformation in the active site. × = C2 atom of the thiazolium ring where binding of a 2-keto acid (decarboxylation) or a donor aldehyde (carboligation) occurs; area A = residues lining the *S*-pocket; A1 = most prominent residues defining the *S*-pocket's size, A2 = residues defining the entrance to the *S*-pocket, a = acceptor binding site; B = bulky residue opposite to the donor binding site, b = binding site for 2-keto acid and donor aldehyde; C = C-terminal α-helix covering the entrance of the substrate channel.

Table 4.7 Residues lining the substrate channel and the active site of various ThDP-dependent enzymes. Corresponding residues are found in similar structural positions, if not otherwise indicated[a].

PfBAL (pdb: 2ag0)	PpBFD (pdb: 1mcz, 1bfd)	ApPDC (pdb: 2vbi)	ZmPDC (pdb: 1zpd)	LlKdcA (pdb: 2vbf)	ScPDC (pdb: 1pyd)	Available variants [references]
Residues lining the S-pocket (area A1 in Figure 4.3)						
L25	(N23)	V24	V24	V23	L25	
H26	P24	A25	A25	P24	P26	
G27	G25	G26	G26	G25	G27	PpBFDG25A [174]
A28	S26	D27	D27	D26	D28	PpBFDS26A [39, 41]; PfBALA28S [48]; ZmPDCD27E,N [184, 185]
F484	F464	I472	I476	I465	I480	PpBFDF464I [53]
(W487)	(Y458)	Y466	Y470	Y459	Y474	
—	L461	E469	E473	E462	E477	PpBFDM365L/L461S [133]; ZmPDCE473I,D,N,T,Q,A [112, 184]
Entrance to the S-pocket (area A2 in Figure 4.3)						
A480	A460	I468	I472	V461	I476	PpBFDA460I, ZmPDCI472A [53]; PpBALA480I [41, 48]
Further important residues lining the donor binding site (area B in Figure 4.3)						
H415	(T380)	W388	W392	F381	A392	ZmPDCW392A [107]; ZmPDCW392M,I [111]
(C414, L398)	C398					
Y397	F397					
Residue stabilizing the V-conformation of ThDP						
M421	L403	I411	I415	I404	I415	
Proton relay system[b]						
H29	H70					PpBFDH70 [39, 41, 174]; PfBALH29A [41]
Q113						
α-Helix covering the entrance to the substrate channel (area C in Figure 4.3)						
H113		H113	H113	H112	H114	ZmPDCH113/H114 [175, 176]; ScPDCH114F/H115F [177]
H114		H114	H114	H113	H115	
aa 550–555	no helix	aa 538–555	aa 544–566	aa 532–547	aa 564–556	ZmPDC C-terminal deletion variants [173]
α-helix,		α-helix,	α-helix,	α-helix,	α-helix,	
1.5 turns		4 turns	>4 turns	4 turns	2–3 turns	

[a]For the positions of the respective areas, see the schematic presentation in Figure 4.3. Residues in parentheses are situated at a similar position but are only partially involved in lining the specific site. pdb = RCBS protein databank code (www.rcbs.org/pdb).
[b]Due to a backbone displacement, the proton relay system in PpBFD and PfBAL is located at different sequence positions compared with PDCs and LlKdcA.

In the following, the different structural regions are discussed comparatively, together with biochemical data, in order to rationalize the structural base for the chemo- and enantioselectivity of the different enzymes.

4.5.2
Substrate Channel

In *Pp*BFD, the channel entrance is clearly visible on the surface of the 3D structure [67]. The situation is different in all other enzyme structures examined as a C-terminal α-helix covers this entrance (area C, Figure 4.3, Table 4.7). The helix is assumed to be flexible in solution in order to allow the substrates to enter the substrate channel and further has some impact on the activity, as was investigated for *Sc*PDC [30]. This is supported by results obtained with *Zm*PDC variants with a C-terminal truncation [173]. Whereas the deletion of the last seven C-terminal amino acids has no effect on the decarboxylase activity, the next few, R561 and S560, are critical not only for the decarboxylase activity resulting in a drastically decrease in k_{cat} for the decarboxylation of the main substrate pyruvate but also for the cofactor binding.

In addition to the uncovered entrance, *Pp*BFD also possesses a very broad and straight substrate channel. In comparison, *Ap*PDC, *Zm*PDC and *Sc*PDC show a narrower and curved substrate channel. The size of the channel of *Pf*BAL and *Ll*KdcA is between those of the PDCs and *Pp*BFD and additionally slightly curved upwards (not educible in Figure 4.3c and d).

4.5.3
Proton Relay System

For *Pf*BAL and *Pp*BFD, catalytically important residues have been identified, operating as proton acceptors and donors in the several protonation and deprotonation steps during the catalytic cycle (Section 4.1.3). In *Pf*BAL, these residues are H29 [41, 67] and Q113, whereas H70 is the proton relay in *Pp*BFD [39, 41, 174]. In *Zm*PDC, H113 and H114 are localized at a similar position (Table 4.7) and might have a proton acceptor/donor function as mutations at this position validate, yielding completely inactive variants [39, 175, 176]. *Ap*PDC, also containing histidine residues at positions 113 and 114, is probably comparable to *Zm*PDC. *Ll*KdcA has H112/H113 [20] and *Sc*PDC H114/H115 [177] in a similar position.

4.5.4
Donor Binding Site

Differences in the substrate range for both the decarboxylase and carboligase ability of the enzymes are directly explainable by the different sizes of the donor binding sites (area b in Figure 4.3). The preference of PDCs for short aliphatic 2-keto acids and also short aliphatic donor aldehydes is mainly a result of the residue located opposite to the C2 atom (area B in Figure 4.3), which is occupied by a bulky tryptophan in the bacterial *Zm*PDC and *Ap*PDC (Table 4.7). However, this binding site in *Ap*PDC is larger than in *Zm*PDC, allowing the binding of at least small amounts of benzaldehyde in this position, which is in line with the observed decarboxylation of benzoylformate (Table 4.3, **11**) [13]. Further, *Ap*PDC shows a very weak benzoin-forming activity (Table 4.5). Although small amounts of aromatic donors can bind, smaller donor aldehydes are the preferred substrates in mixed ligase reactions, yielding predominantly PAC analogues (Table 4.5B, C).

In *Pp*BFD as in *Pf*BAL and *Ll*KdcA (Figure 4.3b–d), this part of the active site is large enough to bind sterically demanding aromatic substrates. However, the size of the binding site is not the only criterion to explain substrate transformation. Optimal stabilization of a molecule in a binding site is also important to allow the rapid formation of an enzyme–substrate complex. This may explain why *Pp*BFD shows only very little decarboxylase activity towards aliphatic 2-keto acids (Table 4.3) and also little carboligase activity with aliphatic donor aldehydes, although there is enough space available in the substrate binding site. In *Ll*KdcA, the shape of the substrate binding pocket allows an optimal fit of the indole moiety, which makes *Ll*KdcA a valuable biocatalyst accepting indole-3-acetaldehyde in carboligation reactions [50] (Table 4.6).

The various constraints in the different enzymes result in an almost selective formation of one predominant ligation product if the two aldehydes to be ligated are sufficiently different. Thus, in the case of an aliphatic and an aromatic aldehyde, PDCs prefer the binding of aliphatic aldehydes as donors, whereas *Pp*BFD prefers aromatic aldehydes. Especially in the case of *Ll*KdcA, the phenomenon of optimal fitting is obvious. With this catalyst in mixed ligase reactions of benzaldehyde and acetaldehyde, both aldehydes can act as donors or acceptors (Table 4.5A). Larger aliphatic aldehydes than acetaldehyde are preferably bound as acyl donors, since steric hindrance with F382 and F542 makes their binding in the acceptor binding site less favorable (for details, see [20]). Therefore, the ligation of larger aliphatic aldehydes with benzaldehyde leads exclusively to the production of PAC analogues (Table 4.5B, C). On the other hand, binding of a substituted benzene ring in the acceptor position would result in several unfavorable interactions with surrounding hydrophobic residues, which also prevents these compounds from acting as acceptors in the carboligation reaction, whereas they fit into the donor binding site. Consequently, the carboligation of substituted aromatic aldehydes with acetaldehyde yields exclusively the HPP-analogues (Table 4.5D).

4.5.5
Acceptor Binding Site

After binding of the donor aldehyde to ThDP, the acceptor aldehyde has to bind in the active site. This can only occur in a strictly defined area of the active site (area a in Figure 4.3). There are predominantly two possibilities for the acceptor aldehyde to approach the ThDP-bound donor: in a parallel mode, with both side-chains pointing into the substrate channel, or in an antiparallel mode, with the acceptor side-chain pointing into the so-called *S*-pocket (Figure 4.2). This step might be decisive not only for the stereoselectivity of the ligation step but also for the chemoselectivity of the reaction, as in principle both aldehydes present in the reaction mixture can also react as acceptors.

4.5.5.1 The S-Pocket Approach

As described in Section 4.2, S-selectivity in *Pp*BFD can be explained by an existing S-pocket which is accessible by acetaldehyde in the wild-type enzyme and was shaped by rational protein design to accept larger acceptor aldehydes. The structural confirmation of the enlarged S-pocket has recently been demonstrated with the variant *Pp*BFDL461A [68].

In contrast to the wild-type enzyme, the S-pocket in this variant offers optimal space for the ethyl group of propanal, providing higher stereoselectivity (Table 4.5B). The data show that enantioselectivity towards the S-product is predominantly a consequence of the optimal stabilization of the acceptor aldehyde side-chain in the S-pocket and is additionally influenced by the size of the donor aldehyde, as larger donor aldehydes, such as substituted benzaldehydes, result in increased selectivity (Scheme 4.5) [68].

4.5.5.2 S-Pockets are Widespread Among ThDP-dependent Enzymes but Not Always Accessible

A superimposition of the crystal structures of *Pp*BFD, *Pf*BAL, *Zm*PDC, *Ap*PDC, *Sc*PDC and *Ll*KdcA revealed S-pockets in all structures except in *Pf*BAL [68]. The S-pocket size increases in the series *Pp*BFD < *Ll*KdcA < *Zm*PDC/*Sc*PDC < *Ap*PDC. In all these enzymes, except for *Pp*BFD, a glutamate residue occupies the position analogous to *Pp*BFDL461 (area A1 in Figure 4.3; Table 4.7). The backbone of the three PDCs is almost identical in this area and the slightly different sizes of the pockets are just a result of different orientations of the side-chains [13].

From this structural comparison, the question arose of why only *Pp*BFD is able to catalyze the formation of (S)-2-HPP from benzaldehyde and acetaldehyde, whereas all other enzymes mentioned are R-selective in mixed reactions (Table 4.5). One explanation is the chemoselectivity of the PDCs. Since small aliphatic aldehydes such as acetaldehyde are exclusively the donor aldehydes and benzaldehyde is the acceptor aldehyde, bulky aromatic aldehydes are usually too large for these S-pockets. A further important factor for the strict R-selectivities are the almost completely blocked entrances of the S-pockets in *Ll*KdcA and in the PDCs by bulky residues such as isoleucine and valine (Table 4.7), preventing the access of even small acceptor aldehydes. Consequently, (S)-2-hydroxy ketones could be formed by improving the access to these pockets. This has been shown successfully for the variant *Zm*PDCI472A, which catalyzes the formation of (S)-HPP (*ee* 70%) whereas exclusively (R)-PAC (*ee* >98%) is formed with the wild-type enzyme using benzaldehyde and acetaldehyde as substrates [53]. Therefore, it is most likely that S-selective variants of *Ap*PDC and *Sc*PDC could be created by opening the access to the S-pockets via site-directed mutagenesis. *Ap*PDC, with the largest S-pocket, has the greatest potential of all of them to accept large acceptor aldehydes [13].

The large *S*-pocket in *Ap*PDC is most probably also the reason for the comparatively low enantioselectivity of the enzyme, yielding for example (*R*)-PAC with an *ee* of just 91%, whereas *Zm*PDC catalyzes this reaction with good enantioselectivity (*ee* 98%) (Table 4.5A). Modeling studies demonstrate that in the case of *Ap*PDC, the *S*-pocket is large enough to bind even benzaldehyde (Figure 4.4). Although the parallel arrangement of the donor and acceptor aldehyde, with the aromatic ring directed towards the substrate channel is preferred (Figure 4.4b), yielding (*R*)-PAC, benzaldehyde fits also into the *S*-pocket (Figure 4.4d). This antiparallel arrangement may yield small amounts of (*S*)-PAC.

(a) (b)

(c) (d)

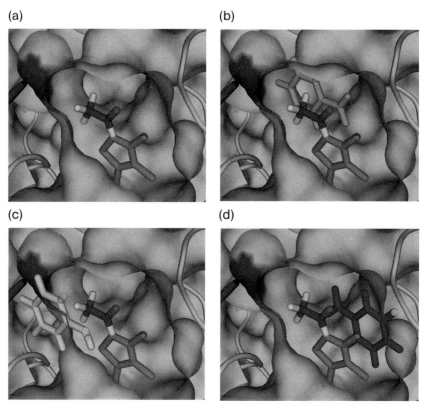

Figure 4.4 Active site of *Ap*PDC with benzaldehyde and acetaldehyde modeled inside. The side-chain of W388 is marked in black (left upper corner). Acetaldehyde (dark gray) is bound in an almost coplanar arrangement to the C2 atom of the thiazolium ring. (a) Without acceptor. (b) Main pathway: benzaldehyde (light gray) approaches towards ThDP-bound acetaldehyde in a parallel mode from the substrate channel, yielding (*R*)-PAC. (c) Due to steric hindrance predominantly with W388 there is no possibility for benzaldehyde (white) to approach from the opposite site, which would theoretically also lead to (*S*)-PAC formation. (d) Explains the formation of low amounts of (*S*)-PAC, as the *S*-pocket in *Ap*PDC is large enough to bind benzaldehyde (dark gray). However, this pathway is less favored than (b). Source: Michael Knoll, University of Stuttgart.

The results show that the biochemical data can be very well explained based on the structure. Further, the modeling studies allowed a reliable prediction of both the chemoselectivity and enantioselectivity of all ThDP-dependent enzymes described here. However, predictions are limited to mixed carboligation reactions of two substrates of significantly different size, for example an aromatic and an aliphatic aldehyde. Only in these cases can preferences for the binding of the small and the large substrate be assigned with sufficient reliability.

4.5.5.3 Carboligation of Two Similar Aldehydes

A special case is the ligation of two small aliphatic aldehydes. The smaller the aldehydes, the more space they have in the active site and the fewer constraints are present. This is especially obvious for the formation of acetoin (16). As demonstrated in Table 4.5E, all enzymes tested show low enantioselectivity for this reaction. The PDCs catalyze the formation of (S)-acetoin in excess, probably making use of their S-pockets. PfBAL, which has no S-pocket, produces (R)-acetoin with an ee of only 40%. This example shows that there must be an alternative route to obtain aliphatic (S)-2-hydroxy ketones, probably according to the arrangement shown in Figure 4.4c. This route is most likely just possible in the case of a large donor binding site in combination with little space-demanding aliphatic aldehydes.

The conformation of the acceptor aldehyde bound in the active site could be demonstrated for LlKdcA by co-crystallization of 2-[(R)-1-hydroxyethyl]deazaThDP [20], a catalytically inactive enantiopure ThDP-analogue mimicking the covalent reaction intermediate 2-[(R)-1-hydroxyethyl]ThDP, which results from the binding of acetaldehyde to ThDP. While a coplanar arrangement of an aromatic donor aldehyde bound to ThDP is most likely, the prediction of the hydroxyethyl-moiety of an aliphatic donor aldehyde is less specific due to fewer constraints. Nevertheless, the crystal structure shows an arrangement of the hydroxyethyl moiety which is very similar to the arrangement predicted for aromatic ThDP-bound aldehydes, with the methyl group being directed into the substrate channel and the α-hydroxy group forming a hydrogen bond to the 4'-amino group of the pyrimidine ring of ThDP [20]. As an alternative conformations of ThDP-bound aliphatic aldehydes can be assumed depending on the size of the active center, more structural data, preferably of co-crystallized donor aldehydes, are necessary to understand the conformation of ThDP-bound aliphatic aldehydes better. It can be assumed that the orientation of the aliphatic donor aldehyde will most probably also be influenced by the acceptor aldehyde.

In summary, the ability of ThDP-dependent enzymes to catalyze regio- and stereoselective C–C bond formation, which in most cases described here are non-physiological transformations, make these enzymes interesting catalysts. Starting from aldehydes or their respective 2-keto acids, a one-pot transformation gives access to 2-hydroxy ketones, valuable chiral building blocks that can undergo diversity-oriented chemoenzymatic modifications. Limitations with respect to substrate range, such as steric hindrance, and stereoselectivity, such as the low tendency towards the formation of S-configured products, have been solved successfully by enzyme engineering, substrate engineering, reaction engineering or a combination of all three techniques. Implementation of other members of the diverse class of ThDP-dependent enzymes and testing their ability to catalyze C–C bond formation starting from different substrates, for example ketones or α,β-unsaturated carbonyls as electrophiles, will broaden even further the scope of possible preparative applications.

References

1 Enders, D., Niemeier, O. and Henseler, A. (2007) *Chemical Reviews*, **107**, 5606–5655.

2 Johnson, J.S. (2004) *Angewandte Chemie-International Edition*, **43**, 1326–1328.

3 Johnson, J.S. (2007) *Current Opinion in Drug Discovery & Development*, **10**, 691–703.

4 Hu, Q.Y. and Kluger, R. (2005) *Journal of the American Chemical Society*, **127**, 12242–12243.

5 Kluger, R. and Yu, D. (2006) *Bioorganic Chemistry*, **34**, 337–344.

6 Kluger, R., Ikeda, G., Hu, Q.Y., Cao, P.P. and Drewry, J. (2006) *Journal of the American Chemical Society*, **128**, 15856–15864.

7 Demir, A.S., Ayhan, P. and Sopaci, S.B. (2007) *Clean*, **35**, 406–412.

8 Frank, R.A.W., Leeper, F.J. and Luisi, B.F. (2007) *Cellular and Molecular Life Sciences*, **64**, 892–905.

9 Pohl, M., Sprenger, G.A. and Müller, M. (2004) *Current Opinion in Biotechnology*, **15**, 335–342.

10 Costelloe, S.J., Ward, J.M. and Dalby, P.A. (2007) *Journal of Molecular Evolution*, **66**, 36–49.

11 Muller, Y.A., Lindqvist, Y., Furey, W., Schulz, G.E., Jordan, F. and Schneider, G. (1993) *Structure*, **1**, 95–103.

12 Dobritzsch, D., König, S., Schneider, G. and Lu, G. (1998) *The Journal of Biological Chemistry*, **273**, 20196–20204.

13 Gocke, D., Berthold, C., Graf, T., Brosi, H., Frindi-Wosch, I., Knoll, M., Stillger, T., Walter, L., Müller, M., Pleiss, J., Schneider, G. and Pohl, M. (2008) *Journal of Molecular Catalysis B: Enzymatic*.

14 Hasson, M.S., Muscate, A., McLeish, M.J., Polovnikova, L.S., Gerlt, J.A., Kenyon, G.L., Petsko, G.A. and Ringe, D. (1998) *Biochemistry*, **37**, 9918–9930.

15 Mosbacher, T.G., Müller, M. and Schulz, G.E. (2005) *FEBS Journal*, **272**, 6067–6076.

16 Versées, W., Spaepen, S., Vanderleyden, J. and Steyaert, J. (2007) *FEBS Journal*, **274**, 2363–2375.

17 Guo, F., Zhang, D., Kahyaoglu, A., Farid, R.S. and Jordan, F. (1998) *Biochemistry*, **37**, 13379–13391.

18 Pang, S.S., Duggleby, R.G., Schowen, R.L. and Guddat, L.W. (2004) *The Journal of Biological Chemistry*, **279**, 2242–2253.

19 Lindqvist, Y., Schneider, G., Ermler, U. and Sundstrom, M. (1992) *The EMBO Journal*, **11**, 2373–2379.

20 Berthold, C.L., Gocke, D., Wood, M.D., Leeper, F.J., Pohl, M. and Schneider, G. (2007) *Acta Crystallographica. Section D, Biological Crystallography*, **63**, 1217–1224.

21 Arjunan, P., Umland, T., Dyda, F., Swaminathan, S., Furey, W., Sax, M., Farrenkopf, B., Gao, Y., Zhang, D. and Jordan, F. (1996) *Journal of Molecular Biology*, **256**, 590–600.

22 Dyda, F., Furey, W., Swaminathan, S., Sax, M., Farrenkopf, B. and Jordan, F. (1993) *Biochemistry*, **32**, 6165–6170.

23 Schütz, A., Sandalova, T., Ricagno, S., Hübner, G., König, S. and Schneider, G. (2003) *European Journal of Biochemistry/FEBS*, **270**, 2312–2321.

24 Pang, S.S., Duggleby, R.G. and Guddat, L.W. (2002) *Journal of Molecular Biology*, **317**, 249–262.

25 Maraite, A., Schmidt, T., Ansorge-Schumacher, M.B., Brzozowski, A.M. and Grogan, G. (2007) *Acta Crystallographica, Section F: Structural Biology and*

Crystallization Communication, **63**, 546–548.

26 Duggleby, R.G. (2006) *Accounts of Chemical Research*, **39**, 550–557.

27 Kutter, S., Wille, G., Relle, S., Weiss, M.S., Hübner, G. and König, S. (2006) *FEBS Journal*, **273**, 4199–4209.

28 Pohl, M., Grötzinger, J., Wollmer, A. and Kula, M.R. (1994) *European Journal of Biochemistry/FEBS*, **224**, 651–661.

29 Raj, K.C., Talarico, L.A., Ingram, L.O. and Maupin-Furlow, J.A. (2002) *Applied and Environmental Microbiology*, **68**, 2869–2876.

30 König, S. (1998) *Biochimica et Biophysica Acta*, **1385**, 271–286.

31 Jordan, F., Nemeria, N.S. and Sergienko, E. (2005) *Accounts of Chemical Research*, **38**, 755–763.

32 Frank, R.A., Titman, C.M., Pratap, J.V., Luisi, B.F. and Perham, R.N. (2004) *Science*, **306**, 872–876.

33 Jordan, F. (2004) *Science*, **306**, 818–820.

34 McCourt, J.A. and Duggleby, R.G. (2005) *Trends in Biochemical Sciences*, **30**, 222–225.

35 Kale, S., Ulas, G., Song, J., Brudvig, G.W., Furey, W. and Jordan, F. (2008) *Proceedings of the National Academy of Sciences of the United States of America*, **105**, 1158–1163.

36 Breslow, R. (1957) *Journal of the American Chemical Society*, **79**, 1762.

37 Kern, D., Kern, G., Neef, H., Tittmann, K., Killenberg-Jabs, M., Wikner, C., Schneider, G. and Hübner, G. (1997) *Science*, **275**, 67–70.

38 Kaplun, A., Binshtein, E., Vyazmensky, M., Steinmetz, A., Barak, Z., Chipman, D.M., Tittmann, K. and Shaanan, B. (2008) *Nature Chemical Biology*, **4**, 113–118.

39 Polovnikova, E.S., McLeish, M.J., Sergienko, E.A., Burgner, J.T., Anderson, N.L., Bera, A.K., Jordan, F., Kenyon, G.L. and Hasson, M.S. (2003) *Biochemistry*, **42**, 1820–1830.

40 Versées, W., Spaepen, S., Wood, M.D., Leeper, F.J., Vanderleyden, J. and Steyaert, J. (2007) *The Journal of Biological Chemistry*, **282**, 35269–35278.

41 Kneen, M.M., Pogozheva, I.D., Kenyon, G.L. and McLeish, M.J. (2005) *Biochimica et Biophysica Acta*, **1753** 263–271.

42 Dünkelmann, P., Kolter-Jung, D., Nitsche, A., Demir, A.S., Siegert, P., Lingen, B., Baumann, M., Pohl, M. and Müller, M. (2002) *Journal of the American Chemical Society*, **124**, 12084–12085.

43 Dünnwald, T. and Müller, M. (2000) *The Journal of Organic Chemistry*, **65**, 8608–8612.

44 Dünnwald, T., Demir, A.S., Siegert, P., Pohl, M. and Müller, M. (2000) *European Journal of ORGANIC Chemistry*, 2161–2170.

45 Müller, M. and Sprenger, G.A. (2004), in *Thiamine: Catalytic Mechanisms in Normal and Disease States* (eds F. Jordan and M.S. Patel), Marcel Dekker, New York, pp. 77–92.

46 Iding, H., Dünnwald, T., Greiner, L., Liese, A., Müller, M., Siegert, P., Grötzinger, J., Demir, A.S. and Pohl, M. (2000) *Chemistry - A European Journal*, **6**, 1483–1495.

47 Demir, A.S., Pohl, M., Janzen, E. and Müller, M. (2001) *Journal of the Chemical Society-Perkin Transactions*, **1** 633–635.

48 Janzen, E., Müller, M., Kolter-Jung, D., Kneen, M.M., McLeish, M.J. and Pohl, M. (2006) *Bioorganic Chemistry*, **34**, 345–361.

49 Dünkelmann, P., Pohl, M. and Müller, M. (2004) *Chimica Oggi-Chemistry Today, Supplement Chiral Catalysis*, **22**, 24–28.

50 Gocke, D., Nguyen, C.L., Pohl, M., Stillger, T., Walter, L. and Müller, M. (2007) *Advanced Synthesis and Catalysis*, **349**, 1425–1435.

51 Kühl, S., Zehentgruber, D., Pohl, M., Müller, M. and Lütz, S. (2007) *Chemical Engineering Science*, **62**, 5201–5205.

52 Domínguez de María, P., Pohl, M., Gocke, D., Gröger, H., Trauthwein, H., Stillger, T. and Müller, M. (2007) *European Journal of ORGANIC Chemistry*, 2940–2944.

53 Siegert, P., McLeish, M.J., Baumann, M., Iding, H., Kneen, M.M., Kenyon, G.L. and Pohl, M. (2005) *Protein Eng. Des. Sel.*, **18**, 345–357.

54 Domínguez de María, P., Stillger, T., Pohl, M., Kiesel, M., Liese, A., Gröger, H. and Trauthwein, H. (2008) *Advanced Synthesis and Catalysis*, **350**, 165–173.

55 Mikolajek, R., Spiess, A.C., Pohl, M., Lamare, S. and Büchs, J. (2007) *Chembiochem*, **8**, 1063–1070.

56 Hildebrand, F., Kühl, S., Pohl, M., Vasic-Racki, D., Müller, M., Wandrey, C. and Lütz, S. (2007) *Biotechnology and Bioengineering*, **96**, 835–843.

57 Stillger, T., Pohl, M., Wandrey, C. and Liese, A. (2006) *Organic Process Research & Development*, **10**, 1172–1177.

58 Kühl, S., Zehentgruber, D., Pohl, M., Müller, M. and Lütz, S. (2006) *Chemical Engineering Science*, **62**, 5201–5205.

59 Domínguez de María, P., Stillger, T., Pohl, M., Wallert, S., Drauz, K., Gröger, H., Trauthwein, H. and Liese, A. (2006) *Journal of Molecular Catalysis B*, **38**, 43–47.

60 Ansorge-Schumacher, M.B., Greiner, L., Schroeper, F., Mirtschin, S. and Hischer, T. (2006) *The Biochemical Journal*, **1**, 564–568.

61 Hischer, T., Gocke, D., Fernandez, M., Hoyos, P., Alcantara, A.R., Sinisterra, J.V., Hartmeier, W. and Ansorge-Schumacher, M.B. (2005) *Tetrahedron*, **61**, 7378–7383.

62 Engel, S., Vyazmensky, M., Berkovich, D., Barak, Z. and Chipman, D.M. (2004) *Biotechnology and Bioengineering*, **88**, 825–831.

63 Sanchez-Gonzalez, M. and Rosazza, J.P.N. (2003) *Advanced Synthesis and Catalysis*, **345**, 819–824.

64 Demir, A.S., Sesenoglu, O., Dünkelmann, P. and Müller, M. (2003) *Organic Letters*, **5**, 2047–2050.

65 Demir, A.S., Sesenoglu, O., Eren, E., Hosrik, B., Pohl, M., Janzen, E., Kolter, D., Feldmann, R., Dünkelmann, P. and Müller, M. (2002) *Advanced Synthesis and Catalysis*, **344**, 96–103.

66 Wilcocks, R., Ward, O.P., Collins, S., Dewdney, N.J., Hong, Y. and Prosen, E. (1992) *Applied and Environmental Microbiology*, **58**, 1699–1704.

67 Knoll, M., Müller, M., Pleiss, J. and Pohl, M. (2006) *Chembiochem*, **7**, 1928–1934.

68 Gocke, D., Walter, L., Gauchenova, E., Kolter, G., Knoll, M., Berthold, C.L., Schneider, G., Pleiss, J., Müller, M. and Pohl, M. (2008) *Chembiochem*, **9**, 406–412.

69 Hossain, M.A., Huq, E., Grover, A., Dennis, E.S., Peacock, W.J. and Hodges, T.K. (1996) *Plant Molecular Biology*, **31**, 761–770.

70 Hossain, M.A., Huq, E., Hodges, T.K. and Hug, E. (1994) *Plant Physiology*, **106**, 799–800.

71 Rivoal, J., Ricard, B. and Pradet, A. (1990) *European Journal of Biochemistry/FEBS*, **194**, 791–797.

72 Kelley, P.M., Godfrey, K., Lal, S.K. and Alleman, M. (1991) *Plant Molecular Biology*, **17**, 1259–1261.

73 Langston-Unkefer, P.J. and Lee, T.C. (1985) *Plant Physiology*, **79**, 436–440.

74 Lee, T.C. and Langston-Unkefer, P.J. (1985) *Plant Physiology*, **79**, 242–247.

75 Mücke, U., Konig, S. and Hübner, G. (1995) *Biological Chemistry Hoppe-Seyler*, **376**, 111–117.

76 Mücke, U., Wohlfarth, T., Fiedler, U., Baumlein, H., Rücknagel, K.P. and König, S. (1996) *European Journal of Biochemistry/FEBS*, **237**, 373–382.

77 Holzer, H., Schultz, G., Villar-Palai, C. and Jüntgen-Schell, J. (1956) *Biochem. Z.*, **327** 331–344.

78 Schmitt, H.D. and Zimmermann, F.K. (1982) *Journal of Bacteriology*, **151**, 1146–1152.

79 Moller, K., Langkjaer, R.B., Nielsen, J., Piskur, J. and Olsson, L. (2004) *Molecular Genetics and Genomics*, **270**, 558–568.

80 Holloway, P. and Subden, R.E. (1993) *Current Genetics*, **24**, 274–277.

81 Krieger, F., Spinka, M., Golbik, R., Hübner, G. and König, S. (2002) *European Journal of Biochemistry/FEBS*, **269**, 3256–3263.

82 Neuser, F., Zorn, H. and Berger, R.G. (2000) *Journal of Agricultural and Food Chemistry*, **48**, 6191–6195.

83 Neuser, F., Zorn, H., Richter, U. and Berger, R.G. (2000) *Biological Chemistry*, **381**, 349–353.

84 Alvarez, M.E., Rosa, A.L., Temporini, E.D., Wolstenholme, A., Panzetta, G., Patrito, L. and Maccioni, H.J. (1993) *Gene*, **130**, 253–258.

85 Lockington, R.A., Borlace, G.N. and Kelly, J.M. (1997) *Gene*, **191**, 61–67.

86 Skory, C.D. (2003) *Current Microbiology*, **47**, 59–64.

87 Dawes, E.A., Ribbons, D.W. and Large, P.J. (1966) *The Biochemical Journal*, **98**, 795–803.

88 Okamoto, T., Taguchi, K., Nakamura, H., Ikenaga, H., Kuraishi, H. and Yamasato, K. (1993) *Archives of Microbiology*, **160**, 333–337.

89 Raj, K.C., Ingram, L.O. and Maupin-Furlow, J.A. (2001) *Archives of Microbiology*, **176**, 443–451.

90 Lowe, S.E. and Zeikus, J.G. (1992) *Journal of General Microbiology*, **138**, 803–807.

91 Wang, Z., Cheng, M., Xu, Y., Li, S., Lu, W., Ping, S., Zang, W. and Lin, M. (2007) *Biotechnology Letters*, **30**, 657–663.

92 Jordan, F., Nemeria, N.S., Zhang, S., Yan, Y., Arjunan, P. and Furey, W. (2003) *Journal of the American Chemical Society*, **125**, 12732–12738.

93 Hübner, G., Tittmann, K., Killenberg-Jabs, M., Schaffner, J., Spinka, M., Neef, H., Kern, D., Kern, G., Schneider, G., Wikner, C. and Ghisla, S. (1998) *Biochimica et Biophysica Acta*, **1385**, 221–228.

94 Hübner, G., Weidhase, R. and Schellenberger, A. (1978) *European Journal of Biochemistry/FEBS*, **92**, 175–181.

95 Dietrich, A. and König, S. (1997) *FEBS Letters*, **400**, 42–44.

96 Talarico, L.A., Ingram, L.O. and Maupin-Furlow, J.A. (2001) *Microbiology*, **147**, 2425–35.

97 Lu, G., Dobritzsch, D., Baumann, S., Schneider, G. and König, S. (2000) *European Journal of Biochemistry/FEBS*, **267**, 861–868.

98 Neuberg, C. and Karczag, L. (1911) *Biochemische Zeitschrift*, **36**, 68.

99 Hildebrandt, G. and Klavehn, W. (1934) US Patent 1 956 950.

100 Gunawan, C., Satianegara, G., Chen, A.K., Breuer, M., Hauer, B., Rogers, P.L. and Rosche, B. (2007) *FEMS Yeast Research*, **7**, 33–39.

101 Satianegara, G., Rogers, P.L. and Rosche, B. (2006) *Biotechnology and Bioengineering*, **94**, 1189–1195.

102 Rosche, B., Sandford, V., Breuer, M., Hauer, B. and Rogers, P.L. (2007) *Applied Microbiology and Biotechnology*, **57**, 309–315.

103 Pohl, M., Lingen, B. and Müller, M. (2002) *Chemistry - A European Journal*, **8**, 5288–5295.

104 Iding, H., Siegert, P., Mesch, K. and Pohl, M. (1998) *Biochimica et Biophysica Acta*, **1385**, 307–322.

105 Baykal, A., Chakraborty, S., Dodoo, A. and Jordan, F. (2006) *Bioorganic Chemistry*, **34**, 380–393.

106 Pohl, M., Mesch, K., Rodenbrock, A. and Kula, M.R. (1995) *Biotechnology and Applied Biochemistry*, **22**, 95–105.

107 Bruhn, H., Pohl, M., Grötzinger, J. and Kula, M.R. (1995) *European Journal of Biochemistry/FEBS*, **234**, 650–655.

108 Candy, J.M. and Duggleby, R.G. (1994) *The Biochemical Journal*, **300**, 7–13.

109 Candy, J.M. and Duggleby, R.G. (1998) *Biochimica et Biophysica Acta*, **1385**, 323–338.

110 Candy, J.M., Koga, J., Nixon, P.F. and Duggleby, R.G. (1996) *The Biochemical Journal*, **315**, 745–751.

111 Pohl, M. (1997) *Advances in Biochemical Engineering/Biotechnology*, **58**, 15–43.

112 Pohl, M., Siegert, P., Mesch, K., Bruhn, H. and Grötzinger, J. (1998) *European Journal of Biochemistry/FEBS*, **257**, 538–546.

113 Bringer-Meyer, S. and Sahm, H. (1988) *Biocatalysis*, 321–331.

114 Rosche, B., Breuer, M., Hauer, B. and Rogers, P.L. (2004) *Biotechnology and Bioengineering*, **86**, 788–794.

115 Goetz, G., Iwan, P., Hauer, B., Breuer, M. and Pohl, M. (2001) *Biotechnology and Bioengineering*, **74**, 317–325.

116 Iwan, P., Goetz, G., Schmitz, S., Hauer, B., Breuer, M. and Pohl, M. (2001) *Journal of Molecular Catalysis B*, **11**, 387–396.

117 Horn, S.J., Aasen, I.M. and Østgaard, K. (2000) *Journal of Industrial Microbiology & Biotechnology*, **24**, 51–57.

118 Rosche, B., Breuer, M., Hauer, B. and Rogers, P.L. (2003) *Biotechnology Letters*, **25**, 847–851.

119 Oku, H. and Kaneda, T. (1988) *The Journal of Biological Chemistry*, **263**, 18386–18396.

120 Amárita, F., Fernández-Esplá, D., Requena, T. and Pelaez, C. (2001) *FEMS Microbiology Letters*, **204**, 189–195.

121 Smit, B.A., Vlieg, J., Engels, W.J.M., Meijer, L., Wouters, J.T.M. and Smit, G. (2005) *Applied and Environmental Microbiology*, **71**, 303–311.

122 Smit, G., Smit, B.A. and Engels, W.J. (2005) *FEMS Microbiology Reviews*, **29**, 591–610.

123 Yep, A., Kenyon, G.L. and McLeish, M.J. (2006) *Bioorganic Chemistry*, **34**, 325–336.

124 Hegeman, G.D. (1970) *Methods in Enzymology*, **17A** 674–678.

125 Tsou, A.Y., Ransom, S.C., Gerlt, J.A., Buechter, D.D., Babbitt, P.C. and Kenyon, G.L. (1990) *Biochemistry*, **29**, 9856–9862.

126 Wendorff, M. (2006) PhD Thesis, Heinrich-Heine-Universität Düsseldorf.

127 Barrowman, M.M. and Fewson, C.A. (1985) *Current Microbiology*. **12**, 235–239.

128 Saehuan, C., Rojanarata, T., Wiyakrutta, S., McLeish, M.J. and Meevootisom, V. (2007) *Biochimica et Biophysica Acta*, **1770**, 1585–1592.

129 Henning, H., Leggewie, C., Pohl, M., Müller, M., Eggert, T. and Jaeger, K.E. (2006) *Applied and Environmental Microbiology*, 7510–7517.

130 Wilcocks, R. and Ward, O.P. (1992) *Biotechnology and Bioengineering*, **39**, 1058–1063.

131 Demir, A.S., Dünnwald, T., Iding, H., Pohl, M. and Müller, M. (1999) *Tetrahedron: Asymmetry*, **10**, 4769–4774.

132 Lingen, B., Grötzinger, J., Kolter, D., Kula, M.R. and Pohl, M. (2002) *Protein Engineering*. **15**, 585–593.

133 Lingen, B., Kolter-Jung, D., Dünkelmann, P., Feldmann, R., Grötzinger, J., Pohl, M. and Müller, M. (2003) *Chembiochem*, **4**, 721–726.

134 Costacurta, Λ., Keijers, V. and Vanderleyden, J. (1994) *Molecular & General Genetics*. **243**, 463–472.

135 Costacurta, A. and Vanderleyden, J. (1995) *Critical Reviews in Microbiology*. **21**, 1–18.

136 Koga, J. (1995) *Biochimica et Biophysica Acta*, **1249**, 1–13.

137 Koga, J., Adachi, T. and Hidaka, H. (1991) *Molecular & General Genetics*, **226**, 10–6.

138 Schütz, A., Golbik, R., König, S., Hübner, G. and Tittmann, K. (2005) *Biochemistry*, **44**, 6164–6179.

139 Okon, Y. and Labandera-Gonzales, C.A. (1994) *Soil Biology & Biochemistry*. **26**, 1591–1601.

140 Somers, E., Ptacek, D., Gysegom, P., Srinivasan, M. and Vanderleyden, J. (2005) *Applied and Environmental Microbiology*. **71**, 1803–1810.

141 Vuralhan, Z., Morais, M.A., Tai, S.L., Piper, M.D. and Pronk, J.T. (2003) *Applied and Environmental Microbiology*. **69**, 4534–4541.

142 Ward, O.P. and Singh, A. (2000) *Current Opinion in Biotechnology*. **11**, 520–526.

143 Guo, Z., Goswami, A., Mirfakhrae, K.D. and Patel, N. (1999) *Tetrahedron: Asymmetry*, **10**, 4667–4675.

144 Guo, Z., Goswami, A., Nanduri, V.B. and Patel, R.N. (2001) *Tetrahedron: Asymmetry*, **21**, 571–577.

145 Spaepen, S., Versées, W., Gocke, D., Pohl, M., Steyaert, J. and Vanderleyden, J. (2007) *Journal of Bacteriology*. 7626–7633.

146 Gonzalez, B. and Vicuna, R. (1989) *Journal of Bacteriology*. **171**, 2401–2405.

147 Hinrichsen, P., Gomez, I. and Vicuna, R. (1994) *Gene*, **144**, 137–138.

148 Demir, A.S., Pohl, M., Janzen, E. and Müller, M. (2001) *Journal of the Chemical Society-Perkin Transactions 1*, 633–635.

149 Demir, A.S., Ayan, P., Igdir, A.C. and Guygu, A.N. (2004) *Tetrahedron*, **60**, 6509–6512.

150 Filho, M.V., Stillger, T., Müller, M., Liese, A. and Wandrey, C. (2003) *Angewandte Chemie-International Edition*. **115**, 3101–3104.

151 Kurlemann, N. and Liese, A. (2004) *Tetrahedron: Asymmetry*, **15**, 2955–2958.

152 Bowen, T.L., Union, J., Tumbula, D.L. and Whitman, W.B. (1997) *Gene*, **188**, 77–84.

153 Chipman, D.M., Duggleby, R.G. and Tittmann, K. (2005) *Current Opinion in Chemical Biology*, **9**, 475–481.

154 Tittmann, K., Vyazmensky, M., Hubner, G., Barak, Z. and Chipman, D.M. (2005) *Proceedings of the National Academy of Sciences of the United States of America*, **102**, 553–558.

155 Pang, S.S., Guddat, L.W. and Duggleby, R.G. (2004) *Acta Crystallographica Section D-Biological Crystallography*, **60**, 153–155.

156 Pang, S.S. and Duggleby, R.G. (2001) *The Biochemical Journal*. **357**, 749–757.

157 Chipman, D.M., Barak, Z., Engel, S., Mendel, S. and Vyazmensky, M. (2004) in *Thiamine: Catalytic Mechanisms in Normal and Disease States* (eds F. Jordan and M.S. Patel), Marcel Dekker, New York, pp. 233–250.

158 Bornemann, S. (2002) *Natural Product Reports*. **19**, 761–772.

159 Engel, S., Vyazmensky, M., Berkovich, D., Barak, Z., Merchuk, J. and Chipman, D.M. (2005) *Biotechnology and Bioengineering*, **89**, 733–740.

160 Engel, S., Vyazmensky, M., Geresh, S., Barak, Z. and Chipman, D.M. (2003) *Biotechnology and Bioengineering*, **83**, 833–840.

161 Swaminathan, A.G. and Venkatesan, L. (2007) *Chimica Oggi-Chemistry Today*, **25**, 54–56.

162 Vinogradov, M., Kaplun, A., Vyazmensky, M., Engel, S., Golbik, R., Tittmann, K., Uhlemann, K., Meshalkina, L., Barak, Z., Hübner, G. and Chipman, D.M. (2005) *Analytical Biochemistry*, **342**, 126–133.

163 Vinogradov, V., Vyazmensky, M., Engel, S., Belenky, I., Kaplun, A., Kryukov, O., Barak, Z. and Chipman, D.M. (2006) *Biochimica et Biophysica Acta*, **1760**, 356–363.

164 Engel, S., Vyazmensky, M., Vinogradov, M., Berkovich, D., Bar-Ilan, A., Qimron, U., Rosiansky, Y., Barak, Z. and Chipman, D.M. (2004) *The Journal of Biological Chemistry*, **279**, 24803–24812.

165 Rosche, B., Sandford, V., Breuer, M., Hauer, B. and Rogers, P. (2001) *Applied Microbiology and Biotechnology*, **57**, 309–315.

166 Chen, G.C. and Jordan, F. (1984) *Biochemistry*, **23**, 3576–3582.

167 Kurniadi, T., Bel Rhlid, R., Fay, L.B., Juillerat, M.A. and Berger, R.G. (2003) *Journal of Agricultural and Food Chemistry*, **51**, 3103–3107.

168 Crout, D.H.G., Dalton, H., Hutchinson, D.W. and Miyagoshi, M. (1991) *Journal of the Chemical Society-Perkin Transactions 1*, 1329–1334.

169 Bornemann, S., Crout, D.H.G., Dalton, H., Hutchinson, D.W., Dean, G., Thomson, N. and Turner, M.M. (1993) *Journal of the Chemical Society-Perkin Transactions 1*, 309–311.

170 Schütz, A., Golbik, R., Tittmann, K., Svergun, D.I., Koch, M.H., Hübner, G. and König, S. (2003) *European Journal of Biochemistry/FEBS*. **270**, 2322–2331.

171 Koga, J., Adachi, T. and Hidaka, H. (1992) *The Journal of Biological Chemistry*, **267**, 15823–15828.

172 Vuralhan, Z., Luttik, M.A., Tai, S.L., Boer, V.M., Morais, M.A., Schipper, D., Almering, M.J., Kotter, P., Dickinson, J.R., Daran, J.M. and Pronk, J.T. (2005) *Applied and Environmental Microbiology*. **71**, 3276–3284.

173 Chang, A.K., Nixon, P.F. and Duggleby, R.G. (2000) *Biochemistry*, **39**, 9430–9437.

174 Siegert, P. (2000) PhD Thesis, Heinrich-Heine-Universität Düsseldorf.

175 Huang, C.Y., Chang, A.K., Nixon, P.F. and Duggleby, R.G. (2001) *European*

Journal of Biochemistry/FEBS. **268**, 3558–3565.

176 Bruhn, H. (1995) Doctoral Thesis, Heinrich-Heine-Universität Düsseldorf.

177 Liu, M., Sergienko, E.A., Guo, F., Wang, J., Tittmann, K., Hübner, G., Furey, W. and Jordan, F. (2001) *Biochemistry*, **40**, 7355–7368.

178 Baykal, A.T., Chakraborty, S., Nemeria, N. and Jordan, F. (2006) *FASEB J.* **20**, A40.

179 Mandwal, A.K., Tripathi, C.K.M., Trivedi, P.D., Joshi, A.K., Agarwal, S.C. and Bihari, V. (2004) *Biotechnology Letters.* **26**, 217–221.

180 Rosche, B., Breuer, M., Hauer, B. and Rogers, P.L. (2005) *Biotechnology Letters.* **27**, 575–581.

181 Rosche, B., Breuer, M., Hauer, B. and Rogers, P.L. (2005) *Journal of Biotechnology.* **115**, 91–99.

182 Mikolajek, R., Spiess, A.C. and Büchs, J. (2007) *Journal of Biotechnology.* **129**, 723–725.

183 Hibbert, E.G., Senussi, T., Costelloe, S.J., Lei, W., Smith, M.E., Ward, J.M., Hailes, H.C. and Dalby, P.A. (2007) *Journal of Biotechnology.* **131**, 425–432.

184 Chang, A.K., Nixon, P.F. and Duggleby, R.G. (1999) *The Biochemical Journal.* **339**, 255–260.

185 Wu, Y.G., Chang, A.K., Nixon, P.F., Li, W. and Duggleby, R.G. (2000) *European Journal of Biochemistry/FEBS.* **267**, 6493–6500.

5
Baeyer–Villiger Monooxygenases in Organic Synthesis

Anett Kirschner and Uwe T. Bornscheuer

5.1
Introduction

In recent decades, biocatalysis and biotransformation using isolated enzymes or whole cell systems became a major alternative technology in organic syntheses [1]. Key factors are the ability of enzymes to catalyze a broad range of synthetically useful reactions, their usually excellent chemo-, regio- and stereoselectivity and the mild reaction conditions. Compared with most chemical reaction conditions, biocatalysis works in aqueous reaction systems, at neutral pH and moderate temperatures. Moreover, enzymatic reactions often do not need a protection or deprotection and hence safe additional reaction steps. Furthermore, the vast progress in microbiology and especially molecular biology nowadays substantially facilitates access to new enzymes, their easy production in recombinant micro-organisms and the ability to tailor-design the biocatalyst by means of rational protein design or directed evolution.

The majority of enzymatic reactions used hydrolases – especially lipases, esterases, proteases and acylases – and other enzyme classes became of interest only more recently. Within the class of oxidoreductases, most applications in organic synthesis were restricted to ketoreductases or alcohol dehydrogenases. However, in the past few years considerable progress has been made using enzymes able to introduce molecular oxygen into non-activated substrates with P450-monooxygenases and Baeyer–Villiger monooxygenases (BVMOs) as the most important biocatalysts.

Whereas the enzymatic Baeyer–Villiger oxidation (BVO) simply requires the enzyme, molecular oxygen and the cofactor NAD(P)H, the chemical reaction uses peracids to convert ketones into the corresponding lactones (or esters). These peracids (hydrogen peroxide, *m*-chloroperbenzoic acid, etc.) are often toxic and difficult to handle, especially at high concentrations and in large-scale synthesis. In addition, enantioselective chemical BVOs have rarely been described and do not yield products with sufficiently high optical purity.

Handbook of Green Chemistry, Volume 3: Biocatalysis. Edited by Robert H. Crabtree
Copyright © 2009 WILEY-VCH Verlag GmbH & Co. KGaA, Weinheim
ISBN: 978-3-527-32498-9

This chapter summarizes the most important biochemical properties of BVMOs and their most recent applications in organic synthesis in comparison with chemical BVOs.

5.2
General Aspects of the Baeyer–Villiger Oxidation

5.2.1
Mechanistic Aspects

The chemical BVO, involving the introduction of an oxygen atom into a carbon-carbon bond adjacent to a keto group, was first described by Adolf von Baeyer and Victor Villiger in 1899 when they published their results on the conversion of cyclic ketones using persulfuric acid [2]. Since then, this reaction has become a powerful tool in organic chemistry. The reaction proceeds via a two-step process as originally proposed by Criegee [3]: the first step represents a nucleophilic attack of the peracid at the carbonyl atom of the substrate ketone forming the tetrahedral 'Criegee intermediate', which subsequently rearranges forming the product ester and releasing the acid (Scheme 5.1). Migration of substituent R^1 and release of the leaving group in reaction step 2 occur in a concerted manner and are usually rate limiting. Additionally, the rearrangement proceeds with strict retention of configuration – a prerequisite for application in stereoselective transformations [4].

The regiochemistry of the BVO is governed by conformational, steric and electronic effects [4]: (i) substituents at the carbonyl carbon exhibiting better stabilization of a positive charge migrate preferentially, (ii) the migrating substituent in the 'Criegee intermediate' has to be antiperiplanar to the O—O bond of the leaving group [5] and (iii) it must be antiperiplanar to a single pair of electrons at the hydroxyl oxygen [6] (Figure 5.1). The choice of the peroxo species and the solvent and addition of catalysts also influence the regiochemistry of oxygen insertion [7].

5.2.2
Chemical Versus Enzymatic Baeyer–Villiger Oxidation

Organic peracids such as m-chloroperoxybenzoic acid and alkyl peroxides are often used in stoichiometric amounts as oxidants in BVOs, producing the corresponding acid or alcohol as side product. In contrast, use of hydrogen peroxide or molecular oxygen as oxidant represents a more environmentally benign approach, but since

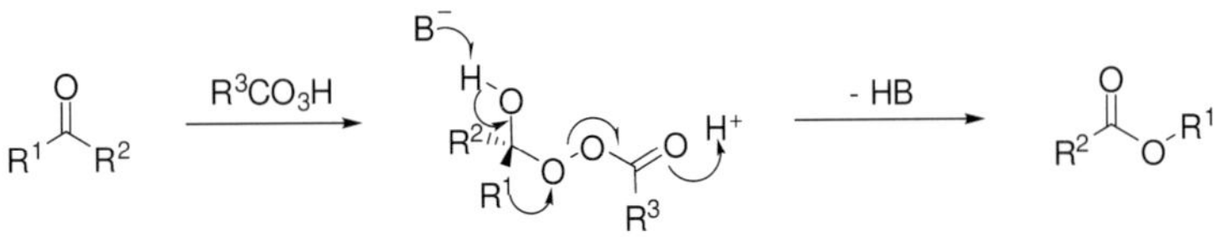

Scheme 5.1 Mechanism of the chemical Baeyer–Villiger oxidation.

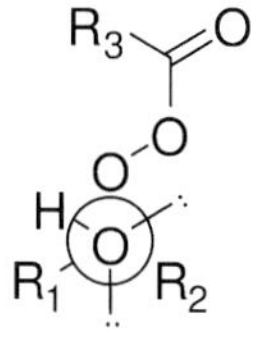

Figure 5.1 Stereoelectronic requirements in the Criegee intermediate for successful rearrangement.

they show less nucleophilicity than peracids, addition of metal catalysts is necessary [8]. Today, both transition metal- and main group element-based complexes are applied [9]. Pioneering work in the field of enantioselective BVOs using hydrogen peroxide and molecular oxygen was done by Strukul and Bolm, respectively. In Strukul's group, platinum complexes with chiral diphosphine ligands and H_2O_2 were used to convert racemic cyclohexanone and cyclopentanone derivatives yielding the corresponding chiral lactones with moderate enantiomeric excess (up to 58% *ee*) [10]. Bolm *et al.* introduced a chiral copper complex for the stereoselective oxidation of 2-arylic cyclohexanones using molecular oxygen and stoichiometric amounts of pivaldehyde giving the corresponding (*R*)-lactones with up to 69% *ee* [11]. Using this catalyst, they were even able to convert racemic bicyclo[4.2.0.]octanone in an enantiodivergent manner yielding the regioisomeric lactone products with 92 and 67% *ee* [12] (Scheme 5.2).

Since then, other chiral metal catalysts have been developed and the influence of different ligands analyzed for their impact on enantioselectivity [13]. However, in view of their substrate spectra, these catalysts are mainly restricted to stereoselective BVOs of cyclobutanone, cyclopentanone and cyclohexanone derivatives and so far in only a few cases could the corresponding product lactones be obtained with high enantiomeric excess ($\geq$ 95% *ee*) [14].

Nature also provides an environmentally benign alternative to these chemical oxidants: enzymes able to perform BVOs. These so-called Baeyer–Villiger monooxygenases (BVMOs) (EC 1.14.13.*x*) belong to the class of oxidoreductases and

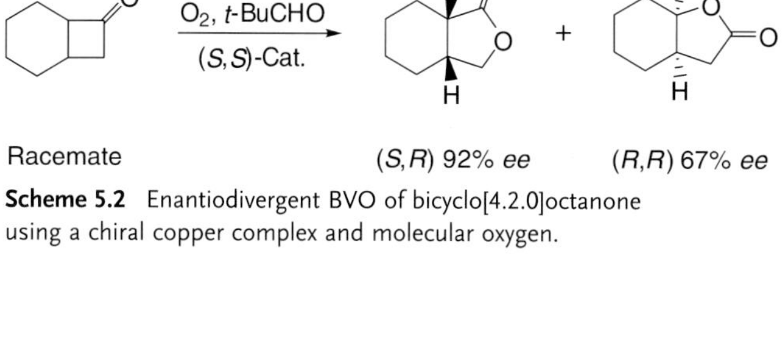

Racemate   (*S*,*R*) 92% *ee*   (*R*,*R*) 67% *ee*

Scheme 5.2 Enantiodivergent BVO of bicyclo[4.2.0]octanone using a chiral copper complex and molecular oxygen.

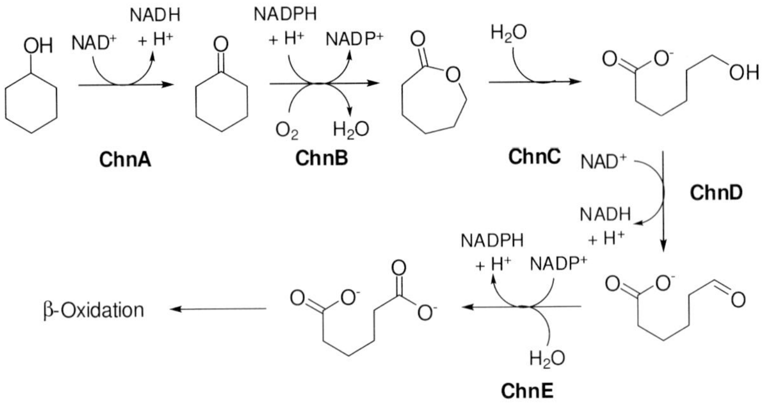

Scheme 5.3 Cyclohexanol degradation by *Acinetobacter calcoaceticus* NCIMB 9871. ChnA, cyclohexanol dehydrogenase; ChnB, cyclohexanone monooxygenase; ChnC, ε-caprolactone hydrolase; ChnD, 6-hydroxyhexanoate dehydrogenase; ChnE, 6-oxohexanoate dehydrogenase.

catalyze the conversion of ketones into esters or lactones using molecular oxygen. Additionally, they are capable of aldehyde and heteroatom oxidation [15–17]. Since the first example of an enzymatic BVO in 1948 [18], intensive research has been performed to explore the biocatalytic potential of BVMOs. So far they have only been found in bacteria and fungi playing important roles in the biodegradation of naturally occurring ketones or the secondary metabolism. Several species, for example, are known to be able to grow on cyclic ketones as sole carbon source [19, 20] whereby the BVMO-catalyzed introduction of an oxygen atom into a ring C–C bond presents the only possibility for aliphatic ring cleavage [21] (Scheme 5.3). *Pseudomonas putida* NCIMB 10007 produces three different BVMOs, two of them being FMN and NADH dependent, allowing the bacterium to grow on (+)- and (−)-camphor as sole carbon source [22, 23]. Several BVO steps were also found to be involved in the biosynthesis of aflatoxin [24] and the anticancer drug mithramycin [25].

In contrast to chiral metal catalysts, BVMOs exhibit remarkably broad substrate spectra including non-natural substrates and often high stereoselectivity. Due to recent advances in genome sequencing projects, many new BVMOs with diverse substrate acceptance and selectivity became available, expanding the pool of oxygenation biocatalysts for synthetic applications. Many of them are encoded in the genomes of pathogenic microorganisms such as mycobacteria, restricting the application of these wild-type isolates in biotransformations. On the other hand, the use of genetically engineered expression strains for recombinant expression of BVMOs facilitates the production of even larger enzyme amounts than could be isolated from wild-type organisms promoting their implementation in organic synthesis.

5.3
Biochemistry of Baeyer–Villiger Monooxygenases

5.3.1
Catalytic Mechanism

BVMOs are flavin-dependent enzymes which require NAD(P)H as cofactor. Most BVMOs known in the literature contain non-covalently bound FAD as prosthetic group and use NADPH, and only a few enzymes with FMN and NADH dependency are known (e.g. 2,5-diketocamphane and 3,6-diketocamphane monooxygenase from *P. putida* NCIMB 10007). Thus, Willets proposed a classification of BVMOs into type I (FAD, NADPH) and type II (FMN, NADH) [26].

The catalytic mechanism of BVMOs was investigated using the type I cyclohexanone monooxygenase (CHMO) from *Acinetobacter calcoaceticus* NCIMB 9871 [27]. The non-covalently bound FAD in the active site of the enzyme is first reduced by NADPH to FADH⁻, which then reacts with molecular oxygen giving flavin-C4a-peroxide (Scheme 5.4). This peroxide species exists in equilibrium with flavin-C4a-hydroperoxide with a pK_S of 8.4, but Sheng et al. showed that only the negatively charged peroxide is able to perform a nucleophilic attack on the carbonyl carbon of the ketone substrate forming the tetrahedral 'Criegee intermediate' [28]. After

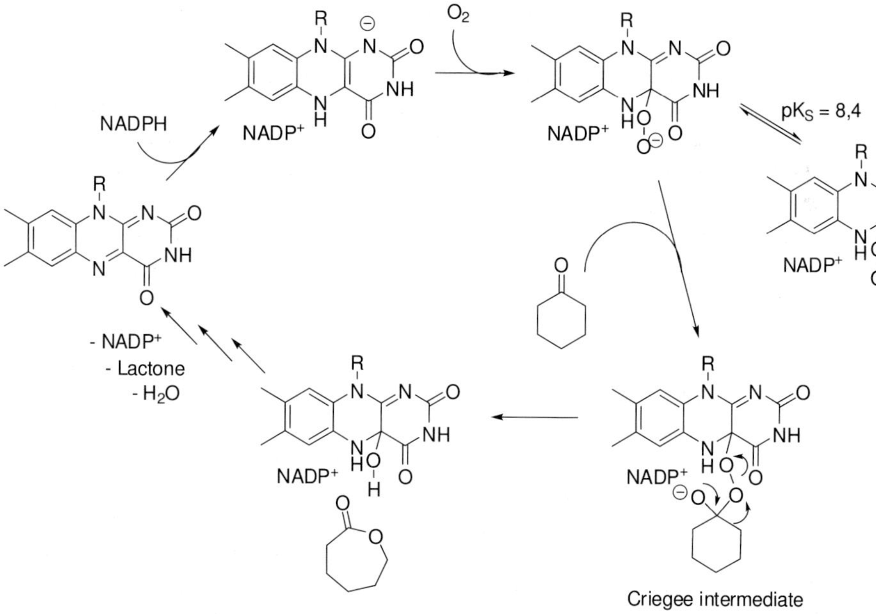

Scheme 5.4 Catalytic mechanism of cyclohexanone monooxygenase from *Acinetobacter calcoaceticus* NCIMB 9871 for BVO of cyclohexanone.

rearrangement of the intermediate, the product ester or lactone is released and the resulting flavin-C4a-hydroxide is converted back to oxidized FAD by elimination of a water molecule. More recently, Ottolina *et al.* postulated that the oxidation of electron-rich compounds such as organic sulfides by BVMOs proceeds through electrophilic attack of the flavin-C4a-hydroperoxide on the heteroatom [29].

In the enzyme-catalyzed BVO, the rearrangement of the 'Criegee intermediate' follows the same rules as in the chemical reaction: only the C−C bond adopting an antiperiplanar configuration with respect to the leaving group can migrate. Due to steric conditions in the enzyme's active site, this can also be the less nucleophilic center.

5.3.2
Structural Features

Type I BVMOs belong to the same superfamily as flavin monooxygenases (FMOs) and *N*-hydroxylating monooxygenases (NMOs) [30]. They possess two GxGxx(G/A) motifs for FAD and NADPH binding and additionally also a highly conserved FxGxxxHxxxW(P/D) fingerprint sequence facilitating the identification of new BVMOs [31].

Elucidation of the first (and so far only) crystal structure of a BVMO (pdb code 1GWI), published in 2004, contributed much to the understanding of the functionality of these enzymes. The three-dimensional structure derived from phenylacetone monooxygenase (PAMO) from *Thermobifida fusca* shows a two-domain architecture consisting of two nucleotide binding domains (Rossman folds) which are connected by a linker region carrying the BVMO sequence motif [32] (Figure 5.2). Hence this fingerprint is not an integral part of the active site.

On the *si*-site of the FAD within the enzyme's active site, several aromatic residues form van der Waals interactions with the isoalloxazin ring, thereby positioning the prosthetic group. A strictly conserved arginine residue (Arg337), which plays an important role in enzyme catalysis, is situated on the *re*-site of the FAD. Mutation of this residue in 4-hydroxyacetophenone monooxygenase (HAPMO) from *Pseudomonas fluorescens* ACB resulted in complete loss of activity [33]. This residue probably stabilizes the negative charge of the flavin peroxide anion during catalysis. In the crystal structure, Arg337 was found to adopt two different conformations, indicating a substantial degree of flexibility throughout the enzymatic reaction [32]. According to the authors, the NADPH-binding domain has to move partly after NADPH binding and reduction of FAD to bring Arg337 into the correct position required for stabilization of the flavin-C4a-peroxide anion. The linker region between the two domains consisting of two antiparallel β-strands thereby acts as hinge, giving an indication of the role of the highly conserved BVMO sequence motif within this linker. Such domain rotations during catalysis had already been reported for other enzymes, such as thioredoxin reductase [34]. However, a more detailed investigation of this model of catalysis is necessary since so far no crystal structure for a BVMO with bound NADPH or substrate is available.

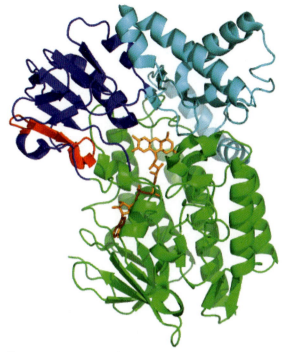

Figure 5.2 3D structure of one monomer of phenylacetone monooxygenase from *Thermobifida fusca*. FAD binding domain in green, NADPH binding domain in blue, BVMO sequence motif in red, FAD in orange.

5.4
Application of Baeyer–Villiger Monooxygenases in Organic Chemistry

5.4.1
Isolated Enzymes Versus Whole Cells

The application of cofactor-dependent enzymes for biocatalyses in organic synthesis is often restricted due to the high costs of the cofactors, which are required in stoichiometric amounts. This also holds true for BVMOs. Especially the need for NADPH, which is approximately 10 times more expensive than NADH, makes a biocatalytic process using isolated BVMOs on a preparative scale far too cost intensive unless a suitable cofactor regeneration system is applied. When using cofactor regeneration, only catalytic amounts of the required cofactor have to be added along with stoichiometric amounts of a cosubstrate, which is converted by a second enzyme simultaneously regenerating the cofactor. Several different NADH-regenerating dehydrogenases have been reported [35], but few enzymes are able to recycle NADPH. A frequently used system consists of glucose-6-phosphate dehydrogenase (GPDH)

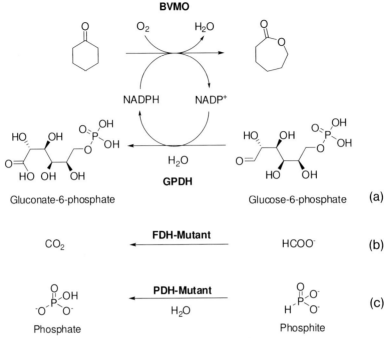

Scheme 5.5 Possible NADPH-regenerating systems: (a) glucose-6-phosphate dehydrogenase (GPDH); (b) mutant of formate dehydrogenase from *Pseudomonas* sp. 101 (FDH); (c) mutant of phosphite dehydrogenase from *Pseudomonas stutzeri* (PDH).

oxidizing glucose-6-phosphate to gluconate-6-phosphate while reducing NADP$^+$ to NADPH (Scheme 5.5a) [36]. However, the auxiliary substrate glucose-6-phosphate itself is comparably expensive. Therefore, attempts have been made to convert NADH-dependent to NADPH-dependent dehydrogenases utilizing cheap substrates by means of directed evolution. Thus, a formate dehydrogenase (FDH) mutant [37] and a phosphite dehydrogenase (PDH) mutant [38] accepting both NADH and NADPH were created (Scheme 5.5b, c). Both are exceptionally attractive cofactor-regenerating enzymes since the cosubstrates required, formate and phosphite, are cheap and the equilibrium reactions are highly in favor of NAD(P)H production. So far only the mutant FDH has been applied in biotransformations with CHMO from *A. calcoaceticus* [39].

Willets *et al.* used an NADP$^+$-dependent alcohol dehydrogenase (ADH) from *Thermoanaerobium brockii* for cofactor regeneration during enzymatic BVO of different cyclic ketones, with the corresponding cyclic alcohols serving as substrates for the ADH (Scheme 5.6) [40].

Furthermore, unusual systems for the reduction of FAD without a NADPH requirement have been reported, for example, application of an organorhodium complex [41] or a light-driven FAD reduction [42], which seem to be proofs of

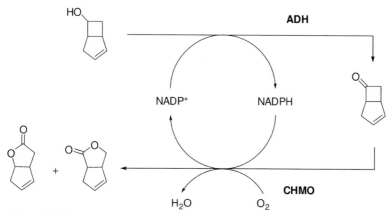

Scheme 5.6 NADPH regeneration in a coupled enzyme system using an alcohol dehydrogenase (ADH) from *Thermoanaerobium brockii* and cyclohexanone monooxygenase from *Acinetobacter calcoaceticus* NCIMB 9871.

principle rather than serious alternatives for cofactor regeneration applicable on a large scale.

There are also problems to be considered when working with isolated BVMOs and a cofactor regeneration system: (i) BVMOs often show low stability in purified form and therefore only comparably short half-lives, (ii) reaction conditions have to be identified which are compatible with both enzymes (e.g. with regard to the buffer system, temperature, possible enzyme inhibition by one of the substrates employed or *in situ*-generated products) and (iii) the required cosubstrate and the product formed therefrom can interfere with the work-up procedure for the desired BVO product.

As an alternative, whole-cell biooxidations can be performed in which advantage is taken of cell-innate NADPH regeneration systems. These are the first steps of glucose oxidation in the pentose phosphate pathway and a $NADP^+$-dependent isocitrate dehydrogenase in the tricarboxylic acid pathway [43]. Using whole cells offers the additional advantage that time- and money-consuming enzyme isolation and purification do not have to be accomplished. On the other hand, unwanted side-reactions can occur, decreasing the yield or optical purity of desired product since whole cells exhibit a large pool of different biocatalysts catalyzing all kinds of reactions. Especially when working with wild-type organisms, the probability of further conversion of the BVO product is increased as BVMOs are often part of metabolic degradation pathways [20, 44, 45]. Also, the substrate can be reduced by ketoreductases or alcohol dehydrogenases, yielding the corresponding alcohol. Nowadays, BVMOs are often recombinantly expressed in fast-growing and easy to handle microorganisms such as *Escherichia coli* (prokaryotic host) [46, 47] or *Saccharomyces cerevisiae* (eukaryotic host) [48, 49] using strong promoters and leading to high overexpression of the recombinant enzyme. Thus, the probability of unwanted side-reactions is minimized and undesirable enzyme activities can even be eliminated using today's genetic

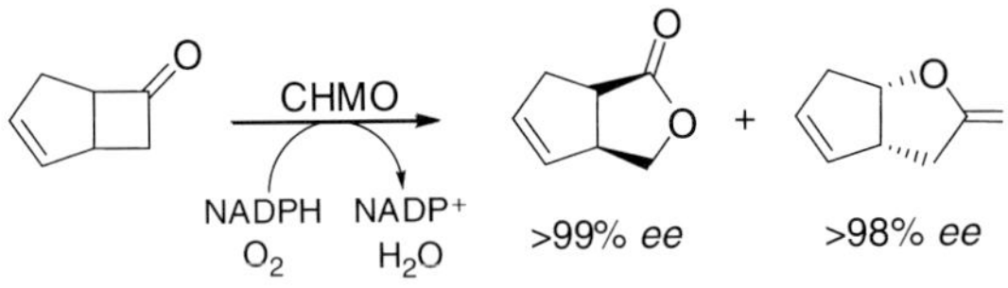

Scheme 5.7 Regiodivergent BVO of bicyclo[3.2.0]hept-2-en-6-one using cyclohexanone monooxygenase from *Acinetobacter calcoaceticus* NCIMB 9871.

engineering techniques. Additionally, enzymes originating from pathogenic organisms can be safely studied in non-pathogenic, recombinant hosts [50, 51].

Common problems observed in whole-cell biotransformations are substrate and product inhibition and toxicity effects exerted by these compounds on the cells, limiting the maximum applicable substrate concentration. As an example, for the BVO of racemic bicyclo[3.2.0]hept-2-en-6-one (Scheme 5.7) in a batch fermentation using recombinant *E. coli* cells expressing CHMO from *A. calcoaceticus* NCIMB 9871, substrate titers of only $0.2–0.4 \, g \, L^{-1}$ were optimal and at lactone concentrations above $4.5\text{-}5 \, g \, L^{-1}$ the reaction stopped due to product inhibition [52]. Performing a fed-batch fermentation, ketone concentrations can be maintained below inhibitory concentrations [53] and an *in situ* product removal approach can be used to increase product yields [54]. A combination of both the so-called *in situ* substrate feeding and product removal strategy (SFPR) makes use of a solid resin binding both substrate and product and therefore keeping the concentrations in the fermentation broth below toxic and inhibitory levels [55].

5.4.2
Baeyer–Villiger Monooxygenases Relevant for Synthetic Applications

The best studied BVMO regarding substrate acceptance and stereopreference is the already mentioned cyclohexanone monooxygenase from *A. calcoaceticus* NCIMB 9871. The enzyme was discovered by Donoghue and co-workers in 1976 and isolated from the wild-type strain for further investigation [44, 56]. However, it took more than 10 years until the corresponding gene was cloned [57] and another decade until recombinant expression systems for this BVMO were available [48, 58]. Since then, many new CHMOs (the names of BVMOs are usually deduced from their natural or preferred substrate) were identified in other bacterial strains and their genes cloned (Table 5.1). Also, BVMOs specialized in the oxidation of larger cycloketones than cyclohexanone are available. Most of the BVMOs described in the literature so far preferentially convert cyclic ketones to their corresponding lactones, for which reason most synthetic applications involve regio- and stereoselective transformations of mono- and bicyclic ketones. As an example, 2,5- and 3,6-diketocamphane 1,2-monooxygenase and 2-oxo-Δ^3-4,5,5-trimethylcyclopentylacetyl coenzyme A monooxygenase from camphor-grown *P. putida* NCIMB 10007 were used as partially purified enzymes [59, 60] as their respective genes had not yet been identified. Furthermore, fungal BVMOs from *Cunninghamella echinulata* [61, 62] and different

Table 5.1 Relevant BVMOs in synthetic applications [64].

Preferred substrates	BVMO/origin	Year of		Examples of biotransformations
		Identification	Gene cloning	
Cyclic ketones	Cyclopentanone monooxygenase/*Comamonas* sp. NCIMB 9872	1976 [65]	2002 [20]	[20, 46]
	Cyclohexanone monooxygenase/*Acinetobacter calcoaceticus* NCIMB 9871	1976 [56]	1988 [57]	[47, 66]
	Cyclohexanone monooxygenase/*Acinetobacter* sp. SE19	2000 [19]	2000 [19]	[67]
	Cyclohexanone monooxygenase/*Arthrobacter* sp. BP2	2003 [68]	2003 [68]	[69]
	Cyclohexanone monooxygenase/*Brachymonas petroleovorans*	2003 [70]	2003 [70]	[69]
	Cyclohexanone monooxygenase 1 and 2 / *Brevibacterium epidermidis* HCU	2000 [71]	2000 [71]	[69]
	Cyclohexanone monooxygenase 1 and 2/*Rhodococcus* sp. Phi1 and Phi2	2003 [68]	2003 [68]	[69]
	Cyclododecanone monooxygenase/*Rhodococcus ruber* SC1	2001 [45]	2001 [45]	[67]
	Cyclopentadecanone monooxygenase/*Pseudomonas* sp. HI-70	2006 [72]	2006 [72]	[72]
	2,5- and 3,6-diketocamphane 1,2-monooxygenase/*Pseudomonas putida* NCIMB 10007	1986 [22]	—	[59, 60]
	2-Oxo-Δ^3-4,5,5-trimethylcyclopentylacetyl coenzyme A monooxygenase/*Pseudomonas putida* NCIMB 10007	1983 [73]	—	[59]
	BVMO Rv3049c/*Mycobacterium tuberculosis* H37Rv	2006 [74]	2006 [74]	[74]
	Fungal BVMO/*Cunninghamella echinulata* NRRL 3655	—	—	[62]
Aryl aliphatic ketones	4-Hydroxyacetophenone monooxygenase/*Pseudomonas fluorescens* ACB	2001 [15]	2001 [15]	[75]
	Phenylacetone monooxygenase/*Thermobifida fusca* YX	2005 [76]	2005 [76]	[77]
Acyclic ketones	BVMO BmoF1/*Pseudomonas fluorescens* DSM 50106	1999 [78]	2007 [79]	[80]
Steroids	Steroid monooxygenase/*Rhodococcus rhodochrous* IFO 3338	1995 [81]	1999 [82]	—
	Steroid monooxygenase/*Cylindrocarpon radicicola* ATCC 11011	1986 [83]	—	[84]

Curvularia and *Drechslera* species [63] were also applied in stereoselective transformations of different cyclic ketones. Since the corresponding genes had not yet been identified and cloned, only wild-type strains or isolated BVMOs therefrom could be used.

Further, few enzymes specialized in aryl aliphatic ketone oxidation have been found, two of them being cloned and recombinantly expressed in *E. coli*: a 4-hydroxyacetophenone monooxygenase (HAPMO) from *P. fluorescens* ACB [15] and phenylacetone monooxygenase from *T. fusca* [76]. These were also explored for their ability to oxidize several aryl aliphatic sulfides. Recently, several BVMOs from different bacterial strains favoring acyclic, short- and medium-chain ketones were identified [51, 79, 85], one of them, BmoF1 from *P. fluorescens* DSM 50106, even showing reasonable enantioselectivity in the kinetic resolution of racemic β-hydroxy ketones [86].

Another very interesting group of BVMOs with a view to synthetic applications in organic chemistry are steroid monooxygenases catalyzing the insertion of an oxygen atom into a ring of the sterol skeleton or the side-chain at position C17 [81, 83]. Although hardly any examples of their biocatalytic application exist [84, 87], they seem to be very interesting catalysts due to their preferred substrates.

5.4.3
Representative Synthetic Applications

Due to their high chemo-, regio- and stereoselectivity, BVMOs are very interesting biocatalysts for applications in organic chemistry. However, until a few years ago, only a handful of well-studied enzymes had been available, which might explain why the majority of synthetic biooxidations were performed with only a few BVMOs, especially CHMO from *A. calcoaceticus* NCIMB 9871. The number of available BVMOs has increased considerably since then and more and more enzymes are being studied for their biocatalytic potential. Comprehensive reviews on enzyme-mediated BVOs were recently published by Mihovilovic and co-workers [64, 88]. Therefore, this chapter will only give an overview of possible synthetic applications of BVMOs in organic chemistry and present the latest findings and results.

5.4.3.1 Kinetic Resolutions of Racemic Ketones
Mainly racemic cyclopentanone and cyclohexanone derivatives bearing a range of different substituents have been applied in kinetic resolutions using BVMOs (Schemes 5.8 and 5.9). The substituent spectrum includes alkyl groups of different

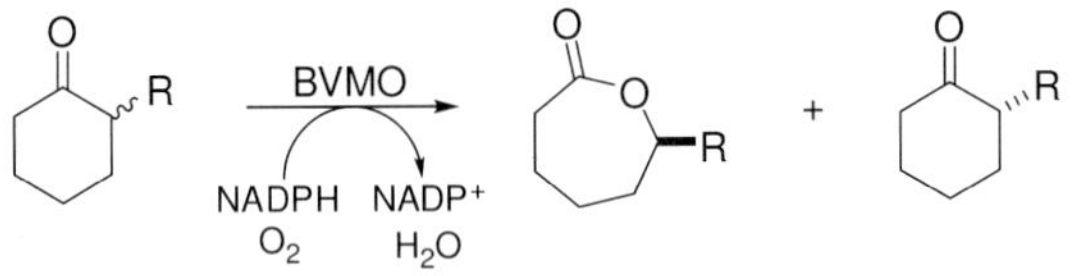

Scheme 5.8 BVMO-mediated kinetic resolution of α-substituted cyclohexanones.

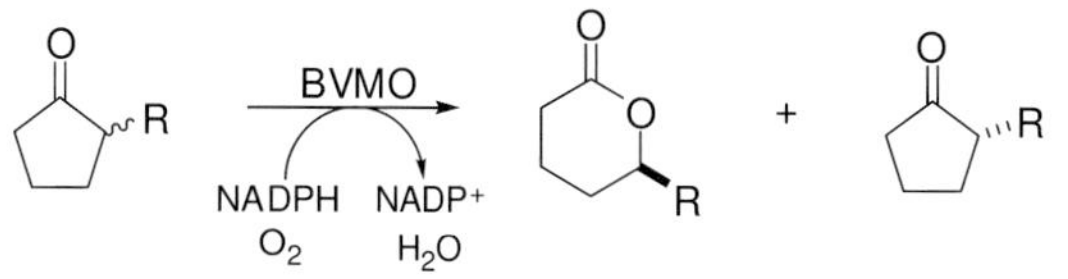

Scheme 5.9 BVMO-mediated kinetic resolution of α-substituted cyclopentanones.

chain length, aromatic groups and substituents with additional functionality. Depending on the enzyme used, stereoselectivity usually increases with increasing chain length until an optimum size reaching synthetically useful E values of >100. The enantiomeric ratio E, corresponding to the ratio of the relative second-order rate constants of the R- and S-enantiomers of the substrate, is used as a measure to compare the enantioselectivities of different enzymes [89]. Interestingly, most of the BVMOs preferentially convert the (S)-ketones to the corresponding (S)-lactones while the R-enantiomers remain unchanged. Depending on the substituent, the priority numbering and therefore also the R/S-assignment may change, but the sense of chirality preferred by the enzymes is retained.

CHMO from *A. calcoaceticus* exhibits high enantioselectivity in the kinetic resolution of α-alkylcyclohexanones with a substituent size ranging from C_2 to C_6 (Table 5.2), while α-methylcyclohexanone is converted with only low preference for the S-enantiomer ($E=6$) [90]. Recently, however, two different bacterial BVMOs were identified, CDMO from *Rhodococcus ruber* and CPDMO from *Pseudomonas* sp., oxidizing the substrate with high selectivity ($E > 200$) [67, 72]. The corresponding cyclopentanones with short-chain alkyl groups in the α-position are generally converted with much lower enantioselectivity [92], probably due to the lower energy difference of the axial and equatorial positions in five-membered ring systems [64] (Table 5.3). As an example, BVO of 2-(2′-acetoxyethyl)cyclohexanone was used to synthesize the corresponding (R)-lactone, which was further converted to (R)-(+)-lipoic acid, a bioactive compound for the treatment of hepatitis, pancreatitis and induced carcinomas (Scheme 5.10) [94, 95].

As an advantage, BVMOs not only exhibit regio- and stereoselectivity but also perform chemoselective oxidations. Thus, cyclic ketones possessing additional

Table 5.2 Examples of BVMO-mediated kinetic resolutions of α-substituted cyclohexanones.

Enzyme	R	% ee$_S$ (% y)	% ee$_P$ (% y)	E	Ref.
CHMO (*A. calcoaceticus*)	Et	>98 (35)	95 (40)	>200	[58]
	Allyl	>98 (29)	>98 (30)	>200	[58]
	iPr	96 (46)	>98 (41)	>200	[58]
	n-Bu	98 (32)	>98 (30)	>200	[58]
	n-Hex	23 (25)	98 (26)	>100	[90]
	Ph	86 (48)	>98 (40)	>100	[90]
	Bn	78 (28)	>96 (22)	>100	[90]
	(CH$_2$)$_2$CN	95 (46)	97 (50)	>200	[91]

(S = Substrate, P = product, y = yield).

Table 5.3 Examples of BVMO-mediated kinetic resolutions of α-substituted cyclopentanones.

Enzyme	R	% ee$_S$ (% y)	% ee$_P$ (% y)	E	Ref.
CHMO (*A. calcoaceticus*)	n-Hex	>98 (42)	>98 (32)	>200	[92]
	n-Undec	>98 (37)	>98 (39)	>200	[92]
	(CH$_2$)$_3$Br	92 (47)	95 (47)	128	[93]
MO$_2$ (*P. putida*)	CH$_2$COOEt	75 (37)	98 (27)	>100	[94]

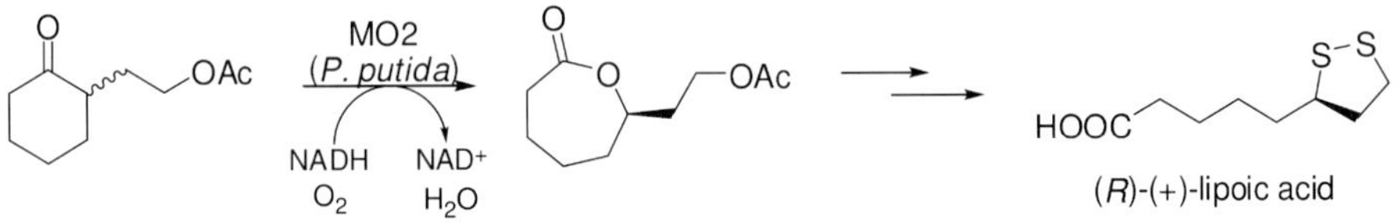

Scheme 5.10 BVO of 2-(2'-acetoxyethyl)cyclohexanone using 2,5-
and 3,6-diketocamphane 1,2-monooxygenase (MO2) from
Pseudomonas putida NCIMB 10007 in the synthesis of (R)-(+)-
lipoic acid.

functional groups such as heteroatoms [96] or double bonds [97] were converted
solely to the corresponding lactones using different BVMOs.

A drawback of kinetic resolutions of racemic substrates is their limitation to 50%
maximum yield, since ideally only one enantiomer is converted by the enzyme.
Through application of *in situ* racemization of the non-converted substrate, the
product yield can be increased to 100%. For such a dynamic kinetic resolution
process, the racemization reaction has to be as fast as or faster than the desired
enzymatic step. The first example of a BVMO-mediated dynamic kinetic resolution of
2-benzyloxymethylcyclopentanone was reported in 2002 (Scheme 5.11) [98]. Only the
substrate shows pH-dependent racemization via keto–enol tautomerization, leading

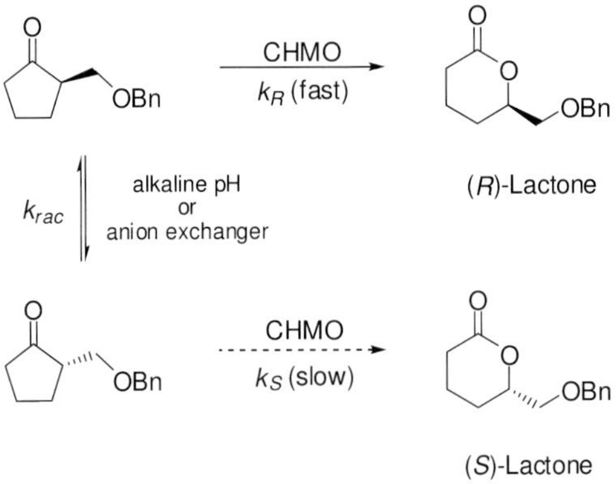

Scheme 5.11 Dynamic kinetic resolution of 2-
benzyloxymethylcyclopentanone using cyclohexanone
monooxygenase from *Acinetobacter calcoaceticus* NCIMB 9871.

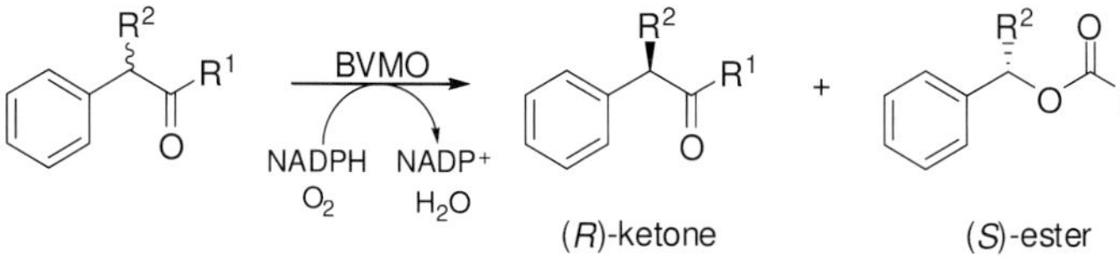

Scheme 5.12 BVMO-mediated kinetic resolution of aryl aliphatic ketones.

to complete racemization within 7 h at pH 9, while the product lactone is not affected. This fact was used in a whole-cell biooxidation using recombinant *E. coli* cells expressing CHMO from *A. calcoaceticus*. After 24 h, 85% (*R*)-lactone was obtained with 96% *ee* when only 1.5 m*M* substrate were employed. Later, a weak anion exchanger (Lewatit MP62) was applied to improve racemization and the reaction conditions were optimized to be able to increase substrate concentrations without compromising the stereoselectivity of the reaction [99].

In addition to cycloketones, racemic aryl aliphatic ketones have also been applied in kinetic resolutions using BVMOs. High *E* values could be achieved in the conversion of different phenyl ketones using PAMO and HAPMO showing *S*-selectivity (Scheme 5.12), but only results of analytical-scale reactions were reported (Table 5.4). In contrast, the corresponding racemic 2-phenylpropionaldehyde was oxidized with only low enantioselectivity ($E \leq 25$) using the same enzymes [100]. CHMO from *A. calcoaceticus* and CPMO from *Comamonas* sp. showed very low selectivity in the conversion of 3-phenylbutan-2-one and 3-phenylpentan-2-one ($E \leq 5$), whereas BmoF1 from *P. fluorescens* DSM 50106 could be successfully applied in the kinetic resolution of 3-phenylbutan-2-one on a preparative scale [80]. The BVMO preferentially converted the *S*-enantiomer to the corresponding phenylethyl acetate with $E = 82$. After enzymatic hydrolysis of the product ester, (*S*)-phenylethanol was obtained in 35% yield with 93% *ee*. Two BVMOs from *P. putida* KT2440 and *P. veronii* MEK700 exhibiting minor *R*-selectivity were also identified [80, 101].

Recently, the first BVMO-mediated kinetic resolution of acyclic 4-hydroxy-2-ketones was reported (Scheme 5.13). Using recombinant whole cells of *E. coli* expressing BmoF1 from *P. fluorescens* DSM 50106, the corresponding (*S*)-hydroxyalkyl acetates

Table 5.4 Examples of BVMO-mediated kinetic resolutions of aryl aliphatic ketones.

Enzyme	R^1/R^2	% ee_S	% ee_P	c (%)	E	Ref.
PAMO (*T. fusca*)	Me/Me	36	98	27	>100 (*S*)	[100]
	Et/Me	27	99	21	>200 (*S*)	[100]
	Me/Et	45	99	32	>200 (*S*)	[100]
	Et/Et	98	95	51	>100 (*S*)	[100]
HAPMO (*P. fluorescens* ACB)	Me/Me	43	97	30	>100 (*S*)	[100]
	Et/Me	98	97	50	>200 (*S*)	[100]
	Me/Et	78	96	45	117 (*S*)	[100]
	Et/Et	98	90	52	87 (*S*)	[100]
BmoF1 (*P. fluorescens* DSM 50106)	Me/Me	80	94	46	82 (*S*)	[80]

(*c* = conversion).

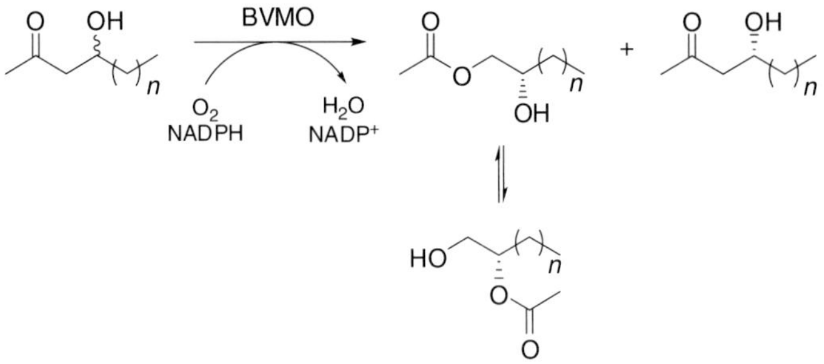

Scheme 5.13 BVMO-mediated kinetic resolution of 4-hydroxy-2-ketones.

were obtained with >90% *ee* ($E \approx 50$) [86]. Migration of the acetyl group from the primary to the secondary hydroxyl group resulted in a 4:1 mixture of the acetylated 1,2-diols. The enantioselectivity could be improved up to $E > 90$ by directed evolution of the wild-type enzyme and by performing the oxidation reactions in shake flasks, probably due to better oxygen supply [102]. Interestingly, a BVMO from *P. veronii* MEK 700 preferentially converted the (*R*)-4-hydroxy-2-ketones but with much lower selectivity ($E \leq 20$) [101].

5.4.3.2 Desymmetrization of Prochiral Ketones

Compared with kinetic resolutions, in desymmetrization reactions of prochiral compounds a 100% (theoretical) yield of optically pure product can be obtained, making such a reaction far more attractive for organic synthesis. Many prochiral cycloketones, both mono- and polycyclic, of diverse structure have been applied in desymmetrizations using BVMOs. A comprehensive compilation of results for biotransformations on a preparative scale can be found in a recent review by Mihovilovic [64]. Here, only selected examples are summarized. Most reactions were again carried out with CHMO from *A. calcoaceticus*, but also more recently discovered BVMOs were investigated, which in some cases led to the formation of enantiocomplementary lactone products.

4-Mono- and disubstituted cyclohexanones with small substituents were oxidized to the corresponding lactones with high activity and stereoselectivity by CHMO from *A. calcoaceticus* whereas lower conversions or enantiomeric excesses were obtained for larger side chains (>C4) (Scheme 5.14 and Table 5.5). For some 4-substituted cyclohexanones, CPMO from *Comamonas* sp. was shown to produce the opposite

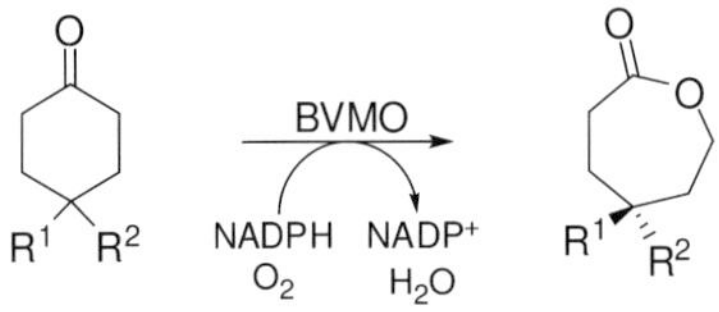

Scheme 5.14 BVMO-mediated desymmetrization of prochiral 4-mono- and 4-disubstituted cyclohexanones.

Table 5.5 Examples of BVMO-mediated desymmetrization of prochiral 4-mono- and 4-disubstituted cyclohexanones.

Enzyme	R^1/R^2	Yield (%)	% ee	Ref.
CHMO (*A. calcoaceticus*)	Me/H	83	>98 (−)	[49]
	Et/H	91	97 (−)	[49]
	Pr/H	80	>98 (−)	[103]
	allyl/H	62	95 (−)	[49]
	CH_2OH/H	80	>98 (−)	[103]
	Cl/H	56	95 (−)	[104]
	Br/H	63	97 (−)	[46]
	I/H	60	97 (−)	[46]
	Et/OH	54	94 (−)	[46]
CHMO (*Brachymonas petroleovorans*)	Me/H	69	>99 (−)	[69]
	Me/OH	48	97 (−)	[69]

lactone enantiomer, but with lower optical purity [64]. In the case of prochiral cyclohexanones with hydroxy functionalities in the 4-position (OH or CH_2OH), the expected seven-membered ring lactones are not obtained as products, but undergo immediate rearrangement to the corresponding five- or six-membered ring lactones, which are thermodynamically more stable (Scheme 5.15).

Mihovilovic's group investigated different BVMOs for their stereoselectivity in the conversion of different prochiral 4-substituted 3,5-dimethylcyclohexanones. For most substrates, enzymes could be identified yielding the corresponding lactones with >99% *ee*; sometimes enantiocomplementary products were also produced [105]. The corresponding lactone products constitute versatile building blocks in the synthesis of different natural compounds [106]. Also ketones bearing heteroatoms within the ring were accepted as substrates and prochiral 2,6-dimethyl- and 2,6-diethylpyran-4-one were oxidized by CHMO from *A. calcoaceticus*, yielding the corresponding lactones in high yield and >99% *ee* [107].

Prochiral 3-substituted cyclobutanones were intensively studied as substrates in BVMO-mediated stereoselective oxidations (Scheme 5.16 and Table 5.6) since the corresponding γ-butyrolactones represent versatile building blocks for the synthesis of various bioactive compounds and natural products such as the lignans entero-

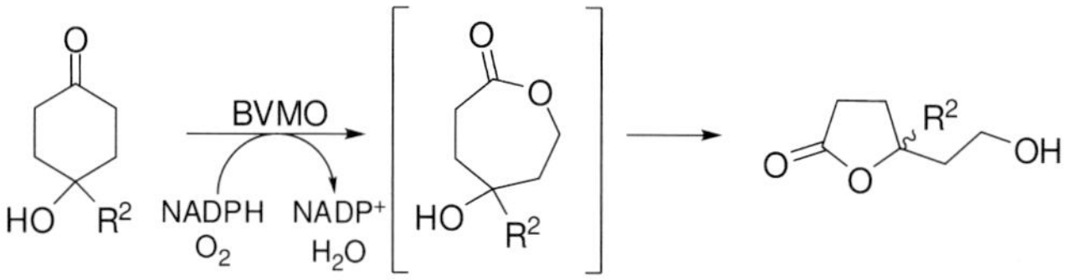

Scheme 5.15 BVO of 4-hydroxycyclohexanones with subsequent rearrangement of the corresponding seven-membered ring lactones.

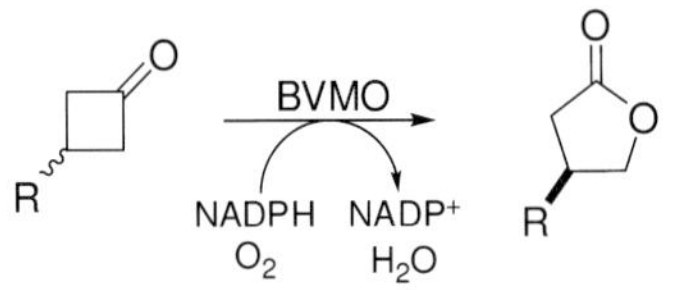

Scheme 5.16 BVMO-mediated desymmetrization of prochiral 3-disubstituted cyclobutanones.

Table 5.6 Examples of BVMO-mediated desymmetrization of prochiral 3-disubstituted cyclobutanones.

R	Enzyme	Yield (%)	% ee	Ref.
n-Bu	CHMO (B. epi. 1)	65	>99 (−)	[108]
i-Bu	CHMO (B. epi. 1)	30	>99 (−)	[108]
Ph	HAPMO (P. fl. ACB)	12	92 (+)	[111]
	CHMO (B. petr.)	45	93 (−)	[108]
	CHMO (B. epi. 1)	73	98 (−)	[108]
	Cunninghamella	65	98 (−)	[62]
4-ClPh	CHMO (R. Phi2)	63	95 (+)	[108]
	Cunninghamella	30	>98 (−)	[62]
Bn	CHMO (B. epi. 1)	30	93 (−)	[108]
	CHMO (A. BP2)	56	93 (−)	[108]
3-MeOBn	CHMO (R. Phi1)	60	98 (−)	[108]
	CHMO (A. calc.)	83	96 (−)	[62]
4-MeO-Bn	CHMO (A. BP2)	89	97 (−)	[108]
3,4,5-(MeO)$_3$Bn	CHMO (A. BP2)	72	94 (−)	[108]
BnOCH$_2$	Cunninghamella	74	98 (−)	[62]

lactone, hinokinin, steganacine and *trans*-burseran, exhibiting diverse pharmacological activities (Scheme 5.17) [108]. 3-(p-Chlorophenyl)-γ-butyrolactone was used as a precursor in the synthesis of the γ-aminobutyric acid (GABA) inhibitor (R)-baclofen [109]. Using whole cells of the fungi *Cunninghamella*, 3-benzyloxymethylcyclobutanone was oxidized to the corresponding (R)-lactone, which was further converted to both enantiomers of β-proline [110]. For most cyclobutanones tested, both antipodal lactones could be produced using different BVMOs. However, in most cases only one enantiomer was obtained in high enantiomeric excess [64].

BVMOs also accepted sterically demanding bridged cycloketones as substrates whereas α, α'-disubstituted cyclohexanones were only poorly converted. Using differently substituted bicyclo[2.2.1]heptan-7-ones, a library of recombinant whole cells expressing various BVMOs of bacterial origin was screened for the production of corresponding lactones in high conversion and stereoselectivity [112]. Thus, for most substrates enzymes could be identified to obtain both lactone enantiomers and corresponding biotransformations were also carried out on a preparative scale (Scheme 5.18 and Table 5.7). The corresponding dicarboxylic acid dimethyl ester-substituted substrate was not accepted by either enzyme.

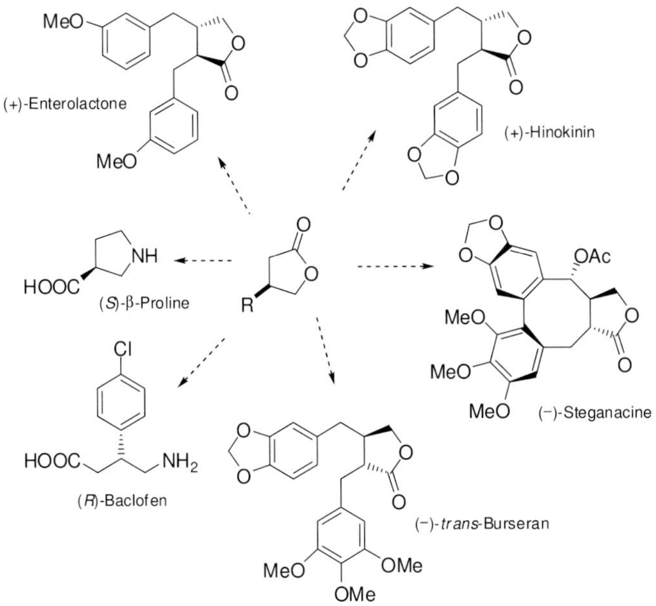

Scheme 5.17 Chiral butyrolactones as key intermediates for total syntheses.

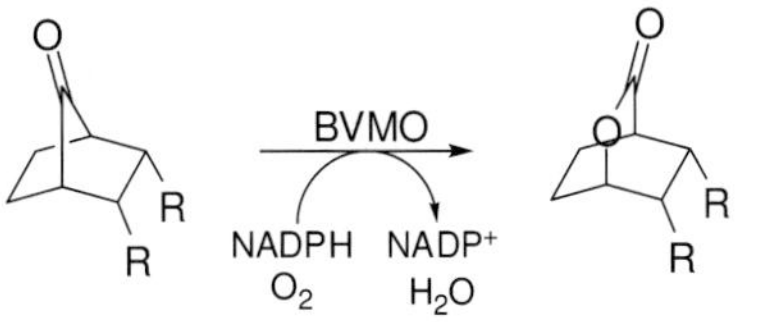

Scheme 5.18 BVMO-mediated oxidation of substituted bicyclo[2.2.1]heptan-7-ones.

Table 5.7 Examples of enzyme-mediated BVO of substituted bicyclo[2.2.1]heptan–7-ones.

R	Enzyme	Yield (%)	% ee	Ref.
$-C_2H_4-$	CHMO (B. epi. 1)	72	96 (−)	[112]
$-CH_2OCH_2-$	CHMO (A. calc.)	53	92 (−)	[112]
$-C_3H_6-$	CHMO (R. Phi2)	63	99 (−)	[112]
	CHMO (B. epi. 2)	67	92 (+)	[112]
$-C_4H_8-$	CHMO (R. Phi1)	58	99 (−)	[112]
	CHMO (B. epi. 2)	78	94 (+)	[112]
Bicyclo[2.2.1]heptane	CHMO (A. calc.)	51	98 (−)	[112]

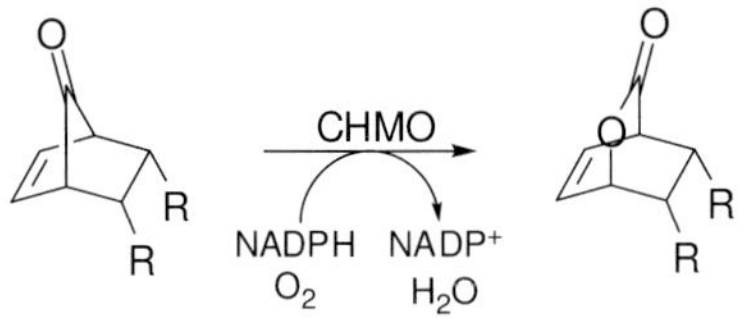

Scheme 5.19 BVMO-mediated oxidation of substituted bicyclo[2.2.1]hepten-7-ones.

Table 5.8 Examples of enzyme-mediated BVO of substituted bicyclo[2.2.1]hepten-7-ones.

Enzyme	R	Yield [%]	% ee	Ref.
CHMO (A. calcoaceticus)	Me	70	>98 (+)	[113]
	Et	83	93 (+)	[113]
	−CH$_2$OCH$_2$−	74	>98 (+)	[113]
	−C$_3$H$_6$−	80	97 (+)	[113]
	−C$_4$H$_8$−	78	>98 (−)	[113]
	−C$_5$H$_{10}$−	57	>98 (+)	[113]

CHMO from *A. calcoaceticus* was also used to oxidize analogous unsaturated bicyclic ketones to the corresponding lactones with high stereoselectivity (Scheme 5.19 and Table 5.8) [113].

In a survey of different fused bicyclic ketones using CHMO from *A. calcoaceticus* and CPMO from *Comamonas* sp., it was observed that the former enzyme exhibits higher stereoselectivity in the conversion of 5,5-bicycloketones whereas the latter enzyme gave the corresponding lactones of prochiral 6,5-bicycloketones with higher *ee* (Scheme 5.20) [114]. Furthermore, it was found that CHMO shows higher conversion of more lipophilic bicyclics than hydrophilic bicyclics and exhibits higher stereoselectivity in the oxidation of 'stretched' *exo* substrates than 'angular' *endo* compounds. The corresponding lactones of prochiral bicyclo[4.3.0]ketones represent key intermediates for the synthesis of indole alkaloids [115].

Recently, Mihovilovic *et al.* analyzed a small library of bacterial BVMOs for their substrate acceptance and stereopreference towards prochiral cyclic ketones of

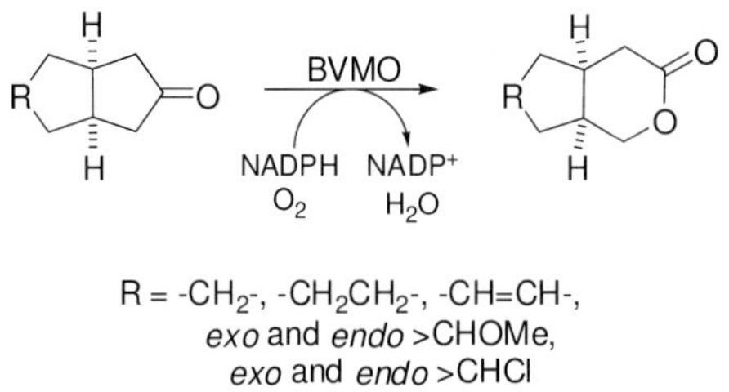

R = -CH$_2$-, -CH$_2$CH$_2$-, -CH=CH-,
exo and endo >CHOMe,
exo and endo >CHCl

Scheme 5.20 Enzyme-mediated BVO of various prochiral bicyclo [4.3.0]ketones and bicyclo[3.3.0]ketones.

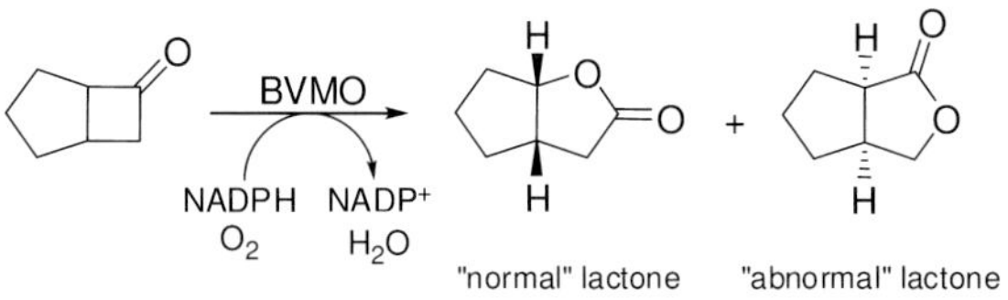

Scheme 5.21 Regiodivergent BVO of a bicyclobutanone.

various structure [69]. Interestingly, they found a correlation of the obtained results with protein sequence data. Using BVMOs showing higher sequence similarity on an amino acid level, similar results regarding substrate acceptance and stereopreference were found. Thus, based on reaction profiles, enzymes could be clustered into two groups, 'CHMO type' and 'CPMO type', which was also supported by phylogenetic tree analysis. This clustering was also confirmed by later biotransformation results using other prochiral ketones [108, 112], possibly facilitating predictions for the stereochemical performance of new BVMOs in the future.

5.4.3.3 Regiodivergent Transformations

In a regiodivergent BVO, both enantiomers of a racemic ketone substrate are converted by one enzyme into different lactone products, ideally with high stereoselectivity (Scheme 5.21). One enantiomer is oxidized to the 'normal' lactone by migration of the higher substituted carbon atom whereas oxidation of the other enantiomer gives the 'abnormal' lactone by migration of the less substituted carbon atom. The reason is probably a different positioning of the antipodal ketones within the active site of the enzyme so that different bonds are situated antiperiplanar to the O-O bond of the leaving group in the Criegee intermediate – a prerequisite for successful rearrangement [116].

Regiodivergent BVO is often found for fused cyclobutanones (Table 5.9). Here again the clustering of BVMOs into 'CHMO-type' yielding approximately a 50 : 50 ratio of both 'normal' and 'abnormal' lactones with good to high enantiomeric excesses and 'CPMO-type' producing predominantly 'normal' lactones in almost racemic form is evident.

The 'normal' oxidation product of bicyclo[3.2.0]hept-2-ene-6-one was used as key intermediate in the synthesis of various prostaglandins [119] whereas the 'abnormal' lactone served as a starting point for the synthesis of cyclosarkomycin, a precursor of the cytostatic (R)-(−)-sarkomycin [120], and also different brown algae pheromones [121]. Racemic bicyclobutanones bearing heteroatoms were also successfully resolved such as N-protected 2-azabicyclo[3.2.0]heptan-6-one, a precursor of the so-called Geisman–Waiss lactone (the corresponding 'normal' BVO product), which serves as an intermediate in the synthesis of necine base pyrrolizidine alkaloids [122].

Regiodivergent behavior of CHMO from A. calcoaceticus was also found in the oxidation of racemic 3-methyl- and 3-ethyl-substituted cyclohexanones [49], while conversion of the corresponding β-substituted cyclopentanones gave both regioisomers in racemic form [92]. Longer tethers in the case of 3-cyclohexanones led to the formation of only proximal regioisomers with moderate ee and the same

Table 5.9 Examples of regiodivergent BVO of different bicyclobutanones.

Ketone	Enzyme	Yield (%)[a]	Ratio[b]	% ee[c]	Ref.
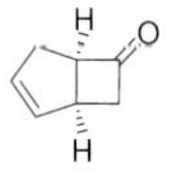	CHMO (A. calc.)	86	51:49	>95 (1S,5R)	[117]
				>95 (1R,5S)	
	CHMO (B. epi. 1)	85	51:49	96 (1S,5R)	[118]
				>99 (1R,5S)	
	CPMO (Com. sp.)	61	97:3	0	[69]
				>99 (1R,5S)	
	CHMO (B. epi. 2)	61	98:2	0	[118]
				>99 (1R,5S)	
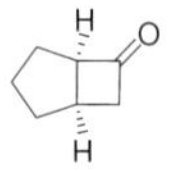	CHMO (B. epi. 1)	78	50:50	>99 (1S,5S)	[118]
				>99 (1R,5R)	
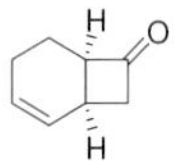	CHMO (A. calc.)	80	54:46	>95 (1R,6S)	[117]
				>95 (1S,6R)	
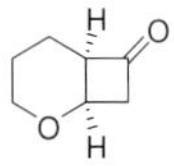	CHMO (B. epi. 1)	64	75:25	97 (1S,6S)	[118]
				>99 (1R,6S)	
	CHMO (B. epi. 2)	71	96:4	0	[118]
				>99 (1S,6R)	

[a]Combined yield of 'normal' and 'abnormal' lactone.
[b]Ratio of 'normal' to 'abnormal' lactone.
[c]Enantiomeric excess of 'normal' and 'abnormal' lactone.

regioselectivity, but no enantioselectivity was found for 3-alkylcyclopentanones with more than four carbon atoms in the side-chain. Also chemical BVO of racemic 3-substituted cycloketones gives product mixtures since the influence of the substituent on the adjacent α-carbon is too weak to make it significantly different from the other. Hence both α-carbon atoms have the tendency to migrate, giving four different lactone products. Using a combination of both chemical and enzymatic catalysis, lactones – also with longer side chains – could be obtained with high enantiomeric excess (Scheme 5.22 and Table 5.10) [104]. Here CPMO from *Comamonas* sp. gave superior results to CHMO. Interestingly, CHMO-catalyzed BVO of 2-cyanomethylcyclohexanone gave the corresponding proximal

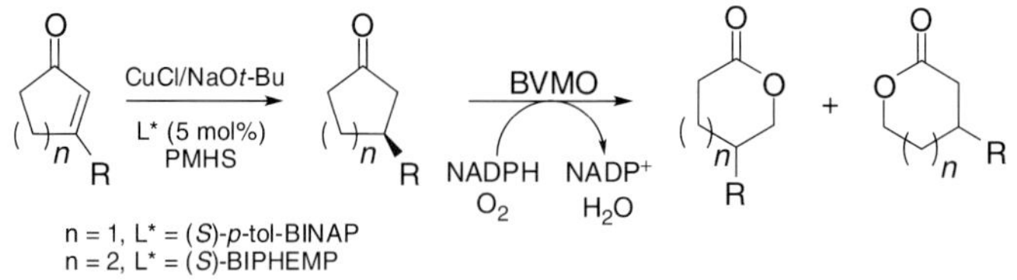

Scheme 5.22 Combination of stereoselective chemical reduction of cyclic enones and BVMO-mediated oxidation of enantiomerically enriched cyclic ketones in the synthesis of optically pure ε-caprolactones and δ-valerolactones.

Table 5.10 Examples of BVMO-mediated oxidation of chiral β-substituted cycloketones.

Enzyme	R (n)	% ee (substrate)	Yield^a (%)	Ratio^b	% ee^c	Ref.
CHMO	Me (1)	>99 (R)	88	10:90	99 (R)	[104]
(A. calcoaceticus)					99 (R)	[104]
	n-Bu (1)	90 (S)	88	95:5	90 (S)	[104]
					—	[104]
	(CH₂)₂Ph (1)	88 (S)	80	>99:1	99 (S)	[104]
	Me (2)	100 (R)	77	>99:1	100 (R)	[104]
		100 (S)	60	99:1	100 (S)	[104]
	Et (2)	81 (R)	89	94:6	94 (R)	[104]
					>99 (S)	[104]
CPMO	Me (1)	>99 (R)	62	100:0	>99 (R)	[104]
(Comamonas sp.)	n-Bu (1)	90 (S)	84	>99:1	92 (S)	[104]
	Me (2)	100 (R)	75	>99:1	100 (R)	[104]

^{a}Combined yield of proximal and distal lactone.
^{b}Ratio of proximal to distal lactone.
^{c}Enantiomeric excess of proximal and distal lactone.

and distal lactones both with >99% *ee* whereas related 2-alkylcyclohexanones were exclusively converted to the proximal caprolactones by this enzyme [91]. Due to the electron-withdrawing nitrile group, the carbon atom bearing the cyanomethyl group is less electronegative than the other α-carbon and therefore both migrate.

Regiodivergent oxidation with high regioselectivity was also found in a series of different terpenones using recombinant *E. coli* strains expressing various BVMOs of bacterial origin [123]. Thus, using different enzymes both 'normal' and 'abnormal' lactones of (−)-*trans*-dihydrocarvone, (−)-carvomenthone and both enantiomers of *cis*-dihydrocarvone could be obtained (Table 5.11). (+)-*trans*-Dihydrocarvone and

Table 5.11 Examples of enzyme-mediated BVO of different optically pure terpenones.

Ketone	Enzyme	Yield (%) (normal lactone)	Yield (%) (abnormal lactone)	Ref.
(+)-*trans*-Dihydrocarvone	CHMO (A. calc.)		70 (−)	[123]
(−)-*trans*-Dihydrocarvone	CHMO (A. calc.)	77 (+)		[123]
	CPMO (Coma.)		18 (+)	[123]
(+)-*cis*-Dihydrocarvone	CHMO (B. epi. 1)	60 (+)		[123]
	CHMO (B. petr.)		7	[123]
(−)-*cis*-Dihydrocarvone	CHMO (A. calc.)	77 (−)		[123]
	CPMO (Coma.)		28 (−)	[123]
(+)-Carvomenthone	CHMO (B. petr.)		76 (−)	[123]
(−)-Carvomenthone	CHMO (B. epi. 1)	71 (+)		[123]
	CPMO (Coma.)		60 (+)	[123]
(+)-Menthone	CHMO (A. calc.)	82 (+)		[123]

(+)-carvomenthone were only converted to the 'abnormal' lactones by all tested BVMOs whereas (+)-menthone gave exclusively 'normal' lactone. Again, classification of BVMOs into 'CHMO type' and 'CPMO type' regarding substrate acceptance and stereopreference was evident. Interestingly, (−)-menthone was not converted by any of the applied enzymes.

5.4.3.4 Large-Scale Application

Although many promising examples of BVMO-catalyzed syntheses of valuable compounds have been reported in the literature, they are rarely used as catalysts in industrial processes. This is probably due to certain characteristics exhibited by BVMOs placing high technical demands on reactor design and downstream processing: (i) they are cofactor dependent, (ii) they need molecular oxygen for catalyzing the reaction, (iii) they are not very stable catalysts either *in vitro* or *in vivo* and (iv) they often exhibit substrate and/or product inhibition [124]. As already mentioned, the need for expensive cofactors can be overcome by applying a cofactor regeneration system or by using whole cells for the reaction process. On an industrial scale, whole-cell biotransformations will be the method of choice since no enzyme isolation step is necessary and the (recombinant) whole cells expressing the desired BVMO can be stored for longer periods with only minor loss of activity [86, 91]. In the first kilogram-scale asymmetric BVO of bicyclo[3.2.0]hept-2-en-6-one (see Scheme 5.7), recombinant *E. coli* cells expressing CHMO from *A. calcoaceticus* were applied [125]. Substrate and product inhibition were overcome by employing a resin-based *in situ* substrate feeding and product removal methodology whereby the substrate ketone and the product lactones formed during the reaction bind to the adsorbent resin (Dowex Optipore L-493), keeping their actual amounts dissolved in the reaction broth very low. Thus, a substrate concentration of about 25 g L^{-1} could be applied without adversely affecting enzyme activity. Additionally, the product lactones could be easily extracted from the resin after complete conversion, simplifying downstream processing. Subsequently, the resin was reusable, lowering the costs of the entire process. Sufficient oxygen supply, which is a demanding factor in terms of reactor design, especially when working with whole cells in high density [126], was made possible by using a specially designed sintered-metal sparger allowing the formation of very small air bubbles and therefore improving oxygenation levels. Thus, regiodivergent biotransformation of bicyclo[3.2.0]hept-2-en-6-one was performed in a 50 L reactor using 900 g of racemic ketone and 6.3 kg of wet resin. After 20 h, reaction was complete, resulting in a space–time yield of 8.2 mmol $L^{-1} h^{-1}$. Both product lactones were obtained in high optical purity (>98% *ee*). Volumetric and biocatalyst productivity were 1 g lactone $L^{-1} h^{-1}$ and 1.37 mmol $g^{-1} h^{-1}$, respectively. However, only 58% isolated product yield was obtained under non-optimized conditions. Nevertheless, this new strategy should facilitate the application of BVMOs in further large-scale (industrial) oxidation processes.

5.4.3.5 Heteroatom Oxidation

As already mentioned, in addition to ketone oxidation, BVMOs are also capable of heteroatom oxidation. Although only a few examples of BVMO-catalyzed oxidations

Table 5.12 Examples of BVMO-mediated oxidation of alkyl aryl sulfides.

Enzyme	Sulfide	c (%)	% ee	Ref.
CHMO (*A. calcoaceticus*)	$C_6H_5-S-CH_3$	88	99 (R)	[16]
	p-F$-C_6H_4-S-CH_3$	91	92 (R)	[16]
	p-F$-C_6H_4-S-C_2H_5$	96	93 (S)	[16]
	o-CH$_3-C_6H_4-S-CH_3$	90	87 (R)	[16]
	p-CH$_3-C_6H_4-S-C_2H_5$	89	89 (S)	[16]
	p-CH$_3-C_6H_4-S-i$Pr	99	86 (S)	[16]
HAPMO (*P. fluorescens ACB*)	$C_6H_5-S-CH_3$	96	99 (S)	[17]
	$C_6H_5-S-C_2H_5$	86	99 (S)	[17]
	$C_6H_5-S-C_3H_6$	85	97 (S)	[17]
	$C_6H_5-S-CH=CH_2$	70	98 (S)	[17]
	$C_6H_5-S-CH_2CH=CH_2$	69	98 (S)	[17]
	$C_6H_5-S-CH_2OCH_3$	63	98 (R)	[17]
	2-Naphthyl$-S-CH_3$	31	95 (S)	[17]
PAMO (*T. fusca*)	$C_6H_5CH_2-S-CH_3$	29	94 (S)	[77]
	$C_6H_5CH_2-S-C_2H_5$	36	98 (S)	[77]

at nitrogen [127], boron [17, 128] or selenium [129] have been published, stereo-selective sulfide and sulfoxide oxidations have been studied in more detail. Chiral sulfoxides are of special interest in synthetic chemistry since they serve as valuable synthons [130] or versatile chiral auxiliaries [131] and are also found in natural products [132]. Several examples of alkyl aryl sulfide oxidation have been reported (Table 5.12). The corresponding sulfoxides could often be obtained with high *ee*; however, the stereochemical outcome depended strongly on the size of the alkyl chain and the presence and chemical nature of substituents at the aromatic ring. In the case of CHMO from *A. calcoaceticus*, enzymatic activity also increased with increasing chain length of the alkyl chain and methyl or fluoride substituents in the *para* position on the aromatic ring [16]. In contrast, high *ees* of the corresponding (S)-sulfoxides in the oxidation of methyl and ethyl benzyl sulfides by PAMO from *T. fusca* were achieved through enantioselective oxidation of the sulfoxides to the corresponding sulfones [77]. Racemic ethyl benzyl sulfoxide was converted with high enantioselectivity ($E = 110$) to ethyl benzyl sulfone, leaving the (S)-sulfoxide behind (Scheme 5.23).

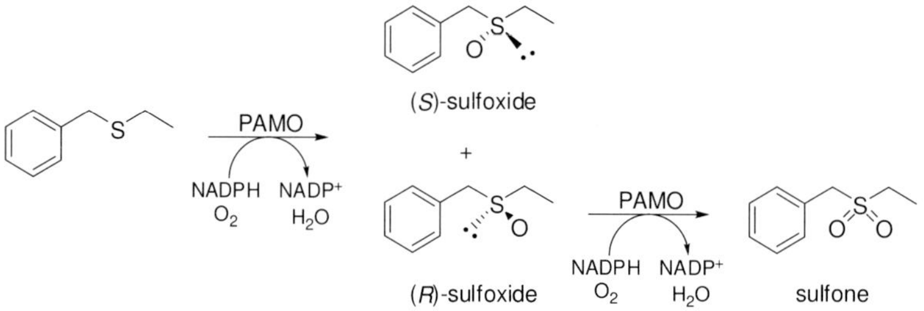

Scheme 5.23 Stereoselective oxidation of ethyl benzyl sulfide by phenylacetone monooxygenase from *Thermobifida fusca*.

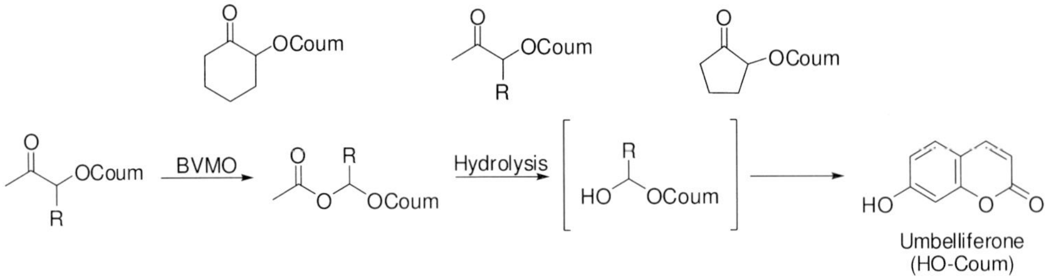

Scheme 5.25 Assay substrates for BVMO fluorescence assays.

on umbelliferone or *p*-nitrophenol release [147], having real potential to facilitate screening of large BVMO mutant libraries (Scheme 5.25).

Another interesting BVMO assay using β-hydroxy ketones as substrates was designed based on the adrenaline assay originally developed for hydrolase activity screening [148]. After hydrolysis of the ester formed by BVO of the substrate ketone, the resulting 1,2-diol is oxidized using $NaIO_4$. Non-converted sodium periodate is subsequently back-titrated using adrenaline, which is oxidized to adrenochrome, a red chromophore showing maximum absorption at 485 nm (Scheme 5.26) [102].

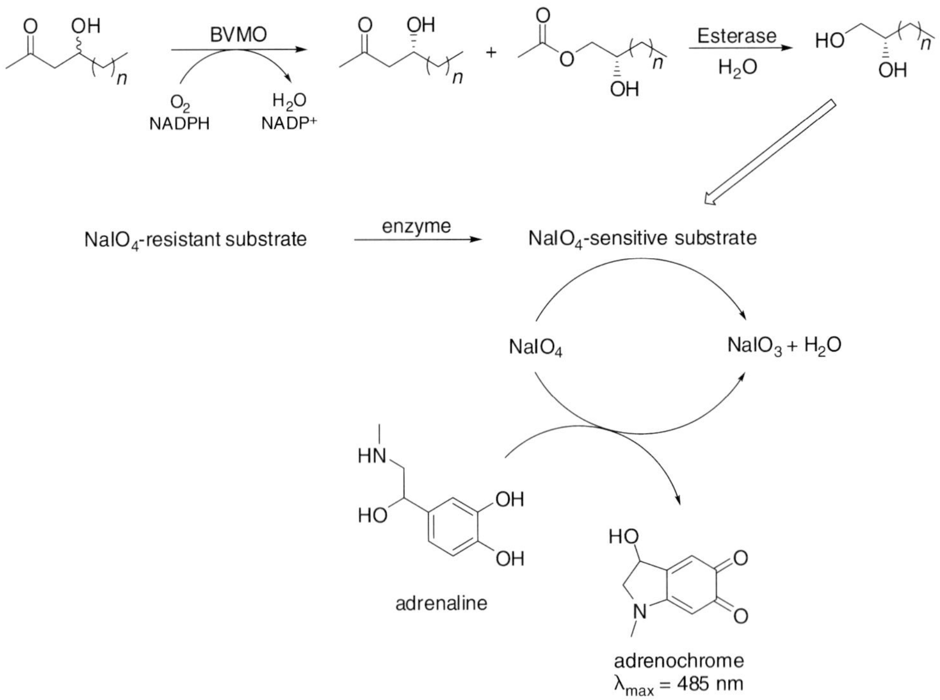

Scheme 5.26 Principle of the adrenaline assay for the detection of BVMO activity in the conversion of 4-hydroxy-2-ketones.

5.6
Conclusions and Perspectives

In this chapter, we have shown that Baeyer–Villiger monooxygenases are useful biocatalysts for the synthesis of (optically pure) lactones and esters from ketones under very mild reaction conditions. In contrast to chemical Baeyer–Villiger oxidations, the enzymatic route can be considered environmentally friendly as the use of toxic, explosive and harmful peracids, transition metal catalysts in the case of enantioselective reactions and the use of solvents can be avoided. Moreover, the broad substrate range and high enantioselectivity make BVMOs very attractive biocatalysts for organic synthesis. Substantial progress was made in the past few years with respect to the number of known BVMOs, the availability of recombinant enzymes and in protein engineering. Still, only a tiny fraction of (putative) BVMOs found in databases of sequenced genomes have been cloned and characterized so far and an even greater diversity can be expected to be present in nature. Hence we are far from assessing the broad potential of these biocatalysts for synthetic applications. Major issues to be addressed reside mostly in the area of process engineering, as the use of whole cell systems appears to be the method of choice for the application of BVMOs. Key factors to be solved here are substrate and product inhibition and stability of the enzymes. With the first known 3D structure of a BVMO and the first examples of successful protein design of BVMOs by rational methods and directed evolution, it can be expected that the near future will see even more examples for the creation of tailor-designed BVMOs and their application in organic synthesis.

Acknowledgments

We thank the Fonds der Chemischen Industrie (Frankfurt, Germany) and the Studienstiftung des Deutschen Volkes (Bonn, Germany) for stipends to Anett Kirschner for her work on the expression, characterization and directed evolution of the BVMO BmoF1 from *P. fluorescens* DSM 50106.

References

1 Liese, A., Seelbach, K. and Wandrey, C. (2006) *Industrial Biotransformations*, 2nd edn, Wiley-VCH Verlag GmbH, Weinheim; Schmid, A., Dordick, J.S., Hauer, B., Kiener, A., Wubbolts, M. and Witholt, B. (2001) *Nature*, **409**, 258–268; Drauz, K. and Waldmann, H. (2002) *Enzyme Catalysis in Organic Synthesis*, Vols 1–3, 2nd edn, Verlag GmbH, Weinheim; Faber, K. (2004) *Biotransformations in Organic Chemistry*, 5th edn, Springer, Heidelberg; Schoemaker, H.E., Mink, D. and Wubbolts, M.G. (2003) *Science*, **299**, 1694–1697.

2 Baeyer, A. and Villiger, V. (1899) *Berichte der Deutschen Chemischen Gesellschaft*, **32**, 3625–3633.

3 Criegee, R. (1948) *Justus Liebigs Annalen der Chemie*, **560**, 127–135.

4 Krow, G.R. (1993) *Organic Reactions*, **43**, 251–798.

5 Crudden, C.M., Chen, A.C. and Calhoun, L.A. (2000) *Angewandte Chemie-International Edition*, **39**, 2851–2855; Crudden, C.M., Chen, A.C. and Calhoun, L.A. (2000) *Angewandte Chemie*, **112**, 2973–2977.

6 Noyori, R., Kobayashi, H. and Sato, T. (1980) *Tetrahedron Letters*, **21**, 2573–2576.

7 Krow, G.R. (1981) *Tetrahedron*, **37**, 2697–2724.

8 Strukul, G. (1998) *Angewandte Chemie-International Edition*, **37**, 1199–1209; Strukul, G. (1998) *Angewandte Chemie*, **110**, 1256–1267; Brink, G.-J. ten., Arends, I.W. C.E. and Sheldon, R.A. (2004) *Chemical Reviews*, **104**, 4105–4123; Punniyamurthy, T., Velusamy, S. and Iqbal, J. (2005) *Chemical Reviews*, **105**, 2329–2363.

9 Del Todesco Frisone, M., Pinna, F. and Strukul, G. (1993) *Organometallics*, **12**, 148–156; Gavagnin, R., Cataldo, M., Pinna, F. and Strukul, G. (1998) *Organometallics*, **17**, 661–667; Corma, A., Nemeth, L.T., Renz, M. and Valencia, S. (2001) *Nature*, **412**, 423–425; Ruiz, J.R., Jiménez–Sanchidrián, C. and Llamas, R. (2006) *Tetrahedron*, **62**, 11697–11703; Murahashi, S.-I., Oda, Y. and Naota, T. (1992) *Tetrahedron Letters*, **33**, 7557–7560.

10 Gusso, A., Baccin, C., Pinna, F. and Strukul, G. (1994) *Organometallics*, **13**, 3442–3451.

11 Bolm, C., Schlingloff, G. and Weickhardt, K. (1994) *Angewandte Chemie-International Edition*, **33**, 1848–1849; Bolm, C., Schlingloff, G. and Weickhardt, K. (1994) *Angewandte Chemie*, **106**, 1944–1946.

12 Bolm, C. and Schlinghoff, G. (1995) *Journal of the Chemical Society, Chemical Communications*, 1247–1248.

13 Frison, J.-C., Palazzi, C. and Bolm, C. (2006) *Tetrahedron*, **62**, 6700–6706; Katsuki, T. (2004) *Russian Chemical Bulletin, International Edition*, **53**, 1859–1870.

14 Bolm, C. (2000) in *Peroxide Chemistry* (ed. W. Adam), Wiley-VCH Verlag GmbH, Weinheim, pp. 494–510.

15 Kamerbeek, N.M., Moonen, M.J. H., van der Ven, J.G. M., van Berkel, W.J. H., Fraaije, M.W. and Janssen, D.B. (2001) *European Journal of Biochemistry*, **268**, 2547–2557.

16 Carrea, G., Redigolo, B., Riva, S., Colonna, S., Gaggero, N., Battistel, E. and Bianchi, D. (1992) *Tetrahedron: Asymmetry*, **3**, 1063–1068.

17 de Gonzalo, G., Torres Pazmino, D.E., Ottolina, G., Fraaije, M.W. and Carrea, G. (2006) *Tetrahedron: Asymmetry*, **17**, 130–135.

18 Turfitt, G.E. (1948) *The Biochemical Journal*, **42**, 376–383.

19 Cheng, Q., Thomas, S.M., Kostichka, K., Valentine, J.R. and Nagarajan, V. (2000) *Journal of Bacteriology*, **182**, 4744–4751.

20 Iwaki, H., Hasegawa, Y., Wang, S., Kayser, M.M. and Lau, P.C. K. (2002) *Applied and Environmental Microbiology*, **68**, 5671–5684.

21 Trudgill, P.W. (1984) in *Microbial Degradation of Organic Compounds*, Vol. **13** (ed. D.T. Gibson), Marcel Dekker, New York, pp. 131–180.

22 Taylor, D.G. and Trudgill, P.W. (1986) *Journal of Bacteriology*, **165**, 489–497.

23 Jones, K.H., Smith, R.T. and Trudgill, P.W. (1993) *Journal of General Microbiology*, **139**, 797–805.

24 Henry, K.M. and Townsend, C.A. (2005) *Journal of the American Chemical Society*, **127**, 3300–3309.

25 Gibson, M., Nur-e-alam, M., Lipata, F., Oliveira, M.A. and Rohr, J. (2005) *Journal of the American Chemical Society*, **127**, 17594–17595.

26 Willets, A. (1997) *Trends in Biotechnology*, **15**, 55–62.

27 Ryerson, C.C., Ballou, D.P. and Walsh, C. (1982) *Biochemistry*, **21**, 2644–2655; Walsh, C.T. and Chen, Y.C. J. (1988) *Angewandte Chemie-International Edition Engl.*, **27**, 333–343; Walsh, C. and Chen, Y.C. J. (1988) *Angewandte Chemie*, **100**, 342–352.

28 Sheng, D., Ballou, D.P. and Massey, V. (2001) *Biochemistry*, **40**, 11156–11167.

29 Ottolina, G., de Gonzalo, G. and Carrea, G. (2005) *Journal of Molecular Structure*, **757**, 175–181.

30 van Berkel, W.J. H., Kamerbeek, N.M. and Fraaije, M.W. (2006) *Journal of Biotechnology*, **124**, 670–689.

31 Fraaije, M.W., Kamerbeek, N.M., van Berkel, W.J. H. and Janssen, D.B. (2002) *FEBS Letters*, **518**, 43–47.

32 Malito, E., Alfieri, A., Fraaije, M.W. and Mattevi, A. (2004) *Proceedings of the National Academy of Sciences of the United States of America*, **101**, 13157–13162.

33 Kamerbeek, N.M., Fraaije, M.W. and Janssen, D.B. (2004) *European Journal of Biochemistry/FEBS*, **271**, 2107–2116.

34 Lennon, B.W., Williams, Jr, C.H. and Ludwig, M.L. (2000) *Science*, **289**, 1190–1194.

35 Van der Donk, W.A. and Zhao, H. (2003) *Current Opinion in Biotechnology*, **14**, 412–426; Eckstein, M., Daussmann, T. and Kragl, U. (2004) *Biocatalysis and Biotransformation*, **22**, 89–96.

36 Wong, C.-H. and Whitesides, G.M. (1981) *Journal of the American Chemical Society*, **103**, 4890–4899.

37 Tishkov, V.I., Galkin, A.G., Fedorchuk, V.V., Savitsky, P.A., Rojkova, A.M., Gieren, H. and Kula, M.-R. (1999) *Biotechnology and Bioengineering*, **64**, 187–193.

38 Woodyer, R., van der Donk, W.A. and Zhao, H. (2003) *Biochemistry*, **42**, 11604–11614.

39 Rissom, S., Schwarz-Linek, U., Vogel, M., Tishkov, V.I. and Kragl, U. (1997) *Tetrahedron: Asymmetry*, **8**, 2523–2526.

40 Willets, A.J., Knowles, C.J., Levitt, M.S., Roberts, S.M., Sandey, H. and Shipston, N.F. (1991) *Journal of The Chemical Society-Perkin Transactions 1*, 1608–1610.

41 de Gonzalo, G., Ottolina, G., Carrea, G. and Fraaije, M.W. (2005) *Chemical Communications*, 3724–3726.

42 Hollmann, F., Taglieber, A., Schulz, F. and Reetz, M.T. (2007) *Angewandte Chemie-International Edition*, **46**, 2903–2906; Hollmann, F., Taglieber, A., Schulz, F. and Reetz, M.T. (2007) *Angewandte Chemie*, **119**, 2961–2964.

43 Walton, A.Z. and Stewart, J.D. (2004) *Biotechnology Progress*, **20**, 403–411.

44 Donoghue, N.A. and Trudgill, P.W. (1975) *European Journal of Biochemistry*, **60**, 1–7.

45 Kostichka, K., Thomas, S.M., Gibson, K.J., Nagarajan, V. and Cheng, Q. (2001) *Journal of Bacteriology*, **183**, 6478–6486.

46 Mihovilovic, M.D., Chen, G., Wang, S., Kyte, B., Rochon, F., Kayser, M.M. and Stewart, J.D. (2001) *The Journal of Organic Chemistry*, **66**, 733–738.

47 Chen, G., Kayser, M.M., Mihovilovic, M.D., Mrstik, M.E., Martinez, C.A. and Stewart, J.D. (1999) *New Journal of Chemistry*, **23**, 827–832.

48 Stewart, J.D., Reed, K.W. and Kayser, M.M. (1996) *Journal of The Chemical Society-Perkin Transactions 1*, 755–757.

49 Stewart, J.D., Reed, K.W., Martinez, C.A., Zhu, J., Chen, G. and Kayser, M.M. (1998) *Journal of the American Chemical Society*, **120**, 3541–3548.

50 Bonsor, D., Butz, S.F., Solomons, J., Grant, S., Fairlamb, I.J. S., Fogg, M.J. and Grogan, G. (2006) *Organic and Biomolecular Chemistry*, **4**, 1252–1260.

51 Fraaije, M.W., Kamerbeek, N.M., Heidekamp, A.J., Fortin, R. and Janssen, D.B. (2004) *The Journal of Biological Chemistry*, **279**, 3354–3360.

52 Alphand, V., Carrea, G., Wohlgemuth, R., Furstoss, R. and Woodley, J.M. (2003) *Trends in Biotechnology*, **21**, 318–323.

53 Doig, S.D., Avenell, P.J., Bird, P.A., Gallati, P., Lander, K.S., Lye, G.J., Wohlgemuth, R. and Woodley, J.M. (2002) *Biotechnology Progress*, **18**, 1039–1046.

54 Simpson, H.D., Alphand, V. and Furstoss, R. (2001) *Journal of Molecular Catalysis B: Enzymatic*, **16**, 101–108.

55 Hilker, I., Alphand, V., Wohlgemuth, R. and Furstoss, R. (2003) *Advanced Synthesis and Catalysis*, **346**, 203–214; Hilker, I., Gutiérrez, M.-C., Alphand, V., Wohlgemuth, R. and Furstoss, R. (2004) *Organic Letters*, **6**, 1955–1958.

56 Donoghue, N.A., Norris, D.B. and Trudgill, P.W. (1976) *European Journal of Biochemistry*, **63**, 175–192.

57 Chen, Y.-C.J., Peoples, O.P. and Walsh, C.T. (1988) *Journal of Bacteriology*, **170**, 781–789.

58 Stewart, J.D., Reed, K.W., Zhu, J., Chen, G. and Kayser, M.M. (1996) *The Journal of Organic Chemistry*, **61**, 7652–7653.

59 Grogan, G., Roberts, S., Wan, P. and Willetts, A.J. (1993) *Biotechnology Letters*, **15**, 913–918.

60 Grogan, G., Roberts, S.M. and Willetts, A.J. (1993) *Journal of the Chemical Society. Chemical Communications*, 699–701.

61 Alphand, V. and Furstoss, R. (2000) *Journal of Molecular Catalysis B: Enzymatic*, **9**, 209–217.

62 Alphand, V., Mazzini, C., Lebreton, J. and Furstoss, R. (1998) *Journal of Molecular Catalysis B: Enzymatic*, **5**, 219–221.

63 Carnell, A. and Willets, A. (1992) *Biotechnology Letters*, **14**, 17–21.

64 Mihovilovic, M.D. (2006) *Current Organic Chemistry*, **10**, 1265–1287.

65 Griffin, M. and Trudgill, P.W. (1976) *European Journal of Biochemistry*, **63**, 199–209.

66 Taschner, M.J. and Black, D.J. (1988) *Journal of the American Chemical Society*, **110**, 6892–6893; Stewart, J.D. (1998) *Current Organic Chemistry*, **2**, 211–232.

67 Kyte, B.G., Rouviere, P.E., Cheng, Q. and Stewart, J.D. (2004) *The Journal of Organic Chemistry*, **69**, 12–17.

68 Brzostowicz, P.C., Walters, D.M., Thomas, S.M., Nagarajan, V. and Rouviere, P.E. (2003) *Applied and Environmental Microbiology*, **69**, 334–342.

69 Mihovilovic, M.D., Rudroff, F., Grötzl, B., Kapitan, P., Snajdrova, R., Rydz, J. and Mach, R. (2005) *Angewandte Chemie-International Edition*, **44**, 3609–3613; Mihovilovic, M.D., Rudroff, F., Grötzl, B., Kapitan, P., Snajdrova, R., Rydz, J. and Mach, R. (2005) *Angewandte Chemie*, **117**, 3675–3679.

70 Bramucci, M.G., Brzostowicz, P.C., Kostichka, K.N., Nagarajan, V., Rouviere, P.E. and Du Pont, 2003, Patent WO/2003/020890A2, *Genes encoding Baeyer-Villiger monooxygenases*, p. 225.

71 Brzostowicz, P.C., Gibson, K.L., Thomas, S.M., Blasko, M.S. and Rouviere, P.E. (2000) *Journal of Bacteriology*, **182**, 4241–4248.

72 Iwaki, H., Wang, S., Grosse, S., Bergeron, H., Nagahashi, A., Lertvorachon, J., Yang, J., Konishi, Y., Hasegawa, Y. and Lau, P.C. K. (2006) *Applied and Environmental Microbiology*, **72**, 2707–2720.

73 Ougham, H.J. and Trudgill, P.W. (1983) *Journal of Bacteriology*, **153**, 140–152.

74 Snajdrova, R., Grogan, G. and Mihovilovic, M.D. (2006) *Bioorganic & Medicinal Chemistry Letters*, **16**, 4813–4817.

75 Kamerbeek, N.M., Olsthoorn, J.J., Fraaije, M.W. and Janssen, D.B. (2003) *Applied and Environmental Microbiology*, **69**, 419–426.

76 Fraaije, M.W., Wu, J., Heuts, D.P. H.M., van Hellemond, E.W., Spelberg, J.H. L. and Janssen, D.B. (2005) *Applied Microbiology and Biotechnology*, **66**, 393–400.

77 de Gonzalo, G., Torres Pazmino, D.E., Ottolina, G., Fraaije, M.W. and Carrea, G. (2005) *Tetrahedron: Asymmetry*, **16**, 3077–3083.

78 Khalameyzer, V., Fischer, I., Bornscheuer, U.T. and Altenbuchner, J. (1999) *Applied and Environmental Microbiology*, **65**, 477–482.

79 Kirschner, A., Altenbuchner, J. and Bornscheuer, U.T. (2007) *Applied Microbiology and Biotechnology*, **73**, 1065–1072.

80 Geitner, K., Kirschner, A., Rehdorf, J., Schmidt, M., Mihovilovic, M.D. and Bornscheuer, U.T. (2007) *Tetrahedron: Asymmetry*, **18**, 892–895.

81 Miyamoto, M., Matsumoto, J., Iwaja, T. and Itagaki, E. (1995) *Biochimica et Biophysica Acta*, **1251**, 115–124.

82 Morii, S., Sawamoto, S., Yamauchi, Y., Miyamoto, M., Iwami, M. and Itagaki, E.

(1999) *The Biochemical Journal*, **126**, 624–631.

83 Itagaki, E. (1986) *The Biochemical Journal*, **99**, 815–824.

84 Itagaki, E. (1986) *The Biochemical Journal*, **99**, 825–832.

85 Rehdorf, J., Kirschner, A. and Bornscheuer, U.T. (2007) *Biotechnology Letters*, **29**, 1393–1398.

86 Kirschner, A. and Bornscheuer, U.T. (2006) *Angewandte Chemie-International Edition*, **45**, 7004–7006; Kirschner, A. and Bornscheuer, U.T. (2006) *Angewandte Chemie*, **118**, 7161–7163.

87 Peterson, J.A., Basu, D. and Coon, M.J. (1966) *The Journal of Biological Chemistry*, **241**, 5162–5164; Fried, J., Thoma, R.W. and Klingsberg, A. (1953) *Journal of the American Chemical Society*, **75**, 5764–5765.

88 Mihovilovic, M.D., Müller, B. and Stanetty, P. (2002) *The European Journal of Organic Chemistry*, 3711–3730; Mihovilovic, M.D., Rudroff, F. and Müller, B. (2004) *Current Organic Chemistry*, **8**, 1057–1069.

89 Chen, C.-S., Fujimoto, Y., Girdaukas, G. and Sih, C.J. (1982) *Journal of the American Chemical Society*, **104**, 7294–7299.

90 Alphand, V., Furstoss, R., Pedragosa-Moreau, S., Roberts, S.M. and Willets, A.J. (1996) *Journal of The Chemical Society-Perkin Transactions 1*, 1867–1872.

91 Berezina, N., Kozma, E., Furstoss, R. and Alphand, V. (2007) *Advanced Synthesis and Catalysis*, **349**, 2049–2053.

92 Kayser, M.M., Chen, A.C. and Stewart, J. (1998) *The Journal of Organic Chemistry*, **63**, 7103–7106.

93 Wang, S., Chen, G., Kayser, M.M., Iwaki, H., Lau, P.C. K. and Hasegawa, Y. (2002) *Canadian Journal of Chemistry*, **80**, 613–621.

94 Adger, B., Bes, M.T., Grogan, G., McCague, R., Pedragosa-Moreau, S., Roberts, S.M., Villa, R., Wan, P.W. H. and Willetts, A.J. (1995) *Journal of the Chemical Society. Chemical Communications*, 1563–1564.

95 Adger, B., Bes, M.T., Grogan, G., McCague, R., Pedragosa-Moreau, S., Roberts, S.M., Villa, R., Wan, P.W. and Willets, A.J. (1997) *Bioorganic and Medicinal Chemistry*, **5**, 253–261.

96 Mihovilovic, M.D., Müller, B., Kayser, M.M., Stewart, J.D., Fröhlich, J., Stanetty, P. and Spreitzer, H. (2001) *Journal of Molecular Catalysis B: Enzymatic*, **11**, 349–353.

97 Bes, M.T., Villa, R., Roberts, S.M., Wan, P.W. H. and Willets, A. (1996) *Journal of Molecular Catalysis B: Enzymatic*, **1**, 127–134; Ottolina, G., de Gonzalo, G., Carrea, G. and Danieli, B. (2005) *Advanced Synthesis and Catalysis*, **347**, 1035–1040.

98 Berezina, N., Alphand, V. and Furstoss, R. (2002) *Tetrahydron: Asymmetry*, **13**, 1953–1955.

99 Gutiérrez, M.-C., Furstoss, R. and Alphand, V. (2005) *Advanced Synthesis and Catalysis*, **347**, 1051–1059.

100 Rodriguez, C., De Gonzalo, G., Fraaije, M.W. and Gotor, V. (2007) *Tetrahedron: Asymmetry*, **18**, 1338–1344.

101 Völker, A., Kirschner, A., Bornscheuer, U.T. and Altenbucher, J. (2008) *Applied Microbiology and Biotechnology*, **77**, 1251–1260.

102 Kirschner, A. and Bornscheuer, U.T. (2008) *Applied Microbiology and Biotechnology*, online; DOI:10.1007/s00253-008-1646-4.

103 Taschner, M.J., Black, D.J. and Chen, Q.Z. (1993) *Tetrahedron: Asymmetry*, **4**, 1387–1390.

104 Wang, S. and Kayser, M.M. (2003) *The Journal of Organic Chemistry*, **68**, 6222–6228.

105 Mihovilovic, M.D., Rudroff, F., Grötzl, B. and Stanetty, P. (2005) *The European Journal of Organic Chemistry*, 809–816; Mihovilovic, M.D., Rudroff, F., Müller, B. and Stanetty, P. (2003) *Bioorganic & Medicinal Chemistry Letters*, **13**, 1479–1482.

106 Taschner, M.J. and Aminbhavi, A.S. (1989) *Tetrahedron Letters*, **30**, 1029–1032; Yokokawa, F., Hamada, Y. and Shioiri, T.

(1996) *Chemical Communications*, 871–872.

107 Mihovilovic, M.D., Rudroff, F., Kandioller, W., Grötzl, B., Stanetty, P. and Spreitzer, H. (2003) *Synlett*, 1973–1976.

108 Rudroff, F., Rydz, J., Ogink, F.H., Fink, M. and Mihovilovic, M.D. (2007) *Advanced Synthesis and Catalysis*, **349**, 1436–1444.

109 Mazzini, C., Lebreton, J., Alphand, V. and Furstoss, R. (1997) *Tetrahedron Letters*, **38**, 1195–1196.

110 Mazzini, C., Lebreton, J., Alphand, V. and Furstoss, R. (1997) *The Journal of Organic Chemistry*, **62**, 5215–5218.

111 Mihovilovic, M.D., Kapitan, P., Rydz, J., Rudroff, F., Ogink, F.H. and Fraaije, M.W. (2005) *Journal of Molecular Catalysis B: Enzymatic*, **32**, 135–140.

112 Snajdrova, R., Braun, I., Bach, T., Mereiter, K. and Mihovilovic, M.D. (2007) *The Journal of Organic Chemistry*, **72**, 9597–9603.

113 Taschner, M.J. and Peddada, L. (1992) *Journal of the Chemical Society. Chemical Communications*, 1384–1385.

114 Mihovilovic, M.D., Müller, B., Schulze, A., Stanetty, P. and Kayser, M.M. (2003) *The European Journal of Organic Chemistry*, 2243–2249.

115 Mihovilovic, M.D., Müller, B., Kayser, M.M. and Stanetty, P. (2002) *Synlett*, 700–702.

116 Kelly, D.R., Knowles, C.J., Mahdi, J.G., Taylor, I.N. and Wright, M.A. (1995) *Journal of the Chemical Society. Chemical Communications*, 729–730; Kelly, D.R., Knowles, C.J., Mahdi, J.G., Wright, M.A., Taylor, I.N., Hibbs, D.E., Hursthouse, M.B., Mish'al, A.K., Roberts, S.M., Wan, P.W. H., Grogan, G. and Willets, A.J. (1995) *Journal of The Chemical Society-Perkin Transactions 1*, 2057–2066.

117 Alphand, V. and Furstoss, R. (1992) *The Journal of Organic Chemistry*, **57**, 1306–1309.

118 Mihovilovic, M.D. and Kapitan, P. (2004) *Tetrahedron Letters*, **45**, 2751–2754.

119 Alphand, V., Archelas, A. and Furstoss, R. (1989) *Tetrahedron Letters*, **30**, 3663–3664.

120 Andrau, L., Lebreton, J., Viazzo, P., Alphand, V. and Furstoss, R. (1997) *Tetrahedron Letters*, **38**, 825–826.

121 Lebreton, J., Alphand, V. and Furstoss, R. (1996) *Tetrahedron Letters*, **37**, 1011–1014; Lebreton, J., Alphand, V. and Furstoss, R. (1997) *Tetrahedron*, **53**, 145–160.

122 Luna, A., Gutiérrez, M.-C., Furstoss, R. and Alphand, V. (2005) *Tetrahedron: Asymmetry*, **16**, 2521–2524.

123 Cernuchová, P. and Mihovilovic, M.D. (2007) *Organic and Biomolecular Chemistry*, **5**, 1715–1719.

124 Law, H.E. M., Baldwin, C.V. F., Chen, J.M. and Woodley, J.M. (2006) *Chemical Engineering Science*, **61**, 6646–6652.

125 Hilker, I., Wohlgemuth, R., Alphand, V. and Furstoss, R. (2005) *Biotechnology and Bioengineering*, **92**, 702–710.

126 Baldwin, C.V. F. and Woodley, J.M. (2006) *Biotechnology and Bioengineering*, **95**, 362–369; Hilker, I., Baldwin, C.V. F., Alphand, V., Furstoss, R., Woodley, J.M. and Wohlgemuth, R. (2006) *Biotechnology and Bioengineering*, **93**, 1138–1144.

127 Ottolina, G., Bianchi, S., Belloni, B., Carrea, G. and Danieli, B. (1999) *Tetrahedron Letters*, **40**, 8483–8486.

128 Latham, Jr, J.A. and Walsh, C. (1986) *Journal of the Chemical Society. Chemical Communications*, 527–528.

129 Latham Jr, J.A., Branchaud, B.J., Chen, Y.-C.J. and Walsh, C. (1986) *Journal of the Chemical Society. Chemical Communications*, 528–530.

130 Nakamura, S., Watanabe, Y. and Toru, T.J. (2000) *The Journal of Organic Chemistry*, **65**, 1758–1766.

131 Carreno, M.C. (1995) *Chemical Reviews*, **95**, 1717–1760.

132 Kyung, K., Han, D.C. and Fleming, H.P. (1997) *Journal of Food Science*, **62**, 406–409.

133 de Gonzalo, G., Ottolina, G., Zambianchi, F., Fraaije, M.W. and Carrea, G. (2006) *Journal of Molecular Catalysis B: Enzymatic*, **39**, 91–97.

134 Colonna, S., Gaggero, N., Bertinotti, A., Carrea, G., Pasta, P. and Bernardi, A. (1995) *Journal of the Chemical Society. Chemical Communications*, 1123–1125.

135 Colonna, S., Gaggero, N., Carrea, G. and Pasta, P. (1996) *Tetrahedron: Asymmetry,* **7**, 565–570.

136 May, O., Nguyen, P.T. and Arnold, F.H. (2000) *Nature Biotechnology*, **18**, 317–320; Reetz, M.T., Zonta, A., Schimossek, K., Jaeger, K.-E. and Liebeton, K. (1997) *Angewandte Chemie-International Edition*, **36**, 2830–2832; Reetz, M.T., Zonta, A., Schimossek, K., Jaeger, K.-E. and Liebeton, K. (1997) *Angewandte Chemie*, **109**, 2961–2963; Schmidt, M., Hasenpusch, D., Kähler, M., Kirchner, U., Wiggenhorn, K., Langel, W. and Bornscheuer, U.T. (2006) *Chembiochem: A European Journal of Chemical Biology*, **7**, 805–809.

137 Reymond, J.-L. (2006) *Enzyme Assays*, Wiley-VCH Verlag GmbH, Weinheim.

138 Reetz, M.T., Brunner, B., Schneider, T., Schulz, F., Clouthier, C.M. and Kayser, M.M. (2004) *Angewandte Chemie-International Edition*, **43**, 4075–4078; Reetz, M.T., Brunner, B., Schneider, T., Schulz, F., Clouthier, C.M. and Kayser, M.M. (2004) *Angewandte Chemie*, **116**, 4167–4170.

139 Reetz, M.T., Daligault, F., Brunner, B., Hinrichs, H. and Deege, A. (2004) *Angewandte Chemie-International Edition*, **43**, 4078–4081; Reetz, M.T., Daligault, F., Brunner, B., Hinrichs, H. and Deege, A.

(2004) *Angewandte Chemie*, **116**, 4170–4173.

140 Bocola, M., Schulz, F., Leca, F., Vogel, A., Fraaije, M.W. and Reetz, M.T. (2005) *Advanced Synthesis and Catalysis*, **347**, 979–986.

141 Torres Pazmino, D.E., Snajdrova, R., Rial, D.V., Mihovilovic, M.D. and Fraaije, M.W. (2007) *Advanced Synthesis and Catalysis*, **349**, 1361–1368.

142 Clouthier, C.M., Kayser, M.M. and Reetz, M.T. (2006) *The Journal of Organic Chemistry*, **71**, 8431–8437.

143 Reetz, M.T., Bocola, M., Carballeira, J.D., Zha, D. and Vogel, A. (2005) *Angewandte Chemie-International Edition*, **44**, 4192–4196; Reetz, M.T., Bocola, M., Carballeira, J.D., Zha, D. and Vogel, A. (2005) *Angewandte Chemie*, **117**, 4264–4268.

144 Clouthier, C.M. and Kayser, M.M. (2006) *Tetrahedron: Asymmetry*, **17**, 2649–2653.

145 Watts, A.B., Beecher, J., Whitcher, C.S. and Littlechild, J.A. (2002) *Biocatalysis and Biotransformation*, **20**, 209–214.

146 Gutiérrez, M.-C., Sleegers, A., Simpson, H.D., Alphand, V. and Furstoss, R. (2003) *Organic and Biomolecular Chemistry*, **1**, 3500–3506.

147 Sicard, R., Chen, L.S., Marsaioli, A.J. and Reymond, J.-L. (2005) *Advanced Synthesis and Catalysis*, **347**, 1041–1050.

148 Wahler, D. and Reymond, J.-L. (2002) *Angewandte Chemie-International Edition*, **41**, 1229–1232; Wahler, D. and Reymond, J.-L. (2002) *Angewandte Chemie*, **114**, 1277–1280.

6
Bioreduction by Microorganisms

Leandro Helgueira Andrade and Kaoru Nakamura

6.1
Introduction

Reduction reactions of organic functional groups, such as nitro, ketone, aldehyde and carbon–carbon double bonds, can be mediated by enzymes and their coenzymes (for some examples, see [1]). These reactions are very important from the viewpoint of green chemistry because, the reductions do not require strong acids, bases and harmful metals. Furthermore, enzymatic reactions are conducted at room temperature whereas chemical reductions usually need high temperatures for activation of catalysts and are conducted at extremely low temperatures for chiral discrimination. The most studied bioreduction reaction is ketone reduction. This reaction is very important due to its wide application in the synthesis of chiral alcohols, in which they are essential intermediates in the preparation of many pharmaceuticals and fine chemicals. The enzymes responsible for the bioreduction of ketones to their corresponding alcohols are ketoreductases/alcohol dehydrogenases with the aid of their coenzymes, NADH (reduced form of nicotinamide adenine dinucleotide) and NADPH (reduced form of nicotinamide adenine dinucleotide phosphate). Reductions of ketones are important in green chemistry since chiral alcohols are obtained from achiral ketones, ideally in 100% chemical and optical yields, whereas kinetic resolutions generally afford a maximum of 50% of the desired isomers. Bioreduction reactions can be performed with the use of whole cells of microorganisms (alcohol/alcohol dehydrogenase producers) or purified enzymes (alcohol/alcohol dehydrogenases).

In this chapter, an outline of and recent progress in bioreduction are presented from the viewpoint of green chemistry.

Handbook of Green Chemistry, Volume 3: Biocatalysis. Edited by Robert H. Crabtree
Copyright © 2009 WILEY-VCH Verlag GmbH & Co. KGaA, Weinheim
ISBN: 978-3-527-32498-9

6.2
Enzymes and Coenzymes

6.2.1
Classification

Dehydrogenases, classified under EC 1.1, are enzymes that catalyze the reduction and oxidation of carbonyl groups and alcohols, respectively [2]. The natural substrates of the enzymes are alcohols such as ethanol, lactate and glycerol and their corresponding carbonyl compounds, but unnatural ketones can also be reduced enantioselectively. To exhibit catalytic activities, the enzymes require a coenzyme; most of the dehydrogenases use NADH or NADPH and a few use flavin, pyrroquinoline quinone and so on. Alcohol dehydrogenases are enzymes responsible for the reduction of carbonyl groups from ketones or aldehydes to the corresponding alcohols. To perform this chemical transformation, these enzymes require a coenzyme NADH or NADPH (hydride source) from which a hydride is transferred to the carbonyl carbon affording the alcohol and NAD^+ or $NADP^+$ (Scheme 6.1) (for reviews on enzymatic reductions, see [3]).

The reaction mechanism of the reduction by dehydrogenases is as follows:

- A holoenzyme (an enzyme with its coenzyme) binds a substrate (a carbonyl compound).
- A hydride on the coenzyme is transferred to the ketone to produce an alcohol.
- The enzyme releases the product alcohol.

In order to achieve 100% yield in alcohol production, a stoichiometric amount of a hydride source is necessary. As the NADH and NADPH are commercially available but are expensive, several approaches have been developed to ensure the maximum yield for this carbonyl reduction (see Section 6.2.2).

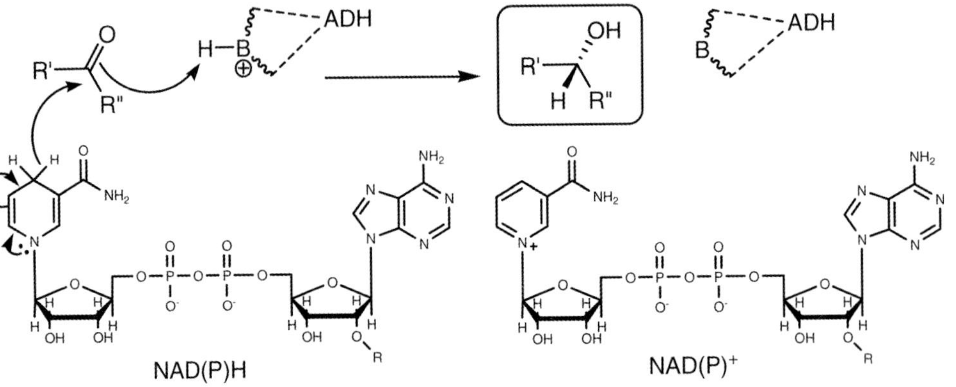

R = H (NADH)
R = PO_3^{2-} (NADPH)
ADH: alcohol dehydrogenase

Scheme 6.1 Mechanism for ketone reduction mediated by ADH and its coenzyme.

6.2.2
Hydrogen Source

Dehydrogenases which perform the reduction of carbonyl groups usually require a coenzyme from which a hydride is transferred to the carbonyl carbon. Since reduction of the substrate is concomitant with oxidation of the coenzyme and the coenzyme is too expensive as a throw-away reagent, it is necessary to recycle and reuse the oxidized form of the coenzyme [3]. Thus, a recycling system which transforms the coenzymes from the oxidized form to the reduced form has to be constructed and hydrogen sources are necessary to perform this reduction reaction. In practice, for biocatalytic reduction, alcohols such as ethanol and 2-propanol, sugars such as glucose, glucose-6-phosphate and glucose-6-sulfate, formic acid and dihydrogen can be used. Some examples are presented in this section.

6.2.2.1 Alcohols as a Hydrogen Source for Reduction
Alcohols such as ethanol and 2-propanol have been widely used to recycle the coenzyme for the reduction catalyzed by alcohol dehydrogenase, since the enzyme catalyzes both reduction and oxidation. Usually, an excess amount of the hydrogen source is used to push the equilibrium to the formation of product alcohols.

There is an interesting example of the use of secondary alcohols in which the enantioselectivity and reaction yield were improved by recycling the coenzyme using secondary alcohols for the reduction of ketones with *Geotrichum candidum* (Scheme 6.2) [4]. Although the enantioselectivity of the reduction of acetophenone with resting cells of the microbe was low, the use of the dried cells as a biocatalyst and addition of a catalytic amount of NAD(P)$^+$ and an excess amount of secondary alcohols such as 2-propanol and cyclopentanol increased the enantioselectivity and also the chemical yield of the reduction; the (S)-alcohol was obtained in >99% enantiomeric excess (*ee*) with high yield.

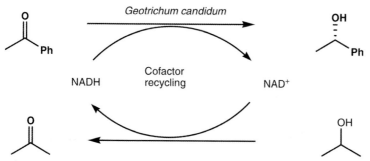

Scheme 6.2 NADH recycling using alcohol as a hydrogen source for reduction.

6.2.2.2 Sugars as a Hydrogen Source for Reduction
Glucose and glucose-6-phosphate have been widely used as reducing agents. In a recent example, a recombinant *E. coli* which coexpresses a carbonyl reductase from yeast (SCR) and a glucose dehydrogenase from *Bacillus megaterium* were used for

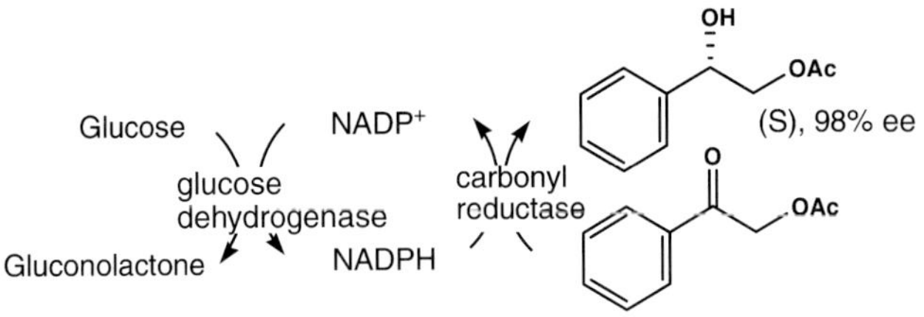

Scheme 6.3 NADH recycling using glucose as a hydrogen source for reduction.

recycling NADPH for the reduction of acetoxyacetophenone to the corresponding (S)-alcohol in 98% *ee* (Scheme 6.3) [5].

6.2.2.3 Formate as a Hydrogen Source for Reduction

Formate is one of the most representative hydrogen sources for biocatalytic reduction because CO_2 formed by the oxidation of formate is easily released from the reaction system. For example, a formate dehydrogenase (FDH) system was applied in the reduction of 6-bromotetralone to (S)-6-bromotetralol, a potential pharmaceutical precursor, with the NADH-dependent ketone reductase from *Trichosporon capitatum* [6]. A resin (XAD L-323) was used to bind the product (Scheme 6.4).

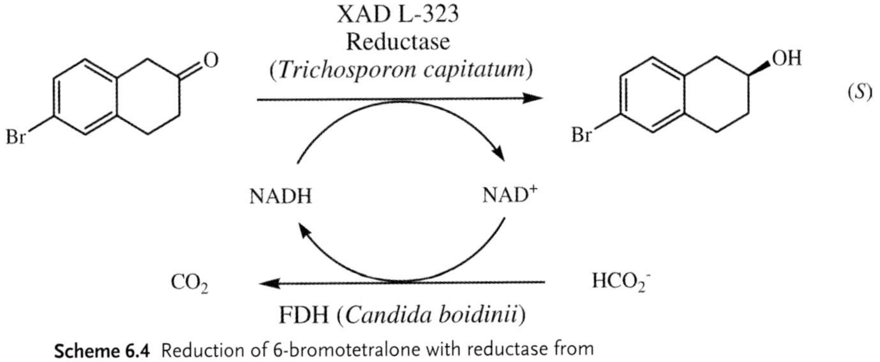

Scheme 6.4 Reduction of 6-bromotetralone with reductase from *Trichosporon capitatum* using formate as a hydrogen source.

6.2.2.4 Molecular Hydrogen as a Hydrogen Source for Reduction

Molecular hydrogen has been used for the recycling of coenzymes [7]. For example, the soluble hydrogenase I (H_2:$NADP^+$ oxidoreductase, EC 1.18.99.1) from the marine hyperthermophilic strain of the archaeon *Pyrococcus furiosus* (PF H2ase I) has been used as a biocatalyst in the enzymatic production and regeneration of NADPH utilizing molecular hydrogen. Utilizing the thermophilic NADPH-dependent alcohol dehydrogenase from *Thermoanaerobium* sp. (ADH M) coupled to the PF H_2ase I *in situ* NADPH-regenerating system, (2S)-hydroxy-1-phenylpropanone was quantitatively reduced to the corresponding (1R,2S)-diol in >98% diastereoisomeric excess (*de*), and a total turnover number (ttn: moles of product/moles of cofactor $NADP^+$ consumed) of 160 could be reached (Scheme 6.5) [7].

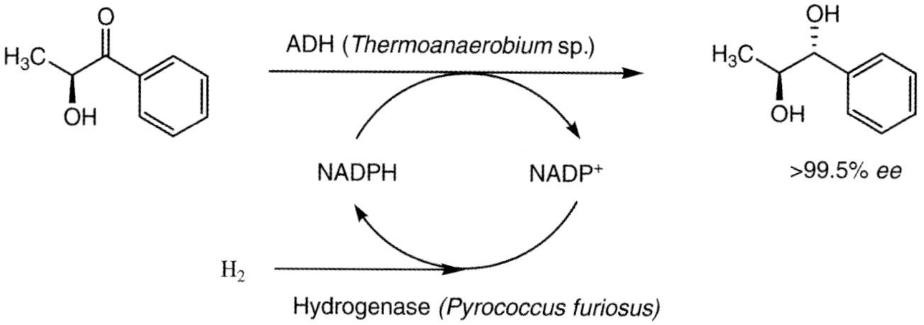

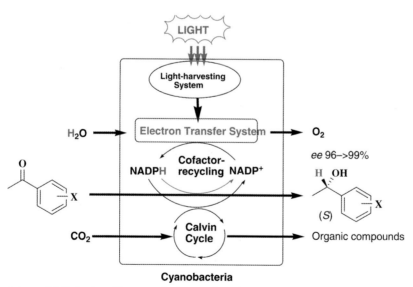

Scheme 6.5 Reduction of ketone with alcohol dehydrogenase from *Thermoanaerobium* sp. using molecular hydrogen as a hydrogen source.

6.2.2.5 Light Energy as a Hydrogen Source for Reduction

Photochemical methods [8] have been developed to provide an environmentally friendly system that employ light energy to regenerate NAD(P)H, for example by the use of a cyanobacterium, a photosynthetic biocatalyst. Using the biocatalyst, the reduction of acetophenone derivatives occurred more effectively under illumination than in the dark (Scheme 6.6) [8b, c]. The light energy harvested by the cyanobacterium is converted into chemical energy in the form of NADPH through an electron transfer system and, consequently, the chemical energy (NADPH) is used to reduce the substrate to chiral alcohol (96 to >99% *ee*). The light energy, which is usually utilized to reduce CO_2 to synthesize organic compounds in the natural environment, was used to reduce the artificial substrate in this case.

Scheme 6.6 Reduction of ketone with photosynthetic biocatalyst using light energy.

6.2.2.6 Electric Power as a Hydrogen Source for Reduction

Electrochemical regeneration of NAD(P)H has long been recognized as a potentially powerful technology for the viewpoint of green chemistry as it would not require a

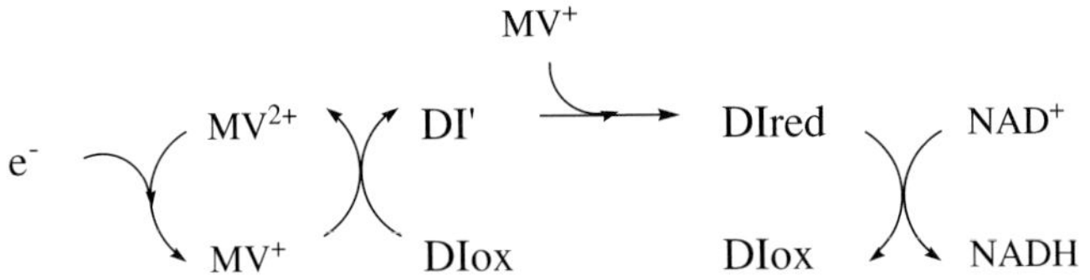

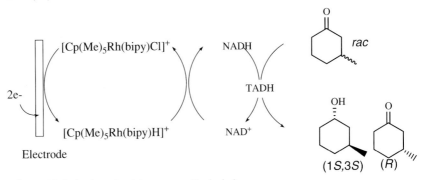

Scheme 6.7 Regeneration of NADH using diaphorase and electric power. DI, diaphorase.

second enzyme and cosubstrate [3b]. However, the method is not effective because of the necessity for high overpotentials with direct reduction of the coenzyme, electrode fouling, dimerization of the coenzyme and the fact that only enzyme in the immediate vicinity of the electrode will be productive. Viologen diaphorase (lipoamide dehydrogenase) was used for the reduction of $NAD(P)^+$, where viologens were used as a mediator of an electron from the electrode to diaphorase (Scheme 6.7) [9a].

Organometallics such as rhodium complexes have also been used for electrochemical regeneration of NAD(P)H from electrodes [9b], as shown in Scheme 6.8. The oxidized form of the coenzyme was reduced by the reduced form of the rhodium complex and thus produced the oxidized form of the rhodium complex, which was reduced electrochemically at an electrode. TADH-catalyzed reduction of racemic 3-methylcyclohexanone afforded (1S,3S)-3-methylcyclohexanol and (R)-3-methylcyclohexanone.

Scheme 6.8 Reduction of cyclohexanone with alcohol dehydrogenase and rhodium complex using electric power.

6.3
Methodologies

6.3.1
Search for the Ideal Biocatalysts

Owing to the wide diversity of carbonyl compounds which can be transformed into chiral alcohols with great synthetic interest, the continuing search for new biocatalysts that can mediate this chemical transformation in an asymmetric fashion is of great relevance. The most commonly employed methodology for searching for enantioselective biocatalysts in carbonyl reduction is screening techniques applied to libraries of microorganisms from culture collections. However, enrichment culture techniques using a ketone compound have been used to isolate microorganisms containing ketone reductase activity [10].

In this section, we will discuss some examples reported in the literature of searches for enantioselective enzymes from screening techniques and recombinant microorganisms.

6.3.1.1 Biocatalysts from Screening Techniques

The main techniques employed to evaluate the enantioselectivity of alcohol dehydrogenases in carbonyl reduction are based on screening of microbial cultures using reactions in test-tubes or small Erlenmeyer flasks and microtiter plates, followed by automated chiral HPLC or GC analysis. For example, for the rapid identification of enantioselective cultures, a microbial screening procedure using reactions in a 24-well microtiter plate format with an automated liquid handling system and automated chiral HPLC analysis was applied by Zaks and co-workers [11]. A collection of about 300 microbes was surveyed for their ability to generate chiral secondary alcohols by enantioselective reduction of a series of alkyl aryl ketones. Microbial cultures demonstrating utility in reducing model ketones were arrayed in multi-well plates and used to identify rapidly specific organisms capable of producing chiral alcohols used as intermediates in the synthesis of several drug candidates. Approximately 60 cultures were shown to reduce selectively various ketones, providing both the R- and S-enantiomers of the corresponding alcohols in 92–99% ee with yields up to 95% at 1–4 g L^{-1}. For example, the bioreduction of 4'-trifluoromethylacetophenone was performed by microorganisms with enantiocomplementary specificity to afford the chiral products in excellent enantioselectivities >99% ee for both enantiomeric forms (Scheme 6.9).

Microbial screening using reactions in Erlenmeyer flasks was employed to evaluate a library composed of several microorganisms [12]. The screening of 310 microbial strains (53 bacteria, 113 yeasts and 144 fungi) yielded eight as suitable biocatalysts for the asymmetric bioreduction of a highly hindered bisaryl ketone to its corresponding alcohol, a chiral precursor for the synthesis of a phosphodiesterase 4 inhibitor (Scheme 6.10). The production of both enantiomers with elevated optical purity (ee >96%) was achieved by different microorganisms. When scaling up the asymmetric bioreduction process mediated by *Rhodotorula pilimanae* ATCC 32762 in laboratory bioreactors (23 L scale), the production of preparative amounts (1.5 g) of the S-enantiomer with elevated optically purity (ee >96%) was achieved.

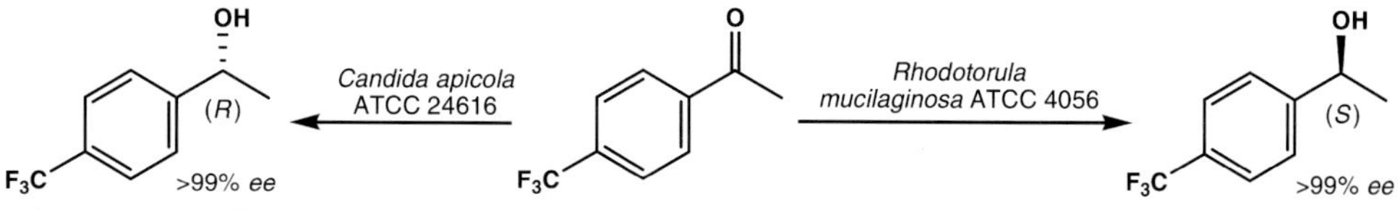

Scheme 6.9 Bioreduction using microorganisms isolated from screening techniques with enantiocomplementary specificity.

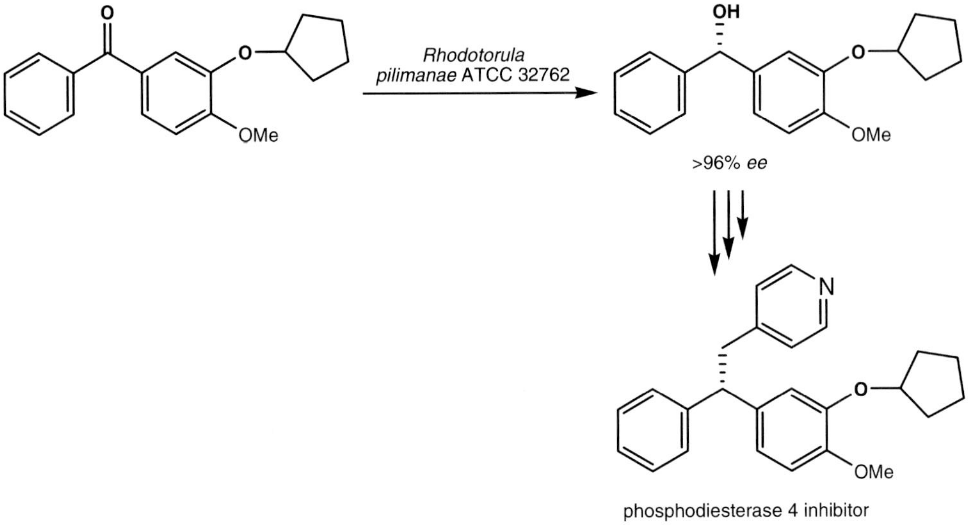

Scheme 6.10 Asymmetric bioreduction of bisaryl ketone using *R. pilimanae* ATCC 32762.

An example of microbial screening using reactions in 96-deep-well plates (well volume 2.0 mL) was reported by Ikunaka *et al.* [13]. This screening technique was employed for the exploration of 757 microbial strains (288 bacteria, 384 yeasts and 85 fungi) to find biocatalysts able to prepare *cis*-4-propylcyclohexanol from 4-propylcyclohexanone. This intensive microbial screening identified *Galactomyces geotrichum* JCM 6359 as the best biocatalyst to deliver hydrides in an equatorial disposition in reducing 4-propylcyclohexanone to *cis*-4-propylcyclohexanol [*cis* : *trans* ratio 99 : 0.5; 74% yield] (Scheme 6.11).

6.3.1.2 Biocatalysts from Recombinant Microorganisms

Although satisfactory results have been observed in many cases in the use of wild-type strains as a source of enzymes responsible for enantioselective reduction of carbonyl compounds, recent development in gene technology have promoted the use of recombinant microorganisms in biocatalytic reductions. The recombinant microorganism strains can overproduce the desired enzymes in higher amounts than the wild-type strains. In addition to this feature, the recombinant strains can also

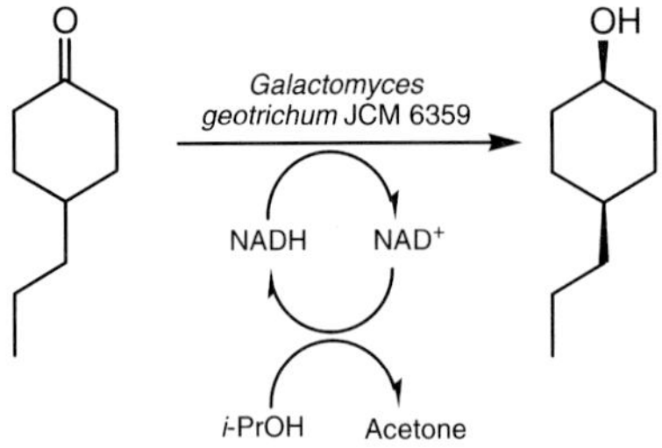

Scheme 6.11 Bioreduction of 4-propylcyclohexanone with *G. geotrichum*.

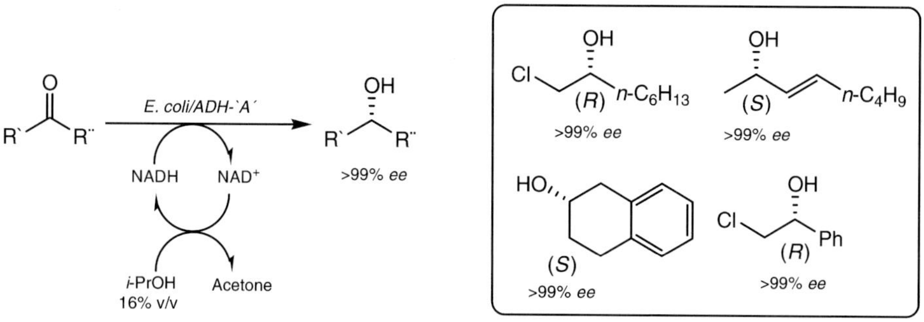

Scheme 6.12 Examples of ketone reduction using lyophilized cells of *E. coli*/ADH-'A'.

improve the enantioselectivity of the carbonyl reduction, because wild-type strains have several reducing enzymes with different enantioselectivities and these enzymes participate in the reduction of a substrate and compete with each other. As a result, the microorganism affords the desired product only in low optical purity. Some recent examples of this approach have been selected for discussion in this section.

E. coli was employed to overproduce a highly organic solvent-tolerant alcohol dehydrogenase ADH-'A' isolated from *Rhodococcus ruber* DSM 44541 [14]. In this work, Kroutil *et al.* carried out several ketone reductions by using lyophilized cells of *E. coli* containing the overexpressed ADH-'A' to prepare versatile building blocks, which were not accessible with the wild-type catalyst, but were obtained in >99% *ee* by this recombinant microorganism (Scheme 6.12).

An interesting example of a tailor-made whole-cell catalyst for a desired asymmetric reduction has been developed by the use of a microorganism that can overproduce an enantioselective alcohol dehydrogenase with *S*- or *R*-enantiopreference and glucose dehydrogenase (cofactor-regenerating) enzyme [15]. The use of a tailor-made whole-cell catalyst in a pure aqueous reaction medium which proceeds at >100 g L^{-1} ketone was developed. This methodology was used to furnish the desired optically active (*R*)- and (*S*)-alcohols with high conversions (>90%) and excellent enantioselectivities (>99% *ee*) (Scheme 6.13).

6.3.2
Reaction Systems for Bioreduction

Bioreduction of carbonyl compounds can be performed using several reaction systems, including different solvents and form of biocatalysts. The most commonly

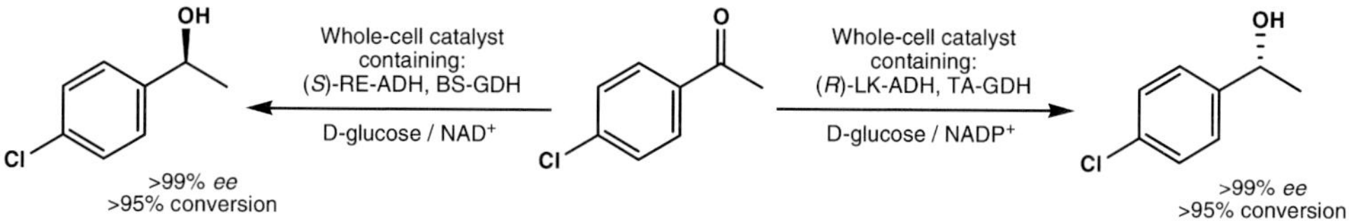

Scheme 6.13 Examples of asymmetric bioreduction using tailor-made whole-cell catalyst with enantiocomplementary specificity.

employed reaction system for bioreduction is in aqueous solution. This is easily explained because in Nature, enzymatic reactions occur in water. As several organic compounds which can be transformed by enzymes have poor solubility in water, important techniques have been developed to overcome this problem and, in some cases, improve the enzymatic activity. Biocatalysts for bioreduction reactions have generally been used in the form of whole cells of microorganisms (ketoreductase/alcohol dehydrogenase producers) and purified enzymes (ketoreductases/alcohol dehydrogenases). Both techniques have several advantages and, of course, some disadvantages depending on the point of view. The biocatalysts can be applied in an immobilized form.

In order to illustrate the available techniques for bioreduction reactions, some recent examples have been selected for discussion in this section.

6.3.2.1 Bioreduction Using Whole-Cell Biocatalysts in an Aqueous Solvent

Bioreduction reactions using whole cells in aqueous solvents can be performed by fermentation technology, immobilized microorganisms and non-immobilized microorganisms. In addition, additives can be added to the reaction system to improve the yield and enantioselectivity of the bioreduction [16].

Chartrain *et al.* employed fermentation technology to produce chiral alcohols as precursors for a large number of pharmaceuticals [17]. Their studies of bioreduction reactions involved microbial screening, process development and scale-up activities. They identified that during the process development, the optimization of the bioreduction environmental conditions (pH, temperature), the timing of ketone addition and the implementation of a nutrient feeding strategy are key factors in achieving increased bioreduction rates. In several examples, the scale-up in pilot-plant bioreactors supported the production of several kilograms of highly optically pure alcohol (up to >99% *ee*; Scheme 6.14).

An important technique employed to reduce inhibition caused by substrate and product in microbial reduction is the use of hydrophobic polymer, Amberlite XAD, as an additive [18]. However, this polymer can also be applied as a material to control the stereochemical course of microbial reductions and increase the chemical yield of the reaction. For example, in the presence of XAD, simple aliphatic and aromatic ketones were reduced by *Geotrichum candidum* IFO 4597 to the corresponding

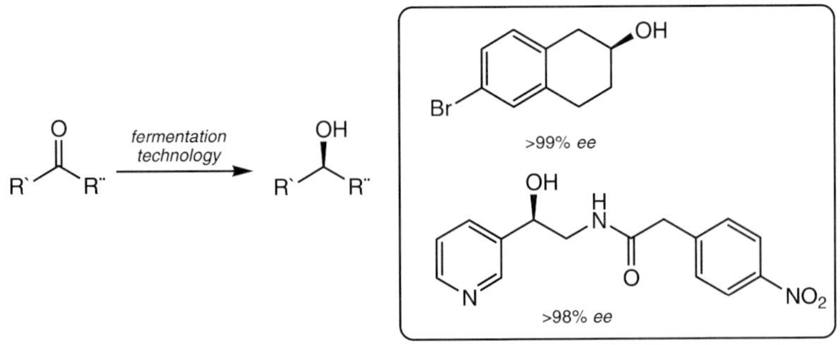

Scheme 6.14 Examples of chiral alcohols prepared by fermentation technology.

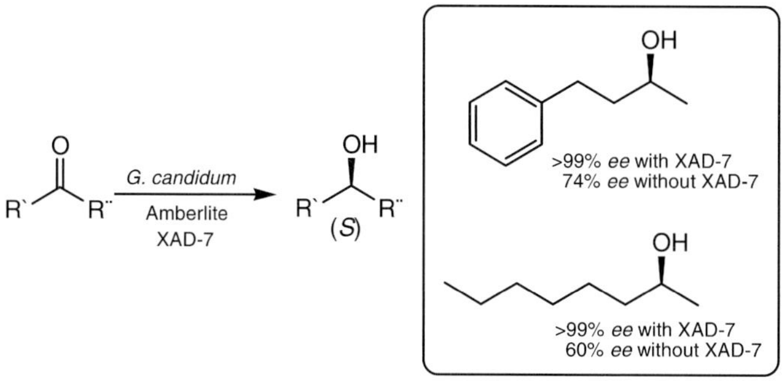

Scheme 6.15 Examples of chiral alcohols prepared using *Geotrichum candidium* IFO 4597 and XAD as additive.

(S)-alcohols in excellent *ee*, whereas low enantioselectivities were observed in the absence of the polymer (Scheme 6.15).

A recent example related to the entrapment of cells with double gel layers to achieve high enantio- and diastereoselectivity in the asymmetric bioreduction of ethyl 3-halo-2-oxo-4-phenylbutanoate was reported by Rodrigues and co-workers [19]. Cells of *S. cerevisiae* were immobilized in calcium alginate beads with double gel layers. The ketone bioreduction performed with this biocatalyst and glucose as electron donor afforded the corresponding alcohols in high chemical yield (90%), *de* (70%) and *ee* (96–99%) (Scheme 6.16).

Microorganism cells can be resuspended in aqueous solution to perform the bioreduction of carbonyl compounds. A recent example reported by Andrade *et al.* made use of this approach (Scheme 6.17) [20]. They employed cells of fungi resuspended in aqueous solution to reduce -substituted acetophenones to give the corresponding chiral alcohols in good yield and high enantioselectivity (up to 99%). These chiral alcohols are important building blocks in the synthesis of a number of chiral drugs, such as sotalol, nifenalol, tomoxetine, norfloxetine and fluoxetine.

6.3.2.2 Bioreduction Using Whole-Cell Biocatalysts in a Conventional Organic Solvent and an Aqueous–Organic Solvent

The use of whole cells for bioreduction in conventional organic solvents is not an easy technique due to the high hydrophilic nature of cells, which tend to agglomerate

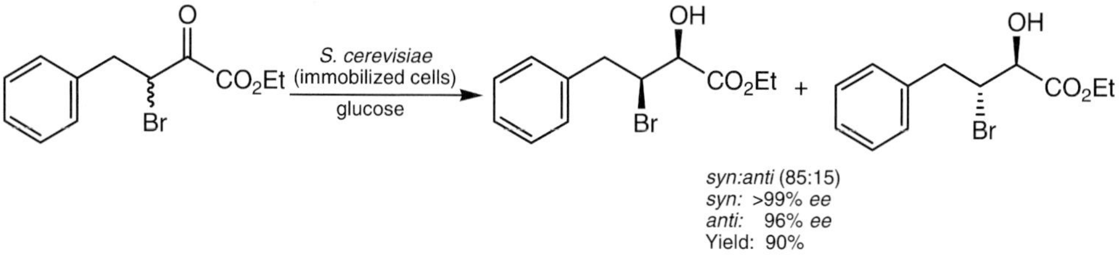

Scheme 6.16 Example of bioreduction mediated by *Saccharomyces cerevisiae* immobilized in Ca alginate beads with double gel layers.

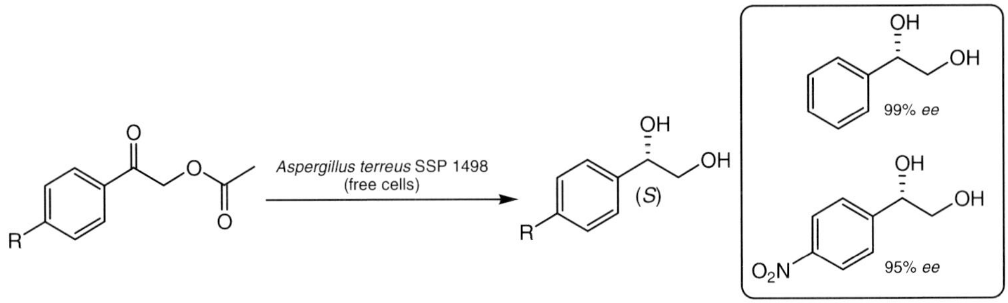

Scheme 6.17 Example of bioreduction by cells of *A. terreus* resuspended in aqueous solution.

when placed directly in organic solvents. This feature can make the contact of the substrate with the enzyme difficult, affording low rates of bioconversion. In addition, whole cell deactivation can be observed. Several studies related to the cells immobilization have been developed to avoid these difficulties (for some examples, see [21]).

An example of cell immobilization for bioreduction in organic solvents was reported by Nakamura *et al.* [21d]. Cells of the fungus *Geotrichum candidum* IFO 4597 were immobilized on a water-absorbing polymer (BL-100) and used for stereoselective reduction in an organic solvent using cyclopentanol or 2-alkanols as additive. Enantiomerically pure (*S*)-1-arylethanols were obtained by the reduction of the corresponding ketones, in contrast to reduction in water by the free cells, in which (*R*)- or (*S*)-1-arylethanols were produced in low *ee* (Scheme 6.18).

Biphasic systems can be applied to dissolve some potentially useful prochiral ketones for bioreductions which are almost insoluble in water. For example, Hu and Xu established an aqueous–organic biphasic system and used it with whole cells of *Oenococcus oeni* to reduce prochiral carbonyl compounds [22]. The conversion increased as the log *P* value of the organic solvent changed from 0.5 to 6.6 in the bioreduction of 2-octanone to (*R*)-2-octanol. Under optimized conditions, the conversion of (*R*)-2-octanol reached 99% from 2-octanone (0.5 *M*) with an optical purity of 99% *ee* using *n*-nonane–aqueous phase (1 : 1 v/v) as the biphasic system (Scheme 6.19). This system also allowed the *anti*-Prelog reduction of aliphatic and aromatic ketones to furnish *R*-configured alcohols in high optical purity.

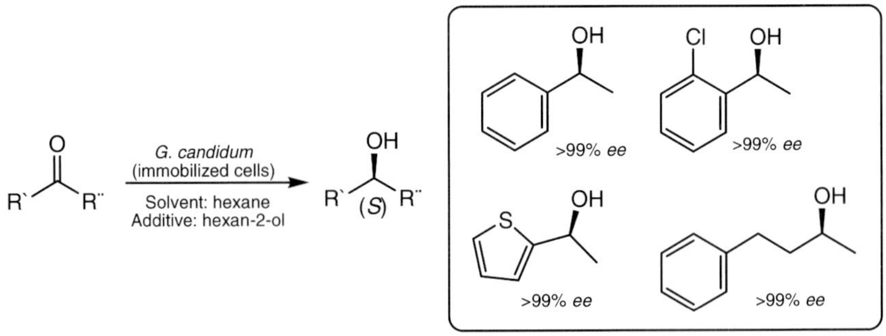

Scheme 6.18 Example of bioreduction by immobilized cells of *Geotrichum candidum* and hexane as solvent.

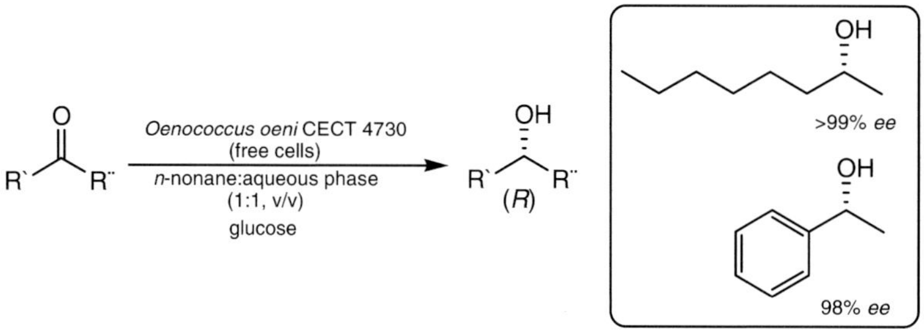

Scheme 6.19 Bioreduction using free cells of O. oeni in a biphasic system.

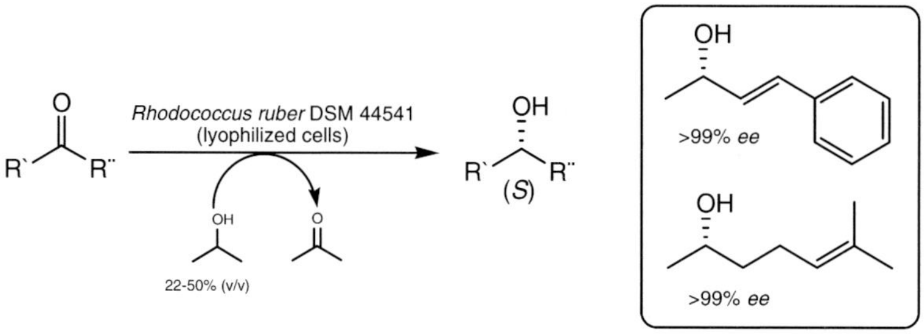

Scheme 6.20 Examples of ketone reduction using lyophilized cells of R. ruber and 2-propanol as hydrogen donor.

Rhodococcus ruber DSM 44541 has been employed as an efficient biocatalyst in ketone bioreduction and enantioselective oxidation of *rac*-alcohols [23]. Chiral *sec*-alcohols were obtained by asymmetric reduction of the corresponding ketone using 2-propanol as hydrogen donor employing lyophilized cells of R. ruber (Scheme 6.20). The microbial reduction system exhibited not only excellent stereo- and enantios-electivity but also a broad substrate spectrum. In addition, the biocatalyst showed exceptional tolerance towards elevated concentrations of organic materials (solvents, substrates and cosubstrates).

Free and immobilized baker's yeast were successfully employed in the asymmetric reduction of prochiral β-keto esters and ketones in glycerol as solvent [24]. The activities with immobilized cells were always higher than with free cells and the enantioselectivity was very high (up to 99%) with both catalysts (Scheme 6.21). The use of glycerol, a non-toxic, biodegradable and recyclable liquid, as an

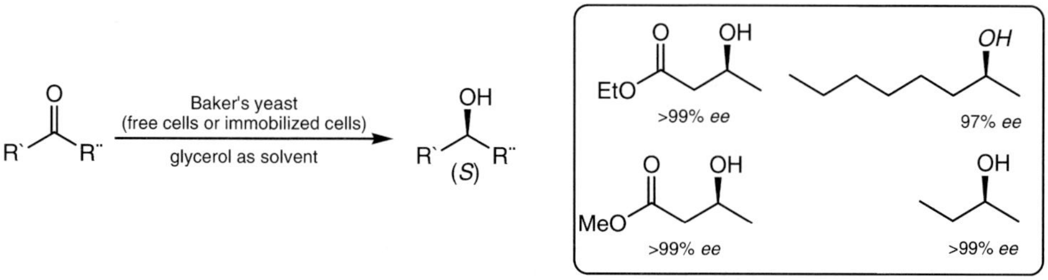

Scheme 6.21 Bioreduction by baker's yeast using glycerol as solvent.

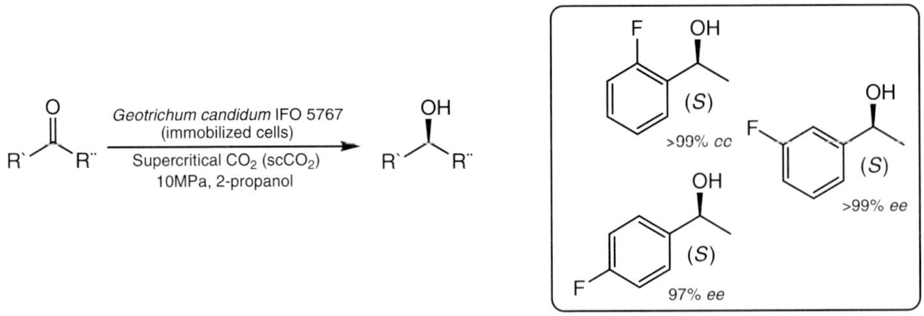

Scheme 6.22 Bioreduction by immobilized resting cells of G. candidum in scCO$_2$ as solvent.

environmentally friendly solvent allowed easy separation of the product by simple extraction with diethyl ether. Both the activity and enantioselectivity were high and competitive with the reactions in water.

6.3.2.3 Bioreduction Using Whole-Cell Biocatalysts in Supercritical Carbon Dioxide, Ionic Liquids and Fluorous Solvents

Recently, solvent systems known as 'green solvents' have also been applied in bioreduction reactions, including supercritical fluids (SCFs), fluorinated solvents and ionic liquids (ILs).

Supercritical Carbon Dioxide (scCO$_2$) scCO$_2$ was employed as a solvent in the bioreduction of aromatic ketones mediated by immobilized resting cells of *Geotrichum candidum* IFO 5767 [25]. For example, this system allowed the bioreduction of fluoroacetophenones to the corresponding chiral alcohols with equivalent enantioselectivity to the system using other solvents (Scheme 6.22).

Fluorous Media Fluorous media such as perfluorinated solvents have been attracting increasing attention in biocatalysis. This interest is due to the immiscibility of fluorous solvents with both organic and aqueous solvents. The main advantage of the use of fluorous media is the ease of separation and purification in the work-up of the reaction.

Immobilized bakers' yeast-mediated reduction of ketones in fluorous media with glucose or methanol as energy sources gives chiral alcohols without loss of stereoselectivity (Scheme 6.23) [26]. The used fluorous solvent was easily and efficiently

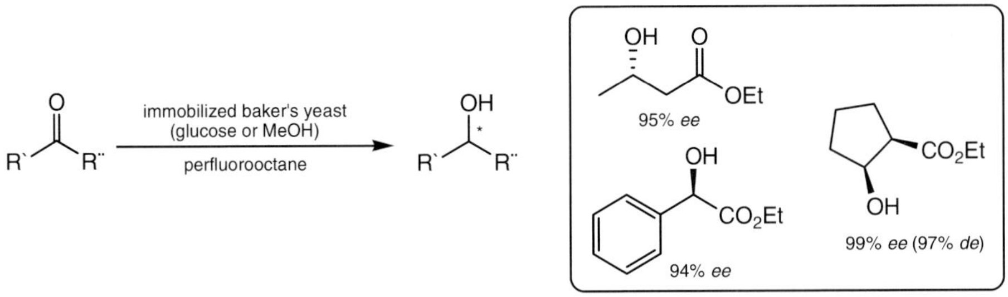

Scheme 6.23 Bioreduction by immobilized baker's yeast in fluorous media as solvent.

recovered by filtration and methanol extraction and was pure enough to be reused without any purification.

Ionic Liquids Reactions using whole-cell biocatalysts in ILs have also been applied to the bioreduction of prochiral ketones with high enantioselectivities [27]. The main interest in performing bioreductions using this solvent is due to its non-volatility and recyclability.

For example, the asymmetric reduction of ketones by an alcohol dehydrogenase preparation from *Geotrichum candidum* (dried cells) in ILs proceeded smoothly with excellent enantioselectivity (>99%) when the cell was immobilized on a water-absorbing polymer containing water, whereas the reaction without the polymer did not proceed (Scheme 6.24) [28].

6.3.2.4 Bioreduction Using Isolated Enzymes

Bioreduction using isolated ketoreductases/alcohol dehydrogenases have attracted great attention, mainly because these enzymes are readily available and cofactor recycling can be easily achieved by the use of different techniques (see Section 6.2.2).

A recent example of the application of isolated ketoreductases in enantioselective ketone reductions was reported by Truppo *et al.* [29]. They performed the bioreduction of a range of substituted benzophenone and benzoylpyridine derivatives with ketoreductase enzymes to afford chiral diarylmethanols in high yield (>90%) and *ee* (up to >99%). As the ketone bioreduction required NADPH cofactor, the hydride source, an NADPH recycling system, was put in place by using glucose and glucose dehydrogenase (GDH) to regenerate the reduced form of the cofactor (Scheme 6.25).

In the search for carbonyl reductases with *anti*-Prelog enantioselectivity, Hua and co-workers examined a carbonyl reductase from *Candida magnoliae* in the bioreduction of various ketones with diverse structures [30]. This carbonyl reductase catalyzed the reduction of a series of ketones and α- and β-keto esters to *anti*-Prelog configured alcohols in excellent optical purity. This reduction was performed with use of an NADPH regeneration system consisting of D-glucose dehydrogenase (GDH) and D-glucose (Scheme 6.26).

Two recent examples involving bioreductions mediated by isolated alcohol dehydrogenase in organic solvents are presented below.

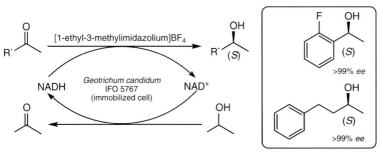

Scheme 6.24 Bioreduction by immobilized *G. candidum* cells and ionic liquid as solvent.

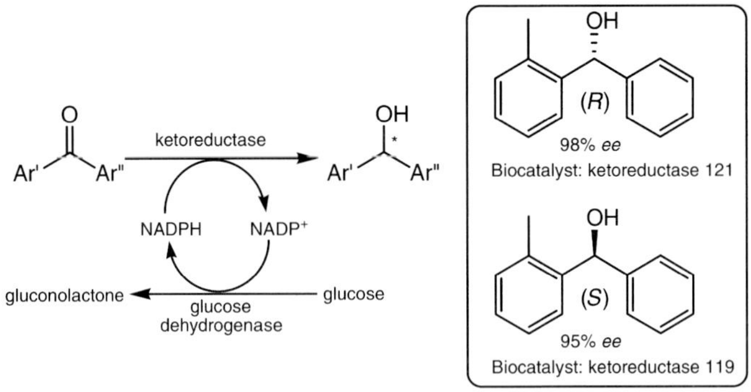

Scheme 6.25 Examples of bioreduction of diaryl ketone by isolated ketoreductases.

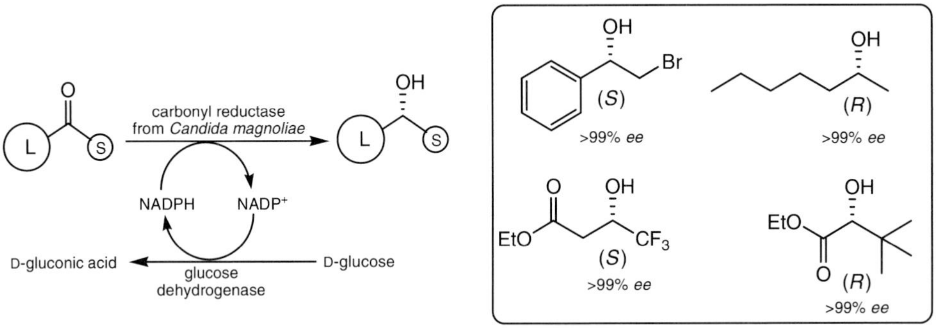

Scheme 6.26 Examples of ketone reduction by carbonyl reductase from *Candida magnoliae*.

Kroutil and co-workers reported that mono- and biphasic aqueous–organic solvent systems (50% v/v) and also micro-aqueous–organic systems (99% v/v) can be successfully employed for the biocatalytic reduction of ketones catalyzed by alcohol dehydrogenase ADH-'A' from *Rhodococcus ruber* DSM 44541 (Scheme 6.27) [31]. A clear correlation between the log *P* of the organic solvent and the enzyme activity (the higher is log *P*, the better is the enzyme activity) was found. The use of organic solvents allowed highly stereoselective enzymatic carbonyl reductions at substrate concentrations close to 2.0 *M*.

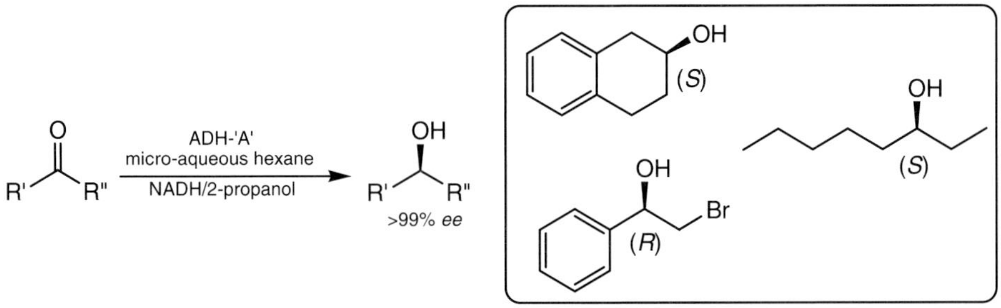

Scheme 6.27 Examples of ketone reduction by alcohol dehydrogenase ADH-'A' from *Rhodococcus ruber* DSM 44541 in micro-aqueous hexane (99% v/v).

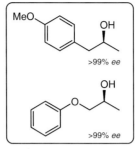

Scheme 6.28 Examples of bioreduction of ketones by xerogel-encapsulated W110A alcohol dehydrogenase from *Thermoanaerobacter ethanolicus* in organic solvents.

Phillips and co-workers reported the application of xerogel-encapsulated W110A alcohol dehydrogenase from *Thermoanaerobacter ethanolicus* for the asymmetric reduction of water-insoluble ketones in organic solvents (Scheme 6.28) [32]. This method allowed the reuse of the xerogel-encapsulated enzyme. In some cases, the use of organic solvents improved the yield and enantioselectivity of the bioreduction.

6.4
Conclusion

Bioreduction of ketones in an asymmetric fashion is, without doubt, one of the most important synthetic methods for preparing chiral alcohols in excellent yield and stereoselectivity. As this chemical transformation is mediated by a natural catalyst, in low amounts, towards a hydrogen source associated with a recyclable system, bioreduction is an excellent example of green technology applied to the synthesis of valuable compounds.

In the future, bioreduction will be even more often employed in industry to produce chiral alcohols, especially after advances in enzyme production technology, the discovery of new versatile enzymes and the creation of tailor-made microorganisms.

References

1 (a) Nitro bioreduction: Pacheco, A. O., Kagohara, E., Andrade, L. H., Comasseto, J.V., Crusius, I.H.-S., Paula, C.R. and Porto, A.L.M. (2007) *Enzyme and Microbial Technology*, **42**, 65–69; (b) Carbonyl bioreduction: Nakamura, K. and Matsuda, T. (2006) *Current Organic Chemistry*, **10**, 1217–1246; (c) Carbon–carbon double bond bioreduction: Stuermer, R., Hauer, B., Hall, M. and Faber, K. (2007) *Current Opinion in Chemical Biology*, **11**, 203–213.

2 Edwin, C. Webb (1992) *Enzyme Nomenclature 1992: Recommendations of the Nomenclature Committee of the International Union of Biochemistry and Molecular Biology on the nomenclature and classification of enzymes*. San Diego, Academic Press, New York.

7
Biotransformations and the Pharma Industry

Hans-Peter Meyer, Oreste Ghisalba, and James E. Leresche

7.1
Introduction

Due to the size and the growth rate of the global population, which is currently estimated at about 6.7 billion, mankind will destroy the biosphere if fundamental changes are not implemented rapidly. As every human being has the same rights to global resources, this destruction will become even faster as many new emerging countries understandably strive for a similar wasteful way of life and resource consumption as in North America, Western Europe and Japan. A change will need sensible adaptations in all kinds of human activity, including the manufacturing industry. Although the principles of green chemistry have been implemented to a large extent in most areas of the chemical industry in the developed world, industry remains under pressure to improve further and still has a negative image in the public perception.

In the discussion on how to improve the environmental compatibility of the organic chemical industry, one gets the impression that the majority of people see the solution in the use of renewable feedstocks to replace petrochemical-based chemical feedstocks. The global consumption of oil, the dominant fossil fuel, is somewhere around a stunning 85 million barrels per day. However, the net contribution capacity of biological renewable materials to replace oil as energy or chemical feedstock will be limited. The different options such as biotechnology, wind, photovoltaic, energy efficiency, intelligent consumption and so on must be carefully balanced before capital investments are made or resources spent in R&D.

Of the Earth's 14.8 billion hectares of land, only one-third, or 5 billion hectares, can be used for agricultural purposes and these arable surfaces are also a non-renewable resource! Due to soil erosion, water limitations and other constraints, agriculture will replace petrochemical feedstocks in only specific cases. In most cases it is ecologically, economically and ethically irresponsible to use 'energy crops' or fractions thereof, such as starch, sugars, lipids and proteins, as raw materials for fuels or feedstock purposes, because this is in direct competition with food production.

Handbook of Green Chemistry, Volume 3: Biocatalysis. Edited by Robert H. Crabtree
Copyright © 2009 WILEY-VCH Verlag GmbH & Co. KGaA, Weinheim
ISBN: 978-3-527-32498-9

if a stereochemical center is present. Thus chiral synthesis will become a state of the art method in organic chemistry. The small-molecule drugs of the future will become structurally more complex and diverse and therefore need new synthetic tools. The method of choice for the derivatization of natural product leads (secondary metabolites) will also be enzyme derivatization.

7.3
The Concept of Green Chemistry

Green chemistry is a concept and guiding principle encouraging the development of manufacturing processes and products with the lowest possible environmental impact or footprint. The goal of the green chemistry concept and principles is to prevent pollution and reduce resource consumptions instead of merely eliminating them. The 12 principles of green chemistry formulated by Anastas and Warner in 1998 are summarized in Table 7.3 [6, 7]. These principles have been followed consciously or subconsciously by synthetic and process organic chemists for a long time for one simple reason: these principles also make a lot of sense from an economic point of view.

After reading the 12 principles, it is clear that biotechnology should be able to make a much greater contribution to green chemistry than is currently the case. The potential contributions of biotechnology are often discussed elsewhere under a number of partly synonymously used terms such as third-wave biotechnology (a term of the Biotechnology Industry Organization), Suschem (used by the European Platform for Sustainable Chemistry), white biotechnology (Europa Bio) or industrial biotechnology. They all have in common the concept of sustainability and the three dimensions of sustainability must always be considered: ecological sustainability, economic sustainability and social sustainability.

Individual chemical companies have all designed specific and refined green chemistry concepts adapted to their specific needs. Figure 7.1 summarizes how

Table 7.3 The 12 principles of green chemistry.

1	Prevent waste
2	Design safer chemicals and products
3	Design less hazardous chemical syntheses
4	Use renewable feedstocks: use raw materials and feedstocks
5	Use catalysts, not stoichiometric reagents
6	Avoid chemical derivatives
7	Maximize atom economy
8	Use safer solvents and reaction conditions
9	Increase energy efficiency
10	Minimize the potential for accidents
11	Design chemicals and products to degrade after use
12	Analyze in real time to prevent pollution

Source: Ref. [6].

Figure 7.1 This summarizes the 'parameters' to consider when challenged with a broad spectrum of organic molecules. The figure contains all the key questions that a process chemist has to address in order to reach 'process excellence' which are in line with the 'green chemistry' concept. Detailed procedures with much more qualitative and quantitative details are available for the synthetic chemist and process engineer.

Lonza, for example, has adapted the green chemistry concept in-house to guide and help chemists to ask the right questions in order to reach optimal ecological and economic results when designing manufacturing processes.

There are four aspects which drive and insure that optimal results for the design of a sustainable process are attained: (1) Innovation, (2) Quality, (3) Ecology and (4) Safety. Innovation is the most important cornerstone from a long-term perspective in process design for NCEs. Innovation makes the right technologies (including biotransformation or fermentation) available and novel creative synthetic routes accessible.

Quality in Figure 7.1 refers also to the quality standards fulfilling current Good Manufacturing Practice (cGMP) requirements when producing APIs or their intermediates. The activities in the process development matching the cGMP guidelines expectations (ICH Q7A) or the so-called FDA SIPQ (Safety–Identity–Purity–Quality) include activities such as finding the critical process parameters, critical scale-up parameters and final qualification and validation procedures. Quality requires that specification and analytical methods for all raw materials, intermediates and the end product, stability of isolated intermediates and products, solubility of lead compounds and cleaning procedures also be addressed.

The ecological benefit is almost guaranteed when following the guidelines and principles summarized in this chapter (Table 7.3) and ecologically critical compounds, for example chlorinated solvents, phosphates and heavy metals, are removed from processes wherever possible. Calculating the E factor ('kilograms in/kilograms out or amount of waste per kilogram of product) is routinely used for benchmarking and continuous improvement, which means that every opportunity for recycling process streams is used. Finally, safety is also an important benchmarking measure and part of the assessment work of the process chemist.

Where is biotechnology in all this? Unfortunately, today, the route selected for the production of an API or intermediate will be via traditional, 'bread and butter' synthetic chemistry. Chemo/enzymatic routes are rarely chosen because the relevant enzymes are either not available in the enzymatic toolbox or the process to develop a robust enzyme suitable for an industrial process will be far too slow. This is a key message with respect to the 'green chemistry' concept in the pharmaceutical and fine chemical industries: first priority must be to invest in novel and commercially available enzymes to fill the toolbox. Another aspect which has not been mentioned up to now is the compression in timeline that has occurred in the development cycle of pharmaceuticals. This, in combination with a missing broad strain and enzyme choice, relegates biotransformation to a second-generation choice at best in the manufacturing of a small-molecule pharmaceuticals. Quick and successful feasibilities using biotransformation under the existing time and quality constraints of the NCE pharma market are very difficult.

In this chapter, we center our discussion on the pharmaceutical and fine chemical industries, where the diversity of chemical structures – which are preferably optically active – requires a particularly diversified reaction and technology toolbox.

7.4
The Organic Chemistry Toolbox

There are many asymmetric tools in organic chemistry – so why should a chemist consider biological reactions catalyzed by enzymes? Textbooks mention that some enzymes are able to speed up a chemical reaction by a factor of 10^{10} and at the same time be very selective. Compared with a simple chemical catalyst such as sulfuric acid, which can also speed up many different reactions, enzymes have a much higher efficiency in terms of rate and selectivity. Building blocks and small-molecule APIs are becoming structurally more complex and asymmetric synthesis is not an exotic academic specialty anymore, but a real industrial requirement for the small-molecule pharmaceutical and fine chemical industries.

The great majority of pharma-relevant applications are related to enantioselective biocatalysis, but there is also an increasing need for regioselective enzyme-catalyzed reactions, for example for the selective esterification of di- and multifunctional compounds, the selective hydrolysis of compounds containing multiple ester functions (dicarboxylic acid esters, sugar esters, etc.) and the selective reduction of compounds with multiple carbonyl groups. Additionally, interest in biocatalysis is growing in other areas where biocatalysts offer clear advantages over purely chemical strategies, such as peptide synthesis, protein modification and the synthesis of oligosaccharides or glycoconjugates. Peptide or oligosaccharide synthesis is difficult and large amounts of waste are produced by classical organic chemistry due to the multifunctionality of these compounds, which calls for sophisticated and laborious protecting group strategies.

7.4.1
Small-Molecule Synthesis

When analyzing the chemical toolbox and the chemical reactions routinely used today (Table 7.4), one realizes that there would still be ample room for improving the ecology of reactions.

Some interesting established classical organic reactions are often not used in industry because safety issues associated with these reactions require expensive investment. Other reasons why these reactions are not used are related to difficulties such as scaling up, too much waste/byproduct formation or that they are not commercially competitive. For example, chemical methods for preparing asymmetric epoxides and their derivatives using the Jacobsen hydrolytic kinetic resolution, Sharpless allylic epoxidation or dihydroxylation are remarkable scientific developments [8–12] but, to the best of our knowledge, these technologies have not been successfully commercialized in the fine chemical industry. The process chemist is forced to shift to other less efficient pathways or accept compromises. A biocatalytic platform should allow access to a single, versatile, scalable, ecologically friendly technology [13].

Although the list in Table 7.5 is neither exhaustive nor exclusive, it gives an additional selection of reactions where biotransformation should be considered as a solution. The list should inspire biotechnologists to investigate and enlarge their toolbox with these opportunities.

7.4.2
Peptide Synthesis

The production methods for peptide APIs (Scheme 7.1), which is characterized by a series of tedious coupling, uncoupling and purification steps using expensive protected precursors (Table 7.6), is still a great challenge today (Table 7.7), even though great progress has already been made. Analogous to small molecules, the molecular structures of peptides are becoming more and more complex (Scheme 7.1)

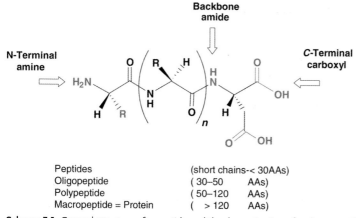

Peptides	(short chains-< 30AAs)
Oligopeptide	(30–50 AAs)
Polypeptide	(50–120 AAs)
Macropeptide = Protein	(> 120 AAs)

Scheme 7.1 General structure of a peptide and the denominations for shorter and longer chains.

Table 7.4 List of readily available reactions from the chemical toolbox: these reactions are examples of well-established chemical reactions routinely used in the organic chemical industry – a 'green assessment' shows that these reactions suffer from ecological and economical drawbacks.

Reaction type	Green assessment	Biotransformation leverage
Addition of an organometallic compound to an epoxide 	⊗ Hazardous chemical ⊗ High salt charge in wastewater/Mg-salts ⊗ CH_4 off-gas ⊗ THF in waste water ⊗ Yield <70% 　Waste for incineration ☺ Modest atom economy	Enzyme alternative?
Chiral addition of a Grignard compound to a ketone 	⊗ Low atom economy due to the high amount of catalyst ⊗ High salt charge, due to the equimolar amount of Mg salts *ex* neutralization	Enzyme alternative to generate tertiary chiral alcohols?

Friedel–Crafts reaction

⊗ Nearly equimolar amount of BF$_3$ or AlCl$_3$
☺ High atom economy
☺ High yield >85%

Could an enzyme control the chiral center with this or other substrates?

Basic ester hydrolysis

☺ High atom economy
⊗ High salt charge in wastewater/Na-salts

Hydrolytic enzymes are commercially available

Nucleophilic substitution

A= H, NH2

⊗ High salt charge in wastewater/KBr

Enzymatic C–C, C–O or C–N bond formation

(Continued)

Table 7.4 (*Continued*)

Reaction type	Green assessment	Biotransformation leverage
Metalation + subsequent acetylation/alkylation	⊗ Bad atom economy during metalation. Often 2 equiv. of base such as BuLi is required (double metalation if NH function present) ⊕ Yield 85% ⊗ High salt charge, due to the high amount of HCl for neutralization	Enzymatic C–C bond formation
Reduction with hydrides such S LiAlH₄, DIBAL	⊗ Al waste ⊗ Low atom economy ⊗ Hazardous reaction	Enzyme solution for reduction to the aldehyde oxidation state
Oxidation	⊗ Oxidation via Cl₂ → high salt charge	Are there enzymes to control the oxidation up to the aldehyde?

Wittig–Horner reaction

⊗ Low atom economy in the Wittig–Horner reaction
⊗ P waste
☺ 79.5% yield

Enzymatic C–C bond formation?

Vilsmeier chloroformylation

☺ Efficient way of functionalization
⊗ P waste
⊗ 55% yield

Enzymes?
Other substrates?

Table 7.5 List of reactions from the chemical toolbox which are less frequently used and assessment of synthetic challenges

Organic reaction	Synthetic challenge	Biotransformation leverage
Aldol reaction	Low atom economy (compromised by self-condensation, regioselectivity and stereoselectivity issues) Good yield Inorganic-waste	Aldolase offers an interesting alternative
Arndt–Eistert synthesis	Safety issue: handling of diazomethane Requires special technology, CAPEX	Probably there is no comparable biotransformation for converting an acid to its next higher homologue
Barbier (-type) reaction	One-step procedure for the preparation of alcohols from organic halides and aldehydes or ketones	Enzymes? See also Grignard

Baeyer–Villiger reaction

RCO_3H

Oxidation of ketones to esters or lactones by peracids

Enzymes are known for performing the same reaction. The open question is whether bio-options can be used on an industrial scale by process chemists. If not, why?

Baylis–Hillmann reaction

EWG + $R\text{–}CHO$ → product with OH and EWG, R

EWG = electron-withdrawing group

Coupling of activated vinyl systems with aldehyde catalyzed by DABCO

Such a reaction is very interesting as it allows the preparation of α-hydroxyalkylated or arylated products. Are there enzymes for such transformations, controlling chirality, increasing reaction rate (reaction is slow)?

Curtius reaction

$RCOOH \longrightarrow RCON_3$

$RNCO \longrightarrow RNH_2$

Stepwise conversion of a carboxylic acid to an amine having one fewer carbon via an acyl azide and isocyanate intermediate. Major issue is the safe handling of azides and therefore special technology is necessary

Often the Hofmann reaction is used as an alternative on an industrial scale. Biotransformation as alternative?

Darzens condensation

$R\text{–}CHO$ + $X\text{–}CHR^1\text{–}COOEt$ $\xrightarrow{EtO^-}$ R,R^2 epoxy $COOEt$ (R^1)

Formation of α,β-epoxy esters by condensation of an aldehyde with esters of α-halo acids

Interesting reaction for the preparation of advanced building blocks. Biotransformation?

(Continued)

Table 7.5 (Continued)

Organic reaction	Synthetic challenge	Biotransformation leverage
Diels–Alder reaction	Interesting reaction for preparing advanced intermediates	Diels–Alderase enzyme?
Fischer indole synthesis	Formation of indoles on heating aryl hydrazones of aldehydes or ketones in the presence of catalysts such as Lewis or proton acids	Can highly functionalized indoles be synthesized by a bio-route?
Fries rearrangement	Rearrangement of phenolic esters to o- and/or p-phenolic ketones on heating with $AlCl_3$ or other Lewis acid catalysts	Would an enzyme be available to perform this rearrangement with high regioselectivity?
Heck reaction	Stereospecific palladium-catalyzed coupling of alkenes with organic halides or triflates	Are there strains or enzymes enabling such C–C bond formation?

Henry reaction

This base-catalyzed aldol-type condensation is an interesting reaction generating advanced intermediates

Is aldolase able to catalyze such a reaction and to control the two chiral centers?

Hunsdiecker reaction

$$RCOOAg + X_2 \longrightarrow RX + CO_2 + AgX$$

Synthesis of organic halides by thermal decarboxylation of carboxylic silver salts in the presence of halogen

Biotransformation avoiding thermal heating and silver salt waste?

Hydroboration reaction and oxidation

Hydroboration on alkenes takes place by *cis* addition followed by oxidation of the boron–carbon bond to form alcohol. Asymmetric version is also established

Could enzymatic asymmetric conversion replace boron chemistry?

Jacobsen epoxidation

Chiral catalyst for asymmetric epoxidation of alkenes. Apparently enantioselectivity and diastereoselectivity depend strongly on the nature of the substrate

Epoxidases?

Krapcho decarboxylation

Decarboxylation of a malonate ester or other derivatives in a dipolar aprotic solvent at high temperature in the presence of water to yield esters, ketones or nitriles

Enzymatic equivalent allowing the same reaction under milder conditions on more complex and sensitive structures?

EWG= COOR, COR, CN, SO$_2$R

(Continued)

Table 7.5 (Continued)

Organic reaction	Synthetic challenge	Biotransformation leverage
Mitsunobu reaction	Reaction allowing the inversion of configuration of alcohol by condensation with acidic component (NuH). Unfortunately the reagents are not ecologically friendly as DEAD and triarylphosphine are used	Dynamic resolution which could be implemented via a chemical or biotech process or a combination of both
Ozonolysis reaction	Efficient cleavage of double bonds with ozone to generate aldehydes, for example. This technology is already very well established on an industrial scale even if safety management of such reactions must not be underestimated	Biotransformation an option (lower investment, milder conditions, increased chemoselectivity on sensitive or highly functionalized substrates?
Sharpless dihydroxylation	Osmium-catalyzed asymmetric cis-dihydroxylation of alkenes Also the Sharpless oxyamination is a very attractive transformation as it provides osmium-mediated cis-addition of nitrogen and oxygen on alkenes	An enzymatic solution would provide valuable solutions to avoid the handling of the prohibitively toxic osmium, reduce the salt charge and CN salts
Sharpless epoxidation	Titanium-catalyzed asymmetric epoxidation of allylic alcohols employing titanium alkoxide, optically active tartrate ester and alkyl hydroperoxide	Are there epoxidases which could provide a broader reactivity spectrum not just limited to allylic alcohols?

Strecker amino acid synthesis

Synthesis of α-amino acids by reaction of aldehydes with ammonia and HCN

Are there enzymes directly combining these 3 substrates by generating the α-amino nitrile optically pure and allowing extension of the application on ketones?

Suzuki–Miyaura reaction

$$R^1\text{-}B(OR)_2 + R^2\text{-}X \xrightarrow[\text{Base}]{\text{Pd}} R^1\text{-}R^2$$

R^1 = aryl, alkenyl, alkyl
R^2 = aryl, alkenyl, alkynyl, benzyl, allyl, alkyl

Low atom economy because multi-step chemistry required for the preparation of the substrates
Good yield|Inorganic-waste, metal waste (leaching)

Are there strains or enzymes enabling C–N bond formation?

Hartwig–Buchwald amination

Low atom economy because multi-step chemistry required for the preparation of the substrates
Good yield|Inorganic-waste, metal waste

Are there strains or enzymes enabling C–C bond formation?

Table 7.6 This technology combination chart shows how the synthesis of peptides can be realized using different methods. These synthetic methods can be applied separately or in combination with each other.

Entry	Solid phase	Liquid phase	Recombinant
Solid phase	• 10-mer to 40-mer • <20-mer fragments • Peptides containing modified/unmodified amino acids • Scale: grams to tons	• Fragment condensation • Peptides containing modified/unmodified amino acids • Scale: grams to tons	
Liquid phase	• Fragments	• <10-mer fragment condensation • All types of amino acids • Scale: grams to tons	
Recombinant		• Fragment condensation • Scale grams to tons	• >30-mer proteins • Complex • All natural amino acids • Scale >100 kg to tons

and the number of chemical steps needed are increasing. For example, the number of amino acids chemically coupled on a large scale has increased from 20 a few years ago to more than 40 natural and unnatural amino acids today.

Peptides need better and 'greener' technologies to provide cleaner production concepts, especially as the predicted demand with respect to quantities is increasing and the standard quality attributes remain high or are increasing due to restrictive regulatory constraints.

Table 7.8 shows the average material consumption for a relatively small 9-mer peptide in conjunction with an annual demand of 20 kg. One realizes that large-scale

Table 7.7 Summary of the advantages of and problems with the different chemical and biological peptide synthetic methods.

Technology	Liquid phase	Solid-phase and hybrid approach	Recombinant and semi-synthesis
Process development	Long	Rapid	Very long
Raw material costs	Medium	High	Low
Advantages	Similar to standard organic chemistry Established technology	Straightforward approach to longer peptides Excellent scalability	Practically no limitations in scale and amounts
Limitations	Very laborious, will remain expensive	High material flows at large scale	Unnatural sequences potentially problematic

Table 7.8 Raw materials needed for the production of 20 kg of a 9-mer peptide [14]: the wasteful peptide synthesis is due to tedious coupling procedures using protected building blocks, the deprotection procedures needed followed by HPLC purification, desalting and lyophilization operations.

Raw material	Amount of material needed to produce 20 kg of peptides (t)
Amino acids	0.4
TCTU[a]	0.3
Piperidine	6.5
NMP[b]	100
DCM[c]	80
Acetonitrile	35
USP[d] water	300

[a]TCTU = [1-Bis(dimethylamino)methylene]-5-chloro-3-oxy-1H-benzotriazol-1-ium tetrafluoroborate.
[b]NMP = 1-Methylpyrrolidin-2-one.
[c]DCM = dichloromethane.
[d]USP = United States Pharmacopeia.

peptide manufacturing is a really wasteful operation, especially if one extrapolates the figures in Table 7.8 for a peptide which is larger than a 9-mer and for which more than 20 kg is produced! The demand for peptides is increasing for different indications, ranging from novel peptide antibiotics to anticancer peptides; hence efficient and cost-effective manufacturing methods are urgently needed. A green chemistry concept will need much higher manufacturability efficiency for peptides as the requirements on raw materials needed for the synthesis, isolation and purification procedures to the isolated final API are excessive.

Table 7.6 suggests that the recombinant bacterial production by fermentation (*de novo* and *in vivo* biosynthesis) of especially larger peptides and proteins is the method of choice. Could biotransformation provide methods and enzymes for *in vitro* and cell-free peptide synthesis and assembly with higher chemoselectivity, quantitative yield and simplified overall production technology? Spirin [15] gives an example of *in vitro* and cell-free translation systems for the synthesis of proteins. Another simpler approach would be to use the reverse reaction of proteases. In this case, individual *exo*-proteases with exclusive specificity for (non-activated or native) amino acids would be needed.

7.4.3
What Should Chemists Consider?

Laird [16] writes in an editorial comment that 'synthesis is not the same as a process' and that the selection of the right synthesis for a new NCE is just the premise of the future industrial process. A number of additional activities must be successfully performed after route selection for a synthesis to become commercially viable. Nevertheless, the key to success is that the ideal route is selected for the synthesis of

Table 7.9 Why will biotransformation play a pre-eminent role in organic chemistry? – there are three main reasons if the enzymes required are commercially available.

Specificity and yield
- High chemo-, regio- and stereoselectivity
- Usually high yields
- Enzymes can be adapted to organic solvents

Working safety and ecology
- Mild reaction conditions and biodegradable catalyst
- No runaway reactions possible
- Reduced health hazards for technical staff

Economic advantages
- No protection deprotection groups needed
- Heterologous protein expression allows competitive production of enzymes
- Low corrosive reaction media result in longer life-span of multipurpose installations

the NCE right from the start. 'Green chemistry' principles should be implemented in the design phase of building blocks and NCEs. In the future, drugs will have more complex structures, which means that the use of biotransformations will become a necessity for their production. Hence it is of paramount importance that the synthetic chemist and the biologist talk to each other very early in the R&D process to implement the nearly unmatched selectivity of enzymatic systems (Table 7.9).

7.5
The Enzyme Toolbox (a Selective Analysis)

In our experience, in over 95% of cases the users of biotransformations are trained chemists. Consequently, in order to have organic chemists consider the use of enzymes in their route selection and later at the bench, two prerequisites are needed. First, the chemists must be aware of the enzymatic options, and second, the enzymes must be commercially available in a ready-to-use form for the laboratory bench and later for production. In other words, the enzyme toolbox or the enzyme shelf needs to be full of different enzymes capable of carrying out all sorts of reactions and accepting all sorts of different molecules. This is exactly where the problem starts: the toolbox is at best only partly filled.

An analysis of one of the main conferences in the area, the International Symposium on Biocatalysis and Biotransformations [17], revealed an old problem (Table 7.10). Of the 365 lectures and posters that were analyzed in 2007, less than 70 dealt with new enzymes. Lipases and esterases still represent 64% of the papers dealing with hydrolytic enzymes or 26% of all papers. On the other hand, there are unknown 'biotransformation' territories to be conquered and radical innovation is needed. In our opinion, both aspects can be best dealt with in an academic environment, where 'risky' and more long-term projects can be performed better

Table 7.10 Overview of the enzyme classes presented as oral or poster presentations at the last three Biotrans Symposia, with an example of the % calculation: 34% of all presentations in 2007 dealt with oxidoreductases; the sum does not add up to 100% because some of the papers presented did not deal with biotransformations.

Enzyme class	2007 (%)	2005 (%)	2003 (%)	Reaction type carried out by the enzyme
1. Oxidoreductases	34[a]	24	28	Redox reactions
2. Transferases	8	6	3	Functional group transfer
3. Hydrolases	41[b]	55	58	Hydrolysis of functional groups
4. Lyases	12	12	10	Non-hydrolytic addition/removal of groups
5. Isomerases	2	2	1	Intramolecular rearrangements
6. Ligases	0	1	0	Formation of $C-O$, $C-S$, $C-N$ or $C-C$ bonds

[a]It seems that the -10% in hydrolytic enzymes are the $+10\%$ in oxidoreductases. This increase in availability of some oxidoreductases (e.g. ketoreductases, ene reductases) in combination with protein e.g. directed evolution have somewhat improved the situation.
[b]64% of all hydrolytic enzymes were papers on lipases/esterases or 26% of all papers.

than in industry. Hydrolytic enzymes, however, do not really represent a prime academic topic any longer, in particular as there are small companies specialized in improving and adapting biocatalysts to industrial needs. There is also a certain risk of price dumping when academic research centers are in direct competition with small- and medium-sized enterprises. Academic research centers are better able to adjust work load imbalances with public funds and academic centers do not always have to calculate on a full-cost basis.

According to this analysis, the other enzyme classes are roughly 10 times less investigated than lipases or esterases and rank in the following order: aldolases, gycosyltransferases, cytochrome P450 enzymes, laccases and ketoreductases. Hence the still half-empty enzyme toolbox for highly selective/specific biotransformations remains problem number one for the pharmaceutical industry! Feasibilities under the existing time and quality constraints of the pharma market are difficult and we have to bridge this to make rapid feasibilities possible.

One success story which should inspire us is the time required today to go from gene to product. This time has been drastically reduced from 5 years in 1988 to 1.5 years in 2000 and to less than 1 year today! This means that once we have enzyme candidates and genes available from the toolbox, the process development and large-scale production should be rapid.

7.5.1
EC 1 Oxidoreductases

Table 7.10 shows that oxidoreductases are the second most investigated class of enzymes. However, oxidoreductases which catalyze oxidations and reductions are a broad and varied group of enzymes reflecting the importance of these biocatalysts in various life-sustaining reactions ranging from respiration to detoxification reaction in the human liver.

7.5.1.1 **EC 1.2.1 Dehydrogenases**

Chiral alcohols are useful intermediates and chemistry has developed asymmetric catalysts such as BINAP for their production. There are two enzymatic approaches possible today, optical resolution or direct synthesis using (cofactor-dependent) dehydrogenases. In fact, the stereoselective reduction of ketones to chiral alcohols using dehydrogenases is a well-developed technology [18]. Ketoreductases (KREDs) for the stereoselective reduction of carbonyl groups represent one of the largest groups of commercially available enzymes. KREDs and engineered KREDs can be obtained from various companies (e.g. DSM, BRAIN, Jülich Chiral Solutions, BioCatalytics, Codexis, Fluka-Sigma-Aldrich). In this section, we will focus on the process technology aspects of dehydrogenase applications.

It is generally accepted that NADH- or NADPH-requiring oxidoreductases can easily be used in whole-cell biocatalysis such as yeast-mediated reductions, where the cofactor recycling step is simultaneously performed within the intact cell, driven by the reduction equivalents introduced via the external carbon and energy source (glucose).

Although several useful biochemical methods for the *in vitro* recycling of cofactors have been very successfully established and used industrially (see below), many chemists still consider the cofactor requirement as a technical and economic obstacle to the more general use of this type of enzyme in their cell-free form. For preparative-scale synthesis, repetitive batch has proved to be an easy to handle technique. The multiple reuse of the enzyme(s) is possible after recovering them from the reacted solution by means of concentration with ultrafiltration equipment. Compared with batch processes, continuous processes often show higher space–time yields. For the application of enzymes in continuous processes and especially for cofactor-dependent systems, the enzyme membrane reactor (EMR) has been developed. The EMR concept has been promoted and successfully applied by Kula and Wandrey [19] and others. It has also been successfully industrialized, with industrial application by various chemical and pharmaceutical companies.

A classical example of the preparative use of an NADH-dependent enzyme is the enantioselective reduction of 3,4-dihydroxyphenylpyruvic acid to D-3,4-dihydroxy-phenyllactic acid catalyzed by D-hydroxyisocaproate dehydrogenase [20]. The essential cofactor PEG–NADH is regenerated *in situ* from PEG–NAD$^+$ by a second enzyme, formate dehydrogenase (FDH), using formate as the hydrogen donor. By coupling to water-soluble poly(ethylene glycol) (PEG) with a molar mass of 20 000, the cofactor can be retained together with the enzymes by an ultrafiltration membrane and the whole process can be performed continuously in an EMR.

As many as 600 000 cycles have been reported for the FDH/formate-driven PEG–NADH recycling system in the case of L-phenylalanine production (Scheme 7.2) by reductive amination with phenylalanine dehydrogenase [21].

At the time when the EMR concept was introduced, NAD and NADH were still very expensive compounds and therefore contributed to a large fraction of the overall process costs. Therefore, for application in continuous processes, molecular weight

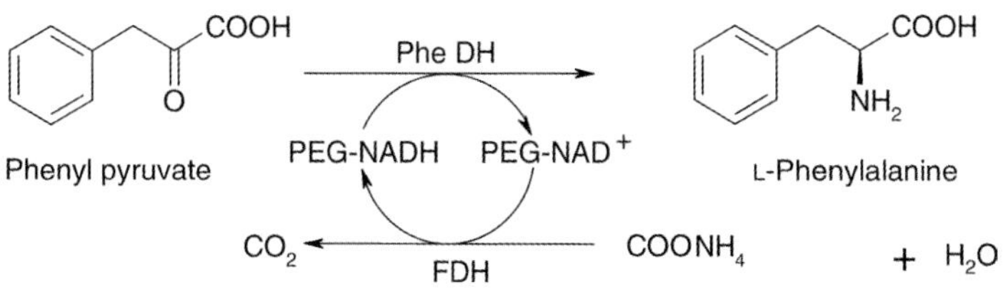

Scheme 7.2 Enzymatic synthesis of L-phenylalanine by reductive amination starting from phenyl pyruvate (using EMR technology).

enhancement by covalent coupling to PEG (in order to allow very high recycling numbers) was an essential condition for the economic viability of such processes. In the meantime, however, thanks to the rapid progress in fermentation technologies, cofactors have become affordable compounds, at least when purchased in bulk quantities.

An EMR process for the enantioselective reduction of 2-oxo-4-phenylbutyric acid (OPBA) to (R)-2-hydroxy-4-phenylbutyric acid (HPBA, ee >99.9%) with NADH-dependent D-lactate dehydrogenase (D-LDH) from *Staphylococcus epidermidis* and FDH/formate for cofactor recycling uses free NADH instead of PEG–NADH [22, 23] (Scheme 7.3). Within the selected residence time of 4.6 h, a cycle number for the cofactor of 1000 can easily be achieved. Instead of NADH, the less expensive NAD^+ is used. A second reason for the application of the native coenzyme is the fact that the activity of D-LDH is reduced to one-tenth when PEG-enlarged coenzyme is used instead of the native coenzyme. HPBA produced by this method with a space–time yield of 165 g L^{-1} d^{-1} is of very high enantiomeric purity and the process shows competitive economy in comparison with chemical 'enantioselective' hydrogenation approaches such as heterogeneous hydrogenation with a modified platinum catalyst (yielding ee 82–91%) or homogeneous hydrogenation with a soluble chiral rhodium–diphosphine complex (yielding ee 96%) [23]. Note that after a second round of process optimization, the space–time yield of this D-LDH/FDH-based process is around 400 g L^{-1} d^{-1} (unpublished data).

On the other hand, the performance, product quality (ee >99.9%) and process economy of the EMR process are comparable to those of an alternative biocatalytic approach using whole cells of reductase-containing *Proteus vulgaris* or *Proteus*

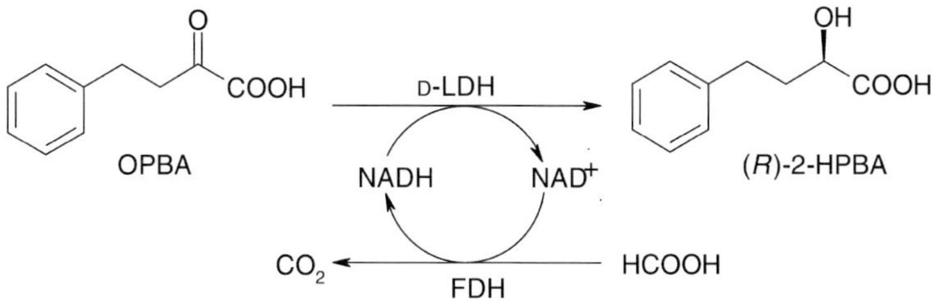

Scheme 7.3 Enantioselective enzymatic reduction of 2-oxo-4-phenylbutyric acid to (R)-2-hydroxy-4-phenylbutyric acid (HPBA) (using EMR technology).

mirabilis in a packed-bed bioreactor [23]. In this case, carbamoylmethylviologen is used as an electron mediator and the oxidized mediator is reduced (recycled) by formate with formate dehydrogenase also present in *Proteus* strains. HPBA is a versatile building block, for example for the angiotension-converting enzyme (ACE) inhibitor benazepril (Cibacen) and related pharmaceuticals.

In a 1996 report on reaction engineering, Wandrey's group reached similar conclusions with respect to the use of native coenzymes instead of PEG-enlarged types in EMR systems [24]: 'With native cofactors quite high cycle numbers of up to 4500 can be reached with charged ultrafiltration membranes. At least in the case of NAD^+ and ATP the cofactor costs are no longer economically limiting. Also for more expensive cofactors (e.g. $NADP^+$), enzyme reaction engineering will help make appropriate processes economically feasible.' An additional 'upgrade' of this technology to improve the cofactor retention further is the use of nanofiltration instead of ultrafiltration membranes [25].

Until fairly recently, NADP(H)-dependent enzymes were not very attractive economically, as no suitable cofactor regeneration system was available. To overcome this limitation, the NAD(H)-dependent FDH from the methylotrophic bacterium *Pseudomonas* sp. 101 was successfully engineered by multipoint site-directed mutagenesis. The finally generated NADP(H)-specific mutant enzyme shows about 60% of the activity of the wild-type FHD with NAD(H). A practical application of this mutant enzyme for cofactor regeneration was described for the enantioselective reduction of acetophenone with the NADPH-dependent alcohol dehydrogenase from *Lactobacillus* sp. [26]. This new option opens up the way to future technical applications of interesting NADP(H)-dependent enzymes. The NADP-specific FDH is now available from Jülich Chiral Solutions (Codexis).

7.5.1.2   EC 1.14.14 P450 Monooxygenases

Monooxygenases (and also dioxygenases) use molecular oxygen as the oxidant. Monooxygenases [cytochrome P450 monooxygenases (CYPs) or flavin monooxygenases (FMOs)] are enzymes inserting an oxygen atom derived from the scission of dioxygen from ambient air into an activated or non-activated C–H bond. The overall reaction can be represented simply by the following equation:

$$RH + O_2 + NADPH + H^+ \rightarrow ROH + H_2O + NADP^+$$

This selective C–H bond activation under mild conditions using oxygen is a good example of green chemistry as it has no equivalent in the chemical synthetic toolbox. Heme-containing P450 monooxygenases are found in all domains of life, but they differ in their performance characteristics. Several thousand bacterial, fungal, plant and animal cytochrome P450 enzymes are known. Bacterial (prokaryotic) P450 monooxygenases are soluble and usually fairly stable but have (with very few exceptions, such as *Nocardia corallina*, *Streptomyces griseus* and *Streptomyces rimosus*) a very narrow substrate specificity in combination with a moderate selectivity. P450 monooxygenases from eukaryotic organisms such as yeasts, fungi, plants and animals are membrane associated and generally have a broader substrate specificity but are much more unstable than their bacterial counterparts. Eukaryotic microbes with broad substrate acceptance and which are frequently used for biotransformations include *Saccharomyces cerevisiae, Candida tropicalis, Aspergillus niger, Beauveria bassiana, Cunninghamella bainieri, Cunninghamella echinulata, Cunninghamella elegans, Curvularia lunata, Penicillium chrysogenum* and *Rhizopus stolonifer*. P450 monooxygenases need cofactors and P450-coupled cofactor recycling systems. This is the reason why they are used as whole-cell systems.

For the enantioselective preparation of chiral synthons, the most interesting oxidations are hydroxylations of unactivated saturated carbons or carbon–carbon double bonds in alkene and arene systems, together with the oxidative transformations of various chemical functions. Of special interest is the enzymatic generation of enantiopure epoxides. This can be achieved by epoxidation of double bonds with cytochrome P450 monooxygenases, ω-hydroxylases or biotransformation with whole microorganisms. Alternative approaches include the microbial reduction of α-halo ketones or the use of haloperoxidases and halohydrin epoxidases [27]. The enantioselective hydrolysis of several types of epoxides can be achieved with epoxide hydrolases (a relatively new class of enzymes). These enzymes give access to enantiopure epoxides and chiral diols by enantioselective hydrolysis of racemic epoxides or by stereoselective hydrolysis of *meso*-epoxides [27, 28]. In the pharmaceutical industry, CYPs and FMOs (recombinant human/mammalian or microbial) are also gaining increasing importance for the whole-cell biocatalyzed synthesis of drug metabolites (on the multi-milligram to gram scale) and for the derivatization of lead compounds (small molecules and natural products, on the multi-gram scale) for biological and pharmacological evaluation [29, 30]. In this context, additional types of P450 (or FMO)-catalyzed reactions such as *N*-, *O*- and *S*-dealkylations, *N*-oxidations/hydroxylations of secondary and tertiary amines, *S*-oxidation of *S*-alkyl derivatives and oxidative ring-closure reactions are also of interest.

7.5.1.3 EC 1.14.13 Baeyer–Villiger Monooxygenases
Baeyer–Villiger monooxygenases (BVMOs) are microbial flavoenzymes that catalyze a wide variety of oxidative reactions such as stereo- and enantioselective Baeyer–Villiger oxidations and sulfoxidations. They are found in biosynthetic pathways in many different organisms and also in microbial degradation pathways [31–33]. BVMOs, like the CYP and FMO enzymes, also use molecular oxygen. Typical substrates for BVMOs are steroids, alkanes and cyclic ketones. The best studied BVMO is cyclohexanone monooxygenase (CHMO) from *Acinetobacter* sp., which can convert >100 different substrates. BVMOs are now available from some enzyme suppliers, but the majority of the described applications are still performed as whole-cell biotransformations, for example the side-chain cleavage of 20-ketosteroids by *Fusarium* sp., the conversion of testosterone/progesterone to testolactone by *Aspergillus* sp. or *Penicillium* sp. and the conversion of 4-hydroxyandrost-4-en-3,17-dione to the corresponding D-ring lactone by *Cylindrocarpon radicicola*.

CYPs, FMOs and BVMOs offer a broad spectrum of synthetic applications. However, to achieve a real breakthrough in this field with these subclasses of enzymes for larger scale biotransformations (preferably with whole cells), in many cases a number of R&D needs must be addressed, such as an increase in expression level, an increase in specific activity and improved enzyme stability and operational lifetime.

7.5.1.4   EC 1.4.3 Oxidases

Laccases are multi-copper oxidases which have been known for long time with lignin-degrading filamentous fungi. Laccases have been used for the removal of the p-methoxyphenyl (PMP) group, which is used for amine and imine group protection [34]. The broad variety of potential and practical laccase applications in many areas of the industrial sector, such as delignification, dye and colorant bleaching and bioremediation, is impressive [35], but the value of this enzyme group for the pharma and specialty chemicals industries still remains to be evaluated and exploited.

7.5.2
EC 2 Transferases

Transferases catalyze the transfer of a specific group, such as methyl, acyl, amino, glycosyl and phosphate, from one substance to another. The transferases are further classified into eight subclasses: EC 2.1 Transfer of one-carbon groups, EC 2.2 Transfer of aldehyde or ketone residues, EC 2.3 Acyltransferases, EC 2.4 Glycosyl-transferases, EC 2.5 Transfer of alkyl or aryl groups (except methyl groups), EC 2.6 Transfer of N-containing groups, EC 2.7 Phosphotransferases and EC 2.8 Transfer of sulfur-containing groups. From this group, transaminases may have the greatest impact for the asymmetric synthesis of chiral amines from the corresponding ketones, followed by sulfotransferases and glucuronyltransferases (both of interest for the preparation of drug metabolites).

7.5.2.1   EC 2.3 Acyltransferases

EC 2.3.2.13 Transglutaminases   Transglutaminases catalyze the acyl transfer reactions of γ-carboxyl groups of glutamine residues to acceptors (e.g. ε-amino groups of lysine) in proteins and they play a role in a number of physiological processes such as coagulation. As transglutaminases crosslink specific proteins into insoluble, protease-resistant high molecular weight complexes, they can be used in the modification of proteinaceous foods to promote gel formation (for example with caseins) or for texture improvements in meat. Activa, a product of Ajinomoto, is currently the only commercially available transglutaminase.

7.5.2.2   EC 2.4 Glycosyltransferases

These are enzymes transferring glycosyl groups used for oligosaccharide synthesis and glycosylation of all sorts of molecules. Some of these enzymes also catalyze hydrolysis, which can be regarded as transfer of a glycosyl group from the donor to water. Also, inorganic phosphate can act as acceptor in the case of phosphorylases. Phosphorolysis of glycogen is regarded as the transfer of one sugar residue from glycogen to phosphate. However, the more general case is the transfer of a sugar from an oligosaccharide or a high-energy compound to another carbohydrate molecule as

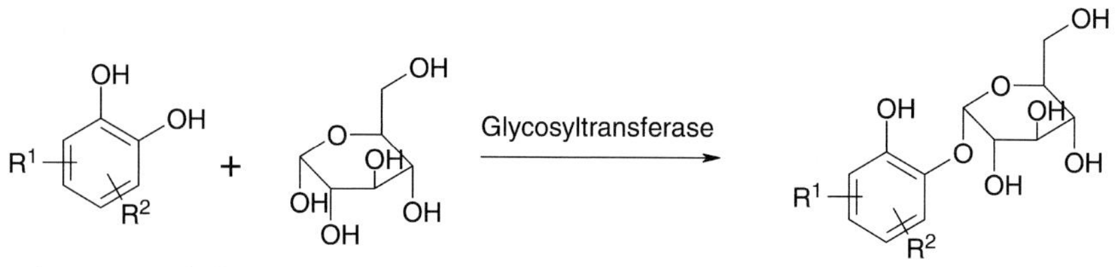

Scheme 7.4 Polyphenolic compounds with beneficial health effects can be improved using glycosyltransferases.

acceptor. The subclass is further subdivided, according to the nature of the sugar residue being transferred, into hexosyltransferases (EC 2.4.1), pentosyltransferases (EC 2.4.2) and those transferring other glycosyl groups (EC 2.4.99).

Carbohydrates have numerous biological roles and more and more drug targets for carbohydrate related drugs are being identified. Glycosyltransferases are an interesting enzyme class which allow the modification of small and large molecules. One example is the biotransformation of natural polyphenols, where sugar moieties are added to the polyphenol (Scheme 7.4) to enhance the physiological and formulation properties of different polyphenols (see www.libragen.com).

7.5.2.3 EC 2.6 Transfer of N-Containing Groups
These include EC 2.6.1 transaminases (Scheme 7.5 and 7.6).

7.5.2.4 EC 2.7 Phosphotransferases
This is a rather large group of enzymes comprising not only those transferring phosphate but also diphosphate, nucleotidyl residues and others. Researchers at Kyowa Hakko (Japan) have developed systems for the large-scale production of UDP-Gal and globotriose (Scheme 7.7) from inexpensive starting materials [38]. Globotriose is a trisaccharide portion of globotriosylceramide, a receptor of verotoxin produced by some bacterial strains. The production system for UDP-Gal combines two metabolically engineered *E. coli* strains with *Corynebacterium ammoniagenes* in a

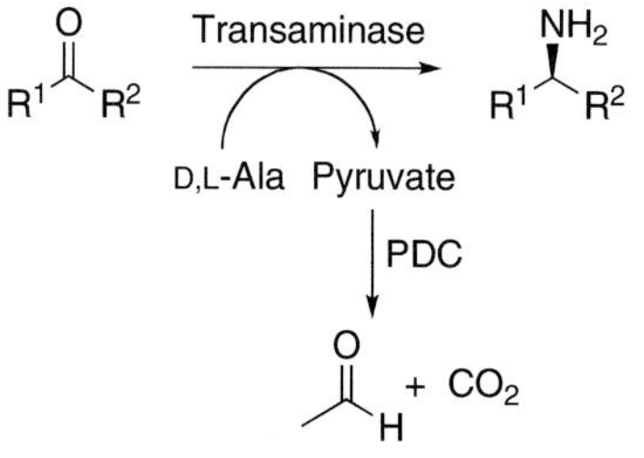

Scheme 7.5 The combination of ω-transaminase with pyruvate decarboxylase (PDC) allowed for efficient asymmetric synthesis of chiral amines as the unfavorable equilibrium could be shifted by removal of pyruvate using the PDC [36].

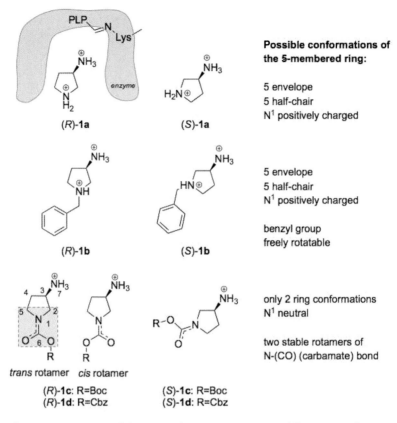

Scheme 7.6 Comparison of the structural features of protected and unprotected 3AP (3-aminopyrrolidine) with respect to stereodiscrimination by the enzyme. The amino groups of both enantiomers are orientated in the same direction to show the possibilities for an enzymatic stereodifferentiation. The two rotamers of **1c** and **1d** are designated as *cis* and *trans* with respect to the orientation of the carbonyl oxygen to the amino group. The atoms in the gray area form a planar geometry. PLP, cofactor pyridoxal-5′-phosphate [37].

defined mixed culture (the three strains were previously grown individually). An amount of 188 g L^{-1} (372 mM) globotriose (Galα1–4Galβ1–4Glc) are produced after 36 h of reaction starting from orotic acid (47 mM), galactose (722 mM), lactose (468 mM) and fructose. Similar approaches using engineered microbes as cooperating 'modules' in mixed cultures were developed by Kyowa Hakko also for the production of CDP-choline and GMP [39].

7.5.3
EC 3 Hydrolases

Hydrolytic enzymes or hydrolases catalyze the hydrolytic cleavage of C–O, C–N and C–C bonds. The idea of using hydrolytic enzymes was first proposed by R—hm in

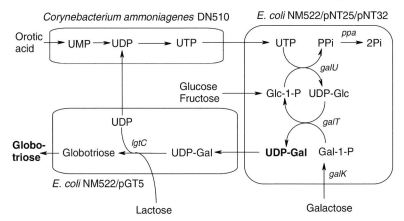

Scheme 7.7 Production of sugar nucleotides by combining metabolically engineered *E. coli* with a nucleoside 5′-triphosphate-producing microorganism, and the concept of producing oligosaccharides by coupling such systems with glycosyltransferases, have great potential for application to the manufacture of other sugar nucleotides and oligosaccharides. Several other bacterial glycosyltransferase genes have already been cloned and overexpressed in *E. coli*. Such recombinant strains would be suitable for the setup of analogous production systems for various sugar nucleotides and oligosaccharides.

1913, who added pancreatic enzymes to detergents. In the meantime, the industrial enzyme market has passed sales of US$2.3 billion, which is generated by a mere 125 products. Practically all of these products are hydrolytic enzymes used in the food, textile, leather and other industries. The existing enzymatic options for protection, deprotection and resolution mainly use hydrolytic enzymes (lipases, esterases, acylases, proteases) for hydroxyl, amino, mercapto and carboxyl groups in the chemistry of peptides, carbohydrates, alkaloids and steroids. The class of hydrolytic enzymes is further divided into 12 subclasses: EC 3.1 Catalyze reactions with ester bonds (esterases and lipases), EC 3.2 Glycosidases, EC 3.3 Reaction with ether bonds, EC 3.4 Peptidases, EC 3.5.Reactions with C–N bonds except peptide bonds, EC 3.6 Reactions with acid anhydrides, EC 3.7 Reactions with C–C bonds, EC 3.8 Reactions with halogen bonds, EC 3.9 Reactions with phosphorus–nitrogen bonds, EC 3.10 Reactions with sulfur–nitrogen bonds, EC 3.11 Reactions with carbon–phosphorus bonds, EC 3.12 Reactions with sulfur–sulfur bonds and EC 3.13 Reactions with carbon–sulfur bonds.

7.5.3.1 EC 3.1.1.1 Esterases and EC 3.1.1.3 Lipases
There are numerous examples of esterase and lipase resolutions (Scheme 7.8 and 7.9) in aqueous and organic solvents (with immobilized and soluble enzymes). These hydrolytic enzymes are widely used and have a relatively long history of use in organic synthesis for the preparation of enantio-pure compounds by resolution. They are probably the only enzymes which are routinely used by chemists because they are simple to use and they are commercially available. Lipases and esterases are also examples of enzymes which have often been used in organic solvents.

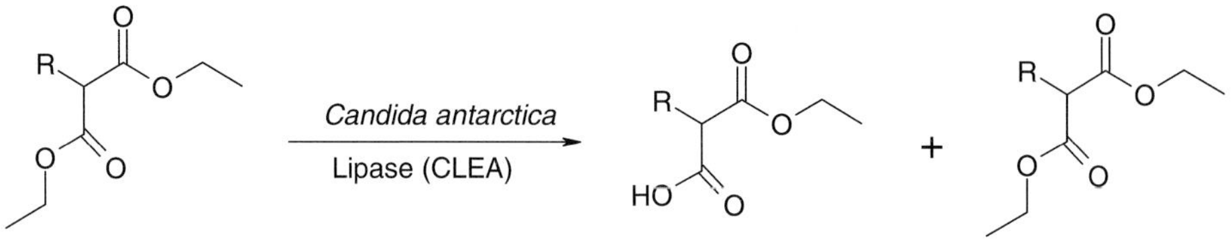

Scheme 7.8 Reaction carried out with a CLEA-immobilized *Candida antarctica* lipase B from CLEA Technologies. The immobilization of the enzyme is necessary to achieve economically competitive costs.

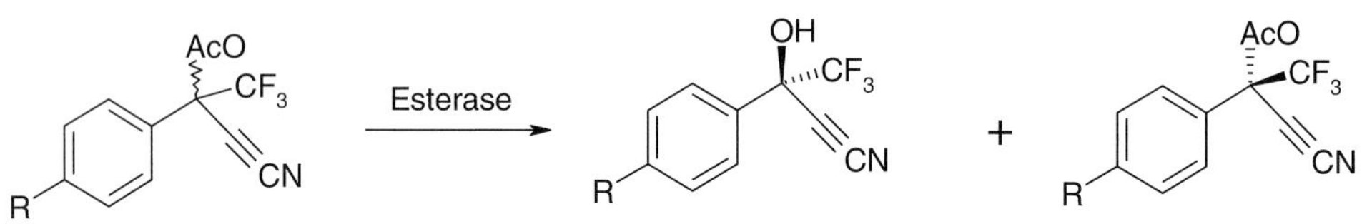

Scheme 7.9 Example of the application of an esterase for the enantioselective synthesis of arylaliphatic tertiary alcohols [40].

7.5.3.2 EC 3.2 Glycosidases

Glycosidases are important and are a widely used enzyme class in the food industry. Glycosidases can be used for the synthesis of oligosaccharides. Bojavora *et al.* reported that water-soluble glycosyl azides can act as a glycosyl donor in transgylcosylation reactions using various glycosidases [41]. The example shown in Scheme 7.10 is a similar product compound to the example discussed under glycosyltransferases. In this case, hydrolysis of the inexpensive glycosylated natural precursor leads to resveratrol, a compound with cardiovascular benefits.

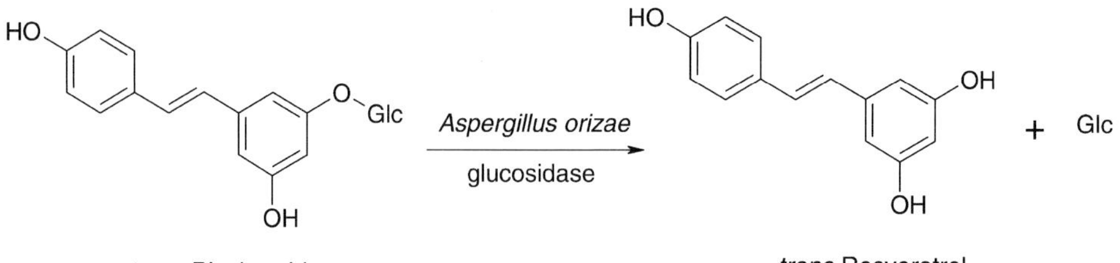

trans-Piceic acid

trans-Resveratrol

Scheme 7.10 Hydrolysis of piceic acid using *Aspergillus orizae* to the corresponding aglycone polyphenol resveratrol, one of the substances believed to responsible for the so-called 'French paradoxon'. The French paradoxon describes the fact that the French have a lower incidence of cardiovascular diseases despite a high red wine consumption. Grape wines contain resveratrol and the corresponding glycoside.

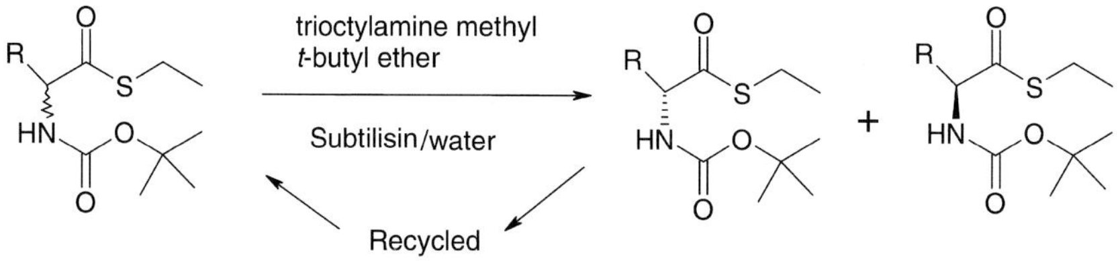

Scheme 7.11 Resolution of N-protected amino acid thioesters using the protease subtilisin; the D-enantiomer was continuously racemized in the presence of an organic base. This kinetic dynamic resolution of the amino acid allowed complete conversion to the desired product [42].

7.5.3.3 EC 3.3 Reaction with Ether Bonds

EC 3.3.2 Epoxide Hydrolases Commercial recombinant epoxide hydrolases are produced by *Aspergillus niger* and used for the hydrolytic kinetic resolution of racemic epoxides.

7.5.3.4 EC 3.4 Peptidases
Numerous proteases such as chymosin, subtilisin (*Bacillus* endopeptidase) and savinase (alkaline protease) are used in such diverse areas as cheese making and detergents. EC 3.4.21.61 Subtilisin is a protease which has been used for fine chemical purposes. Scheme 7.11 gives an example of the subtilisin-catalyzed enantioselective hydrolysis of racemic α-amino acid thioesters. It is important to overcome the intrinsic yield limitation of racemic resolutions (maximally 50%) by some means of *in situ* racemization.

7.5.3.5 EC 3.5 Reactions with C–N Bonds Except Peptide Bonds

EC 3.5.5 Nitrilase Nitrile-converting enzymes such as nitrilases, nitrile hydratases (see Section 7.5.4.1, EC 4.2.1) and amidases have become preferred enzyme tools in chemistry (Scheme 7.12) because cyanide can be easily introduced chemically to provide the required educts for these enzymes. Nitrilases have been reported for desymmetrization, including the hydrolytic desymmetrization of prochiral bis(cyanomethyl) sulfoxide [43].

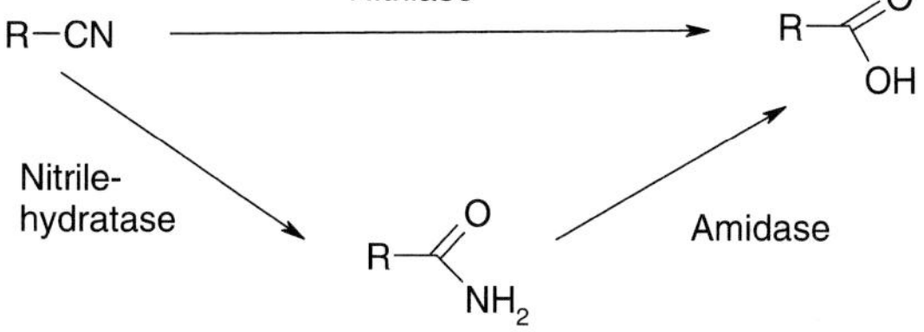

Scheme 7.12 The use of nitrilases with a typically broad substrate specificity but low enantioselectivity or the use of nitrile hydratase and enantioselective amidase combinations for the enzymatic conversion of nitriles.

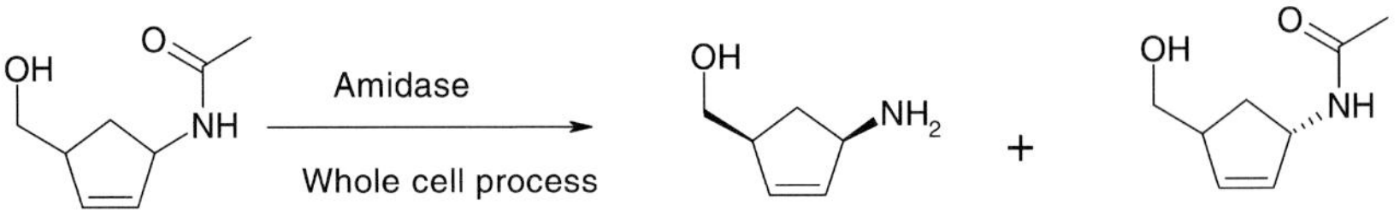

Scheme 7.13 Process scheme for the production of (1*R*, 4*S*)-1-amino-4-hydroxymethylcyclopent-2-ene.

EC 3.5.1 Amidases Scheme 7.13 shows an example of the use of an amidase or amidohydrolase for the production of (1*R*, 4*S*)-1-amino-4-hydroxymethylcyclopent-2-ene using *Rhodococcus erythropolis* CB101 [44].

7.5.3.6 EC 3.7 Reactions with C–C Bonds

EC 3.7.1 β-Diketone Hydrolases β-Diketone hydrolases catalyze the hydrolytic cleavage of carbon–carbon bonds. They are as simple to use as any other hydrolytic enzyme. Therefore, they are of interest for potential applications in organic synthesis such as the desymmetrization of prochiral substrates. One such example is the resolution of heteroannular bicyclic β-diketones by a 6-oxocamphor hydrolase from *Rhodococcus* sp. [45].

7.5.3.7 EC 3.8 Reactions with Halogen Bonds

EC 3.8.1 Dehalogenases The production of enantiopure epoxides has been reported using a cascade reaction of alcohol dehydrogenase and halohydrin dehalogenase [46].

7.5.4
EC 4 Lyases

Lyases catalyze the hydrolytic cleavage of C–C, C–O, C–N and other bonds. C–C bond formation with the simultaneous formation of a new stereocenter can be carried out using the reverse reaction of lyases.

7.5.4.1 EC 4.2.1 Nitrile Hydratases

$$R-C \equiv N + H_2O \rightarrow R-C(O)NH_2$$

Nitrile hydratases are usually not stereoselective enzymes, but there are exceptions such as the proposed route by Bioverdant to the drug levetiracetam. Nitrilase hydratases hydrolyze nitriles to the corresponding acetic acid amides. Nitrile hydratases from *Rhodococcus rhodochrous* (J1) is an enzyme used by Lonza for the production of several thousand tons of nicotinamide in a continuous three-step immobilized enzyme process [47] (Scheme 7.14).

7.5.4.2 EC 4.3.1.5 Phenylalanine Ammonia Lyase (PAL)
PAL is a widely distributed enzyme which deaminates the aromatic acid L-phenylalanine. Heterologous expression of fungal PAL in *E.coli* has been reported for the production of *p*-hydroxycinnamic acid [48].

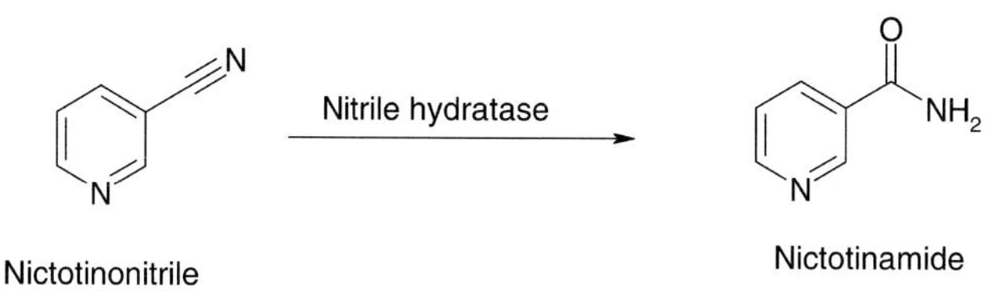

Nictotinonitrile

Nictotinamide

Scheme 7.14 Example of an industrial process for the biotechnological production of several thousand tons of nicotinamide using immobilized cells of wild-type *Rhodococcus rhodochrous* J1 originally developed by Nagasawa and Yamada for the production of acrylamide. The advantages of the process are that no acid or base is required for hydrolysis, no byproducts are formed and the immobilized biocatalyst can be used in a continuous process at low temperature and low pressure.

7.5.4.3 EC 4.1.2 Aldolases

Aldolases catalyze C–C bond formation and generate new stereogenic centers. This stereocontrolled enzymatic formation of C–C bonds is a very attractive option, due to the general interest in this fundamental reaction in organic chemistry [49]. However, compared with other types of enzymatic reactions, the number of reports on asymmetric C–C bond formation is still small and the number of practical examples on the preparative scale is very limited. One of the most classical reactions for the formation of C–C bonds is aldol condensation. In Nature, such stereocontrolled reactions are catalyzed by enzymes of the class of lyases (EC 4). The majority of these enzymes can be found in the biosynthesis of carbohydrates and are used for the synthesis of natural and unnatural carbohydrates. Aldolases (and transketolases) have been intensively investigated and their scope of applications has been evaluated and reviewed by the groups of Whitesides and Wong [50–54]. Aldolases accept a wide range of aldehydes in place of their natural substrates and permit the synthesis of carbohydrates such as aza sugars, deoxy sugars, deoxythio sugars, fluoro sugars and C_8- and C_9-sugars. In the case of D-fructose-1,6-diphosphate aldolase (FDP aldolase, type A), more than 75 aldehydes have been identified as substrates [53].

The baker's yeast-mediated acyloin condensation, which was discovered some 80 years ago, has found many practical synthetic applications. However, this seems mainly due to the ready availability of this cheap biocatalyst and also to the operational simplicity of its application.

Many enzymes can catalyze condensation reactions, which finally lead to the formation of a C–C bond. A recent example is the chemo-enzymatic synthesis of aromatic 1,2-amino alcohols using threonine aldolase in combination with a decarboxylase [55].

N-Acetylneuraminic acid aldolase (a type B aldolase) catalyzes the cleavage of *N*-acetylneuraminic acid (Neu5Ac) to *N*-acetylmannosamine (ManNAc) and pyruvate (Pyr). The reverse reaction can be applied to synthesize Neu5Ac (Scheme 7.15), which plays an important physiological role as the terminal sugar residue in mammalian glycoproteins and glycolipids [56]. This reverse reaction has been known for a long time and some research groups have used it for the preparation of Neu5Ac on a small scale. The careful kinetic analysis of the system and the use of EMR allowed scaling-

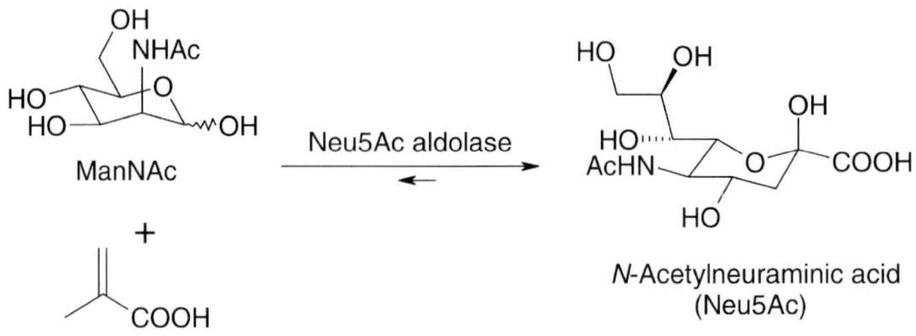

Scheme 7.15 Aldolase-catalyzed production of
N-acetylneuraminic acid (using EMR technology).

up to the multi kilogram scale (with a space–time yield of 650 g L^{-1} d^{-1}) in a collaboration between academic research and industry (Ciba-Geigy and Research Center Jülich) [57, 58]. This process was later transferred to Jülich Chiral Solutions (now owned by Codexis) and further developed up to the multi-hundred kilogram scale. The reaction can also be performed with immobilized Neu5Ac aldolase [59]. A number of companies have patented or are developing Neu5Ac-based drugs, e.g., sialidase inhibitors or antiallergic agents.

As Neu5Ac aldolase accepts a range of substrates (more than 60 are known) in place of ManNAc, this enzyme can also used to synthesize Neu5Ac derivatives [59]. For example, if ManNAc is replaced by D-mannose, Neu5Ac aldolase can be used for the preparative-scale synthesis of KDN (3-deoxy-β-D-glycero-D-galacto-2-nonulosonic acid). [60].

L-Dopa, a metabolic precursor of dopamine, is a very important drug for the treatment of parkinsonism, but is also of interest in other therapeutic indications (annual production >250 t). A very interesting industrialized bioprocess (Ajinomoto) [61] is the production of L-dopa using β-tyrosinase (tyrosine phenol lyase, TPL) in a resting cell system (see also Table 7.11). This enzyme catalyzes a variety of reactions: α,β-elimination (7.1), β-replacement (7.2) and the reverse of α,β-elimination (7.3):

$$RCH_2CHNH_2COOH + H_2O \rightarrow RH + CH_3COCOOH + NH_3 \tag{7.1}$$

$$RCH_2CHNH_2COOH + R'H \rightarrow R'CH_2CHNH_2COOH + RH \tag{7.2}$$

$$R'H + CH_3COCOOH + NH_3 \rightarrow R'CH_2CHNH_2COOH + H_2O \tag{7.3}$$

where R,R$'$ = phenyl, hydroxyphenyl, dihydroxyphenyl or trihydroxyphenyl.

L-Tyrosine and related amino acids can be synthesized in very high yields through the reverse of α,β-elimination (7.3). In the case of L-dopa synthesis by the resting cells of *Erwinia herbicola*, high concentrations of L-dopa are obtained in the reaction mixture (Scheme 7.16). This new access to L-dopa was first described in 1969 [62]. After a first round of reaction engineering and strain development, titers of around 50 g L^{-1} were achieved in 1975 [63]. The now commercialized fed batch process [64], with a final product concentration of 110 g L^{-1}, has many advantages over the classical chemical process (Table 7.11) and can serve as an excellent example of a sustainable production process. This process is also a very good example of a successful

Table 7.11 Comparison between the enzymatic and the chemical processes for the production of L-dopa. The enzymatic process (60 m3 scale) shows a fivefold improved productivity in comparison with the chemical process and with a significant reduction in time required to complete the process.

	Enzymatic synthesis	Chemical synthesis
Main staring materials	Pyrocatechol, pyruvic acid, and ammonia	Vanillin, hydantoin, hydrogen, acetic anhydride
Number of individual reactions	1	8
Reaction byproducts	Water	Ammonia, carbon dioxide, acetic acid
Optical separation	Not necessary	Separation of reaction intermediates (acetyl-D/L-veratroylglycine) with the enzyme acylase and racemization of the D-compound
Production facilities (reaction and isolation)	Versatile equipment can be used for both the enzymatic reaction and the isolation process	Special plant is required for both the synthesis reactions and the separation process
Time required for production	~3 days	~15 days
Downstream processing	Simple → crystallization, ultrafiltration, recrystallization; no organic solvent extraction	Complex, stepwise
Types of amino acids present in the product as impurities in minor amounts	L-Tyrosine (traces of added inducer) (→ this problem can be eliminated by using a production strain engineered for constitutive expression of TPL [65])	L-Tyrosine, 3-methoxy-L-tyrosine, 3,4, 6-trihydroxyphenylalanine

Source: Ajinomoto Co. Ltd [64].

collaboration between academia and industry and for straightforward technology transfer, as the complex reverse reaction was developed to practical applicability by an academic group (Kumagai, Yamada and co-workers), who also carried out the bioreaction engineering and the strain development work up to the pilot scale. The academic group further supported the project during the industrial implementation phase by creating genetically engineered production strains not requiring the induction of TPL by addition of L-tyrosine [65].

7.5.4.4 EC 4.1.2 Hydroxynitrile Lyases (Oxynitrilases)
A highly interesting option is oxynitrilase-catalyzed enantioselective cyanohydrin formation by addition of cyanide to aldehydes. The well-known reaction catalyzed by

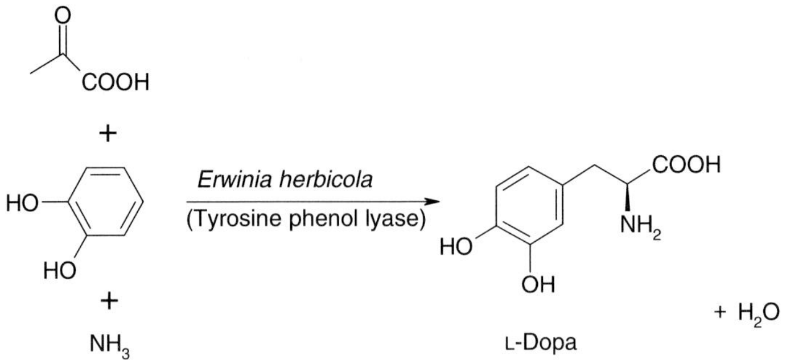

Scheme 7.16 The new synthetic process for L-dopa.

the almond oxynitrilase (EC 4.1.2.10) has been reinvestigated [66–70] and applied to various preparative purposes directed to the synthesis of chiral α-hydroxy acids and β-amino alcohols. (*R*)-Cyanohydrins are formed by addition of HCN to aldehydes or ketones under the catalysis of almond (*R*)-oxynitrilase in organic solvents. The use of (*S*)-oxynitrilase from *Sorghum bicolor* allows the formation of the corresponding (*S*)-cyanohydrins [68, 70]. Asano and co-workers described a considerable number of interesting new plant oxynitrilases found in a dedicated screening program [71, 72]. Thanks to this effort, the size of the oxynitrile toolbox and the application possibilities are significantly enhanced.

7.5.4.5 EC 4.1.1 Decarboxylases

Pyruvate decarboxylase, benzoylformate decarboxylase, phenylpyruvate decarboxylase, enzymes are capable of acyloin-type condensation reactions leading to the formation of chiral α-hydroxy ketones, which are versatile building blocks in the pharmaceutical and fine chemical industries. In addition, α-acetolactate decarboxylase and arylmalonate decarboxylase can be used in non-condensation reactions to produce asymmetric carbon centers [73]. New subclasses of 'broad spectrum' lyases capable of enantioselectively adding water or ammonia to C–C double bonds in various types of compounds, for example in analogy with the fumarase hydratase, aspartate ammonia lyase or methylaspartate ammonia lyase reactions, would be of great interest to pharmaceutical chemists. Unfortunately, the known enzymes are very substrate specific. More versatile new enzymes for such reactions remain to be discovered.

7.5.5
EC 5 Isomerases

Isomerases catalyze geometric and structural rearrangements within a molecule. Isomerases are used in the food industry and on a large scale to produce fructose on a very large scale. Glucose isomerase (EC 5.3.1.5) especially from *Streptomyces* species is used to produce fructose for non-alcoholic beverages. Isomerases are classified into five groups: EC 5.1 Racemases and epimerases, EC 5.2 *cis–trans*-Isomerases, EC 5.3

Intramolecular oxidoreductases, EC 5.4 Mutases or intramolecular lyases and EC 5.5 Intramolecular lyases. However, not many applications reported of isomerases for biotransformation purposes have been reported. One recent example is one-pot two-enzyme biotransformation for the production of N-acetylneuraminic acid by Zimmermann and Kragl using an epimerase to improve reaction equilibria [74].

7.5.6
EC 6 Ligases

Ligases catalyze the linking of two molecules, coupled with the hydrolysis of an energy-rich pyrophosphate bond in ATP or another nucleoside triphosphate. They are classified in the following order, most of which are of more clinical than synthetic interest: EC 6.1 Formation of carbon–oxygen bonds, E.E.6.2 Formation of carbon–sulfur bonds, EC 6.3 Formation of carbon–nitrogen bonds, EC 6.4 Formation of carbon–carbon bonds and EC 6.5 Formation of phosphate–ester bonds. So far, ligases have been found to be of very little interest to the biocatalysis community (see also Table 7.11).

7.6
Outlook and Conclusions

The demand for enantiomerically pure chiral drugs and building blocks, for which novel enantioselective biocatalysis approaches are needed, will continue to grow. Solutions are also required for laborious and wasteful chemical protection group strategies needed for the synthesis of the multifunctional molecules. The interest in biocatalysis is also growing in areas such as peptides, glycoconjugates, oligosaccharide synthesis and protein and small-molecule modifications (for example, glycosylations). Although enzymes are real alternatives for these needs, most of these tools are not yet available for use on the bench by the synthetic chemist. Chemists often still have an old-fashioned opinion on what biology can do for them and biologists do not understand the intrinsic problems of chemical route selection for synthesis. However, this common understanding is an important prerequisite to set the course right for the future so that the correct priorities for the search for novel enzymes can be set.

Three aspects must be dealt with in order to make biotransformations commercially successful and which will introduce at the same time ecologically friendly processes to organic chemistry: (1) close ranks between chemistry and biology for route selection and priority setting; (2) find and improve novel enzymes ready for use on the bench; and (3) introduce innovative process engineering.

Novel enzymes are needed, but which ones are the most important for synthetic organic chemistry? The Swiss Industrial Biocatalysis Consortium (SIBC) analyzed the needs of seven global companies (Ciba, Givaudan, Hoffmann-La Roche, Lonza, Novartis, SAFC and Syngenta), all of which use biotechnology for their chemical syntheses, and the conclusions are summarized in Table 7.12 [75].

Table 7.12 The following enzyme needs were identified and are listed according to their importance by the SIBC.

Oxidoreductases

Dehydrogenases

1. NADH-dependent dehydrogenases for the asymmetric reduction of ketones, keto acids and alkenes
2. Oxidation of alcohols with dehydrogenases have second priority

Oxygenases

1. Monohydroxylations, especially hydroxylations of non-activated centers and of non-natural substrates are important reactions. Improve the practicability and robustness of the *in vitro* P450 systems, and develop an FMO-based alternative system
2. Peroxidases
3. Other mentioned reactions were transformation of ribonucleotides, stereospecific epoxidations and the oxidation of ketones to esters and lactones (Baeyer–Villiger)

Lyases

1. Synthetically useful enzymes for C–C bond formation (preferably asymmetric) using aldolases and hydroxynitrile lyases
2. C–N (aminolyases) and C–O (hydratases) bond formations. Find lyases with a broad substrate acceptance

---

Epoxidases, amidases, nitrilases and nitrile hydratases are other hydrolytic enzymes of some industrial interest. With respect to transferases, transaminases may have the highest impact, followed by sulfotransferases and glucuronosyltransferases. Racemases and isomerases have limited industrial applications in the pharmaceutical and specialty sectors. Hydrolases in general are of lower priority, because the technology has already been implemented in industry and no longer represents an academic challenge. Among the hydrolytic enzymes, the lipases/esterases are the most popular enzymes used in chemical synthesis, followed by proteases and acylases. Generally, a larger number of available enzymes and strains are needed and these must be more robust with a broader substrate acceptance and higher selectivity. However, this is a domain of specialized enzyme suppliers today.

Modern protein engineering methods such as directed evolution allow the exploration of enzyme functions for synthetic purposes which would hardly ever be required in the natural environment. However, we need to continue to explore the biodiversity for the discovery of completely new, so far unknown, subclasses of enzymes. We need to include the screening from natural sources again and not rely too much on the metagenomic approach, which is excluding radical innovation as known sequences or functionalities must be used as references. The Laboratory of Applied Microbiology at Kyoto University with Yamada (now retired) and his co-workers and colleagues Kumagai, Shimizu, Asano, Nagasawa and others serve as an example of the success stories that result from the screening of the natural environment for novel enzyme activities. More than 30 industrialized processes have been realized; one of them is the nicotinamide fed batch process with the highest end product concentration ever reported for a biotransformation: 1465 g per liter of reactor volume. This immobilized catalyst is now used by Lonza (Scheme 7.15) for a highly productive three-step continuous process. German groups have also achieved good screening results, e. g. Kula and Hummel in Jülich and Düsseldorf (new amino acid and hydroxy acid dehydrogenases with high industrial potential) and Wagner and Syldatk in Braunschweig and Stuttgart (hydantoinases, carbamoylases).

An excellent updated compilation of existing industrial biotransformations (with process data and scale of operation) was recently edited by Liese, Seelbach and Wandrey [76]. This publication clearly demonstrates that the application spectrum of biocatalysis for industrial processes is much broader than generally perceived. Thus, the hope seems justified that the contribution of biotechnology in medium- to large-scale production processes can be significantly increased even more in the future, provided that the necessary efforts to enlarge the biocatalytic toolbox are made by academia and industry.

Acknowledgment

The authors greatly appreciate the appraisal of the manuscript and comments of our colleague Karen T. Robins of Lonza.

References

1 Caldwell, J. (1996) *Chimica Oggi-Chemistry Today*, (10), 65–66.

2 Stinson, S.C. (1993) *Chemical & Engineering News*, **71** (23), 38–64.

3 Thayer, A.M. (2007) *Chemical & Engineering News*, **85** (32), 11–19.

4 Rouhi, A.M. (2004) *Chemical & Engineering News*, **82** (24), 47–62.

5 Grimley, J. (2006) *Chemical & Engineering News*, **84** (49).

6 Anastas, P. and Warner, J. (1998) *Green Chemistry: Theory and Practice*, Oxford University Press, New York.

7 Mestres, R. (2004) *Green Chemistry*, **6**, G10–G12.

8 Tokunaga, M., Larrow, J.F., Kakiuchi, F. and Jacobsen, E.N. (1997) *Science*, **277**, 936.

9 Aouni, L., Hemberger, K.E., Jasmin, S., Kabir, H., Larrow, J., Le-Fur, I., Morel, P. and Schlama, T. (2004) in *Asymmetric Catalysis in Industrial Scale: Challenges and Approaches and Solutions* (eds H.-U. Blaser,

E. Schmidt) Wiley-VCH Verlag GmbH, Weinheim, pp. 165–199.

10 Pfenninger, A. (1986) *Synthesis*, 89–116.

11 Kolb, H.C., Van Nieuwenzhe, M.S. and Sharpless, K.B. (1994) *Chemical Reviews*, **94**, 2483–2547.

12 Lohray, B.B. (1999) *Tetrahedron: Asymmetry*, **3**, 1317.

13 Short, P. (2005) *Chemical & Engineering News*, **83** (24), 27.

14 Scarso, A., The landscape for the peptide market: investements, pipeline and technology, presented at the Tides Conference, Las Vegas, NV, 21 May 2007.

15 Spirin, A.S. (2004) *TIBTECH*, **22**, 538–545.

16 Laird, T. (2007) *Organic Process Research & Development*, **11**, 783.

17 *Proceedings of the 8th International Symposium on Biocatalysis and Biotransformation*, Oviedo, Spain, 8–13 July 2007 http://pubs.acs.org/cen/coverstory/84/8449cover.html.

18 Hummel, W. (1999) *TIBTECH*, **17**, 487–492.

19 Kula, M.-R. and Wandrey, C. (1987) *Methods in Enzymology*, **136**, 9–21.

20 Pabsch, K., Petersen, M., Rao, N.N., Alfermann, A.W. and Wandrey, C. (1991) *Recueil des Travaux Chimiques des Pays-Bas*, **110**, 199–205.

21 Hummel, W., Schütte, H, Schmidt, E., Wandrey, C. and Kula, M.-R. (1987) *Applied Microbiology and Biotechnology*, **26**, 409–416.

22 Schmidt, E., Ghisalba, O., Gygax, D. and Sedelmeier, G. (1992) *Journal of Biotechnology*, **24**, 315–327; (1993), **31**, 233.

23 Schmidt, E., Blaser, H.U., Fauquex, P.F., Sedelmeier, G. and Spindler, F. (1992) in *Microbial Reagents in Organic Synthesis* (ed. S. Servi), Kluwer, Dordrecht, pp. 377–388.

24 Kragl, U., Kruse, W., Hummel, W. and Wandrey, C. (1996) *Biotechnology and Bioengineering*, **52**, 309–319.

25 Seelbach, K. and Kragl, U. (1997) *Enzyme and Microbial Technology*, **20**, 389–392.

26 Seelbach, K., Riebel, B., Hummel, W., Kula, M.-R., Tishkov, V.I., Egorov, A.M., Wandrey, C. and Kragl, U. (1996) *Tetrahedron Letters*, **37**, 1377–1380.

27 Archelas, A. and Furstoss, R. (1997) *Annual Review of Microbiology*, **51**, 491–525.

28 Archelas, A. and Furstoss, R. (1998) *TIBTECH*, **16**, 108–116.

29 Ghisalba, O. and Kittelmann, M. (2007) in *Modern Biooxidations. Enzymes, Reactions and Applications* (eds R.D. Schmid and V.B. Urlacher), Wiley-VCH Verlag GmbH, Weinheim, pp. 211–232.

30 Hanlon, S.P., Friedberg, T., Wolf, C.R., Ghisalba, O. and Kittelmann, M. (2007) in *Modern Biooxidations. Enzymes, Reactions and Applications* (eds R.D. Schmid and V.B. Urlacher), Wiley-VCH Verlag GmbH, Weinheim, pp. 233–252.

31 Sima Sariaslani, F. and Rosazza, J.P.N. (1984) *Enzyme and Microbial Technology*, **6**, 242–253.

32 Kamerbeek, N.M., Janssen, D.B., van Berkel, W.J.H. and Fraaije, M.W. (2003) *Advanced Synthesis and Catalysis*, **345**, 667–678.

33 Fraaije, M.W. and Janssen, D.B. (2007) in *Modern Biooxidations. Enzymes, Reactions and Applications* (eds R.D. Schmid and V.B. Urlacher), Wiley-VCH Verlag GmbH, Weinheim, pp. 77–97.

34 Verkade, J.J.M., van Hemert, L.J.C., Quaedflieg, P.J.L.M., Schoemaker, H.E., Schürmann, M. and van Delft, F.L. (2007) *Advanced Synthesis and Catalysis*, **349**, 1332–1336.

35 Xu, F., Damhus, T., Danielsen, S. and Ostergaard, L.H. (2007) in *Modern Biooxidations. Enzymes, Reactions and Applications* (eds R.D. Schmid and V.B. Urlacher), Wiley-VCH Verlag GmbH, Weinheim, pp. 43–75.

36 Höhne, M., Kühl, S., Robins, K. and Bornscheuer, U.T. (2008) *ChemBioChem*, **9**, 363–365.

37 Höhne, M., Robins, K. and Bornscheuer, U.T. (2008) *Advanced Synthesis and Catalysis*, **360**, 807–812.

38 Koizumi, S., Endo, T., Tabata, K. and Ozaki, A. (1998) *Nature Biotechnology*, **16**, 847–850.

39 Hashimoto, S. and Ozaki, A. (1999) *Current Opinion in Biotechnology*, **10**, 604–608.

40 Kourist, R., Bartsch, S. and Bornscheuer, U.T. (2007) *Advanced Synthesis and Catalysis*, **349**, 1393–1398.

41 Bojavora, P., Petrásková, L., Ferrandi, E. E., Monti, D., Pelantová, H., Kuzma, M., Simersk, P. and Křen, V. (2007) *Advanced Synthesis and Catalysis*, **349**, 1514–1520.

42 Arosio, D., Caligiuri, A., D'Arrigo, P., Pedrocchi-Fantoni, G., Rossi, C., Saraceno, C., Servi, S. and Tessaro, D. (2007) *Advanced Synthesis and Catalysis*, **349**, 1345–1348.

43 Kiełbasinki, P., Rachwalski, M., Mikolajczyk, M., Szyrej, M., Wieczorek, M. W., Wijtmans, R. and Rutjes, F.P.J.T. (2007) *Advanced Synthesis and Catalysis*, **349**, 1387–1392.

44 Shaw, N., Robins, K. T. and Kiener, A. (2004) in *Asymmetric Catalysis on Industrial*

Scale (eds H.U. Blaser and E. Schmidt), Wiley-VCH Verlag GmbH, Weinheim, pp. 105–115.

45 Hill, C. L., Chiang-Hung, L., Smith, D.J., Verma, C.S. and Grogan, G. (2007) *Advanced Synthesis and Catalysis*, **349**, 1353–1360.

46 Seisser, B., Lavandera, I., Faber, K., Spelberg, J.H.L. and Kroutil, W. (2007) *Advanced Synthesis and Catalysis*, **349**, 1399–1404.

47 Rohner, M. and Meyer, H.-P. (1995) *BIOPROCESS Engineering*, **13**, 69–78.

48 Vannelli, T., Zhixiong, Z., Breinig, S., Qi, W.W. and Sariaslani, F.S. (2007) *Enzyme and Microbial Technology*, **41**, 413–422.

49 Dean, S. M., Greenberg, W.A. and Wong, C.-H. (2007) *Advanced Synthesis and Catalysis*, **349**, 1308–1320.

50 Whitesides, G.M. and Wong, C.-H. (1990) *Enzymes in Organic Synthesis*, Pergamon Press, Oxford.

51 Wong, C.H. and Whitesides, G.M. (1994) *Enzymes in Synthetic Organic Chemistry*, Tetrahedron Organic Chemistry Series, Vol. 12, Pergamon Press, New York.

52 Bednarski, M.D., Simon, E.S., Bischofberger, N., Fessner, W.-D., Kim, M.-J., Lees, W., Saito, T., Waldmann, H. and Whitesides, G.M. (1989) *Journal of the American Chemical Society*, **111**, 627–636.

53 Bednarski, M.D. (1991) in *Applied Biocatalysis*, Vol. 1 (eds H.W. Blanch and D.S. Clark), Marcel Dekker, New York, pp. 87–116.

54 Takayama, S., McGarvey, G.J. and Wong, C.-H. (1997) *Annual Review of Microbiology*, **51**, 285–310.

55 Steinreiber, J., Schürmann, M., van Assema, F., Wolberg, M., Fesko, K., Reisinger, C., Mink, D. and Griengl, H. (2007) *Advanced Synthesis and Catalysis*, **349**, 1379–1386.

56 Schauer, R. (1985) *Trends in Biochemical Sciences*, **10**, 357–360.

57 Kragl, U., Gygax, D., Ghisalba, O. and Wandrey, C. (1991) *Angewandte Chemie*

(International Edition in English), **30**, 827–828.

58 Kragl, U., Kittelmann, M., Ghisalba, O. and Wandrey, C. (1995) *Annals of the New York Academy of Sciences*, **750**, 300–305.

59 Liu, J.L.-C., Shen, G.J., Ichikawa, Y., Rutan, J.R., Zapata, C., Vann, W.F., W.F. and Wong, C.-H. (1992) *Journal of the American Chemical Society*, **114**, 3901–3910.

60 Sugai, T., Kuboki, A., Hiramatsu, S., Okazaki, H. and Ohta, H. (1995) *Bulletin of the Chemical Society of Japan*, **68**, 3581–3589.

61 Yamada, H. (1998) in *New Frontiers in Screening for Microbial Biocatalysts* (eds K. Kieslich, C.P. van der Beek, J.A.M. de Bont and W.J.J. van den Tweel), Studies in Organic Chemistry, Vol. 53, Elsevier, Amsterdam, pp. 13–17.

62 Kumagai, H., Matsui, H., Ohgishi, H., Ogata, K., Yamada, H., Ueno, T. and Fukami, H. (1969) *Biochemical and Biophysical Research Communications*, **34**, 266–270.

63 Yamada, H. and Kumagai, H. (1975) *Advances in Applied Microbiology*, **19**, 249–288.

64 Enei, H., Nakazawa, E., Tsuchida, T., Namerikawa, T. and Kumagai, H. (1996) *Japan Bioindustry Letters*, **13** (1), 2–4.

65 Katayama, T., Suzuki, H. and Kumagai, H. (2000) *Applied and Environmental Microbiology*, **66**, 4764–4771.

66 Smitskamp-Wilms, E., Brussee, J., van der Gen, A., van Scharrenburg, G.J.M. and Sloothaak, J.B. (1991) *Recueil des Travaux Chimiques des Pays-Bas*, **110**, 209–215.

67 Effenberger, F. (1994) *Angewandte Chemie (International Edition in English)*, **33**, 1555–1564.

68 Kragl, U., Niedermeyer, U., Kula, M.-R. and Wandrey, C. (1990) *Annals of the New York Academy of Sciences*, **613**, 167–175.

69 Niedermeyer, U. and Kula, M.-R. (1990) *Angewandte Chemie*, **102**, 424–425.

70 Effenberger, F., Hörsch, B., Förster, S. and Ziegler, T. (1990) *Tetrahedron Letters*, **31**, 1249–1252.

71 Nanda, S., Kato, Y. and Asano, Y. (2005) *Tetrahedron*, **62**, 10908–10916.

72 Asano, Y., Tamura, K., Doi, N., Ueatrongchit, T., Kittikun, A.H. and Ohmiya, T. (2005) *Bioscience, Biotechnology, and Biochemistry*, **69**, 2349–2357.

73 Ward, O.P. and Singh, A. (2000) *Current Opinion in Biotechnology*, **11**, 520–526.

74 Zimmermann, V. and Kragl, U., in *Proceedings of the 8th International Symposium on Biocatalysis and Biotransformation*, 8–13 July 2007, Oviedo, Spain, poster 249, http://www.uniovi.es/biotrans2007/.

75 Meyer, H.-P. and Münch, T. (2005) *BioWorld Europe*, **1**, 14–16.

76 A. Liese, K. Seelbach and C. Wandrey (eds) (2006) *Industrial Biotransformations*, 2nd edn, Wiley-VCH Verlag GmbH, Weinheim.

8
Hydrogenases and Alternative Energy Strategies

Olaf Rüdiger, António L. De Lacey, Victor M. Fernández, and Richard Cammack

8.1
Introduction: The Future Hydrogen Economy

With the increasing uncertainty about the availability of fossil fuels and concerns about CO_2 emissions and their effects on climate, a sustainable energy source is essential. Hydrogen is widely considered to be a viable convertible form of energy for the future. It is readily transported through pipelines and can be stored, with some energy loss, in the form of compressed or liquefied gas. A great advantage is the high conversion efficiency into electricity in fuel cells, in comparison with the inevitable thermodynamic losses in heat engines. At present, only limited amounts of hydrogen are produced commercially, mainly by steam conversion of fossil fuels [1, 2]. However in a future sustainable economy, hydrogen could be produced by electrolysis of water (Figure 8.1). A major limitation to be overcome is the cost of the catalysts. New catalysts are being developed for the consumption and production of hydrogen [3] but those used in fuel cells currently rely on precious metals such as platinum or their alloys [4]. The known reserves of the noble metals in the lithosphere represent a limitation to their continued use [5]. Moreover these catalysts have problems of side-reactions with O_2 and sensitivity to gases such as carbon dioxide, carbon monoxide and sulfide [6, 7].

In the biosphere, hydrogen has been an energy carrier for billions of years. It is efficiently consumed and produced by microorganisms, in electron-transfer reactions catalyzed by the hydrogenases. These enzymes harbor iron- and nickel-containing clusters with the special property of reacting with or producing H_2, according to the following simplified general reaction:

$$2H^+ + 2e^- \rightarrow H_2 \qquad (8.1)$$

where e^- represents reducing equivalents, usually from a suitable electron donor protein. They are highly efficient catalysts. They catalyze the oxidation and production of hydrogen at rates comparable to those with metals such as platinum, and,

Sustainable Hydrogen Economy

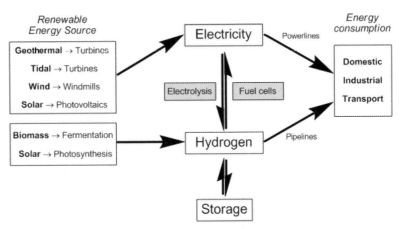

Figure 8.1 Outline of renewable energy resources involving hydrogen. Shaded boxes indicate processes that currently require precious metals such as platinum and could be substituted by hydrogenases.

moreover, with a low overpotential [8, 9]. Where necessary, hydrogenases can be resistant to inhibition by oxygen and carbon monoxide [10].

The hydrogenases, combined with photosynthesis, offer the tantalizing prospect of carbon-neutral H_2 production from renewable resources and conversion to electricity [11]. It has been known for many years that some green algae can produce hydrogen and oxygen by photolysis of water using only sunlight [12], the ultimate clean energy technology. This has spurred the study of the chemistry, mechanism and applications of hydrogenases. Over the past 40 years, a great deal has been learned about their structure and chemistry, how they are made and how they are used in biological processes. Several different lines of approach have been made for the application of hydrogenases for hydrogen energy:

- The harnessing of hydrogen-producing microorganisms to produce hydrogen. This is somewhat equivalent to the *domestication*, selective breeding for human exploitation, a process which for plants and animals took place over thousands of years. There are two major approaches: by anaerobic fermentation of organic wastes and as a byproduct of oxygenic or anaerobic photosynthesis.

- The design of biologically inspired catalysts or *biomimicry*, based on the chemical structure and mechanism of hydrogenases.

- The application of hydrogenases as catalysts in the reversible interconversion of H_2 and electricity. There has been recent success in connecting or 'wiring' hydrogenases to carbon electrodes as a possible substitute for precious metal electrodes. This application will be a major topic of this chapter.

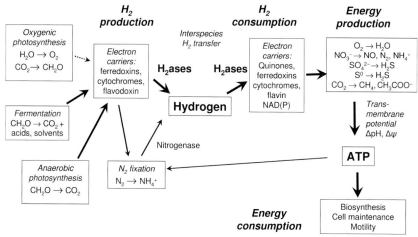

Figure 8.2 Hydrogen metabolism in the biosphere. CH_2O represents fermentable substrates from biomass, such as carbohydrates.

8.1.1
Biological Hydrogen Energy Metabolism

The outline scheme of hydrogen metabolism in the biosphere is shown in Figure 8.2. Eukaryotes that can produce H_2 include certain algae, protozoa and fungi, but not higher animals or plants. Unlike the hydrogen energy technologies outlined in Figure 8.1, in biological systems hydrogen gas is not allowed to accumulate, since it is usually assimilated, either by the organism that produces it or by other microorganisms, a process known as interspecies hydrogen transfer [13]. H_2-producing microorganisms are not restricted to normal temperatures or pH; nor do they need pure water. Hyperthermophiles can grow at temperatures up to $100\,^{\circ}C$ and their hydrogenases operate optimally at these temperatures. Species can be found that grow in strongly saline environments and over a pH range from strongly acidic to alkaline (the internal pH of the cell remaining near 7, however). They can also be tolerant to heavy metals.

Hydrogen is produced in anaerobic fermentations, in organisms such as *Clostridium* species and *Escherichia coli* [14]. When there is an excess of organic reductant, dissolved H_2 may be released as a metabolic byproduct.

In biological systems, H_2 is produced and used locally; the transport distance is often no further than the micrometre distance between neighboring cells. In biological systems H_2 occurs not as a gas, but in solution and at atmospheric pressure the maximum concentration is only 2 mM. If reducing equivalents are stored, it is in the form of more stable metabolites such as polysaccharides or lipids, which are readily produced and interconverted by enzymic systems. Poly-3-hydroxybutyrate, another storage compound, is interesting as a potentially valuable, biodegradable material ('bio-plastic'), which is produced as an alternative to H_2 [15].

Nitrogenases are another source of biogenic hydrogen. They are microbial enzymes used in nitrogen fixation, that catalyze the six-electron reduction of nitrogen to ammonia, using adenosine triphosphate, according to the reaction

$$N_2 + 8e^- + ATP \rightarrow 2NH_4^+ + H_2 + 16ADP + 16phosphate \tag{8.2}$$

where e^- represents an electron donated by either ferredoxin or flavodoxin. The production of at least one H_2 appears to be an obligatory part of the reaction, but nitrogenases also produce further H_2 depending on the conditions, even when N_2 is excluded.

In contrast to nitrogenase, hydrogenases operate almost at thermodynamic equilibrium, requiring only a sufficiently low redox potential to convert protons to H_2. In hydrogen-oxidizing bacteria such as *Ralstonia eutropha* they are able to oxidize H_2, even at low concentrations, in air.

H_2 produced by anaerobic fermentation is reoxidized by autotrophic organisms. The oxidants are compounds such as O_2, nitrate, sulfate and CO_2, which are reduced to H_2O, nitrogen oxides or N_2, H_2S or CH4, respectively (Figure 8.2). In each case a hydrogenase is required to react with H_2, according to Equation 8.2, producing H^+ and e^-. The reducing equivalents, e^-, are oxidized in membrane-bound protein complexes, generating transmembrane gradients, which are then dissipated by the F0/F1 ATPase to generate ATP [16].

8.2
Chemistry of Hydrogenase Catalytic Sites

H_2 is inherently an extremely stable molecule [17]. Despite the low redox potentials that are generated in the metabolic pathways of photosynthetic and fermentative bacteria, only hydrogenases, nitrogenase and a few enzymes that have some characteristics in common with hydrogenases (notably, the nickel-iron-sulfur protein carbon monoxide dehydrogenase and the complex iron-sulfur protein pyruvate: ferredoxin reductase [18]) have been reported to reduce protons to H_2. Of these, the hydrogenases are by far the most efficient [8]. The biological role, properties and applications of hydrogenases have been described in an edited volume [19] and recently in a thematic issue of *Chemical Reviews* on hydrogen [17, 20–24].

As can be surmised from Figure 8.2, hydrogenases in different organisms are able to transfer electrons to and from many different substrates, depending on the conditions of growth. An organism may contain several different hydrogenases, which are expressed under different conditions and which have different specificities for electron donors and acceptors. The reactions with acceptors and donors are all thermodynamically reversible reactions, but some hydrogenases catalyze the reaction much more rapidly in one direction than others, reflecting their cellular function. Hence there are uptake, production and bidirectional hydrogenases. Many hydrogenases have a modular construction, with domains that resemble electron-transfer proteins from other redox enzymes [25, 26], resulting in electron transfer chains for different electron donors and acceptors (Figure 8.3).

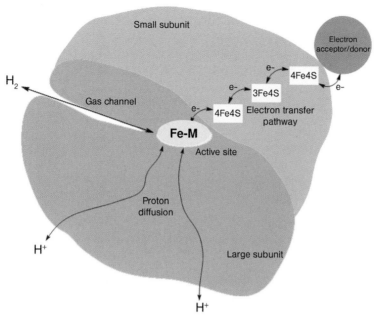

Figure 8.3 Hydrogenase general working scheme, where M is Ni for the [NiFe] hydrogenase or Fe for the [FeFe] hydrogenase.

Hydrogenases have evolved for efficiency in hydrogen production or consumption and a number of common features have been selected. There are three types, based on the organometallic complexes in hydrogenases. All of them contain iron with CO ligands, almost all contain iron–sulfur clusters and some of the most efficient enzymes also contain nickel. The type of active site that is employed in any biological environment may be partly a consequence of the availability of nickel as a trace metal in the environment and partly a consequence of evolution. Nickel–iron-containing hydrogenases are found in the Bacteria and Archaea, whereas iron-containing hydrogenases are found in Bacteria and Eukaryotes such as green algae, fungi and protozoa.

8.2.1
NiFe

The active sites of NiFe-hydrogenases comprise a dinuclear center, with an Ni ion with cysteine sulfur coordination, bridged to a low-spin Fe(II) which has CO and CN^- ligands (Scheme 8.1a). This structure was revealed by X-ray crystallography of the hydrogenase of *D. gigas* [27, 28] and is representative of many nickel-containing hydrogenases. The dinuclear center is stable in different redox states, some of which are paramagnetic (Scheme 8.2). The oxidized states of the hydrogenase contain an interesting bridging ligand, which in the first crystallographic structure was somewhat disordered [27], but which is now believed to represent an oxygen-derived species [29]. Current mechanisms favor the view that this bridging position is the location for a hydride species in the catalytic cycle [23, 30].

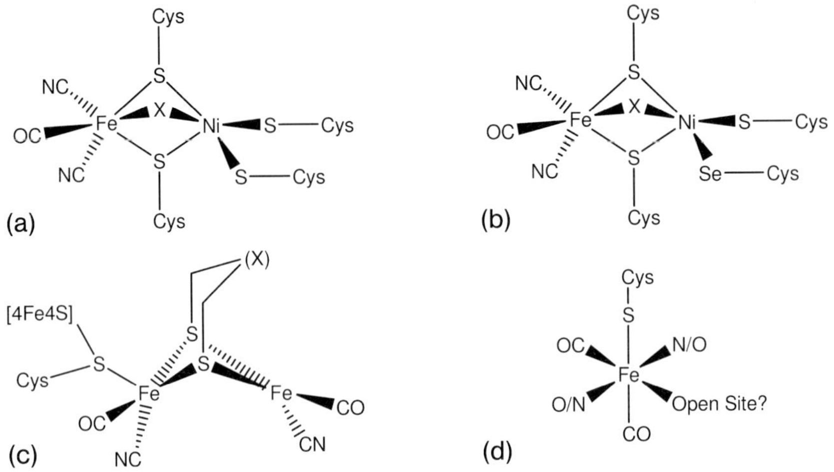

Scheme 8.1 Schematic representation of the different hydrogenase active sites. (a) NiFe-hydrogenase; the active site is coordinated to the enzyme through two bridging and two terminal cysteines. The X ligand can be O for the oxidized species, CO for the inhibited forms or H in the catalytic cycle. (b) NiFeSe- hydrogenase has selenocysteine instead of the equivalent cysteine to Cys530 of *D. gigas* NiFe-hydrogenase. (c) H cluster from the FeFe-hydrogenases. X is still unknown, but N or C is suspected. (d) Predicted structure of the active site iron of the iron-sulfur cluster-free hydrogenase (Hmd).

Most NiFe-hydrogenases lose activity in the presence of O_2. In the case of *D. gigas* hydrogenase and similar hydrogenases, two paramagnetic oxidized forms are produced, which contain nickel in the formal oxidation state Ni(III). These were named the Ni-A nor Ni-B states, after their EPR signals. Both are inactive towards H_2, but they can be reduced to EPR-silent states, Ni-SU and Ni-SI$_1$, respectively, which are still unreactive with H_2, but which can then recover their activity. The Ni-A state is also called the *unready* state, since it required prolonged reduction to become activated, whereas Ni-B is a *ready* state, which can be activated rapidly by reducing agents. On reduction a third paramagnetic state, Ni-C, is formed, which is considered to be an intermediate in the reaction cycle. A likely assignment of these states is as follows: Ni-A and Ni-SU are μ-peroxo, Ni-B and Ni-SI$_1$ are μ-hydroxo and Ni-C and Ni-R$_1$ are μ-hydrido species. The paramagnetic states, indicated with an asterisk, are Ni(III), whereas the EPR-silent states are Ni(II). The bridging position in the Ni-SI$_1$ state can bind H_2, CO or other gaseous ligands.

The oxygen-insensitive soluble hydrogenases, such as the NAD-reducing enzyme from *R. eutropha*, are more complex proteins. Although their sequences indicate the presence of typical NiFe centers, their NiFe center does not normally show nickel EPR signals in any state. This appears to reflect additional features to avoid inhibition by O_2, as will be discussed later.

Some other proteins with hydrogenase-like NiFe active sites function in the sensing of hydrogen concentrations and regulation of hydrogenase expression. These 'sensor hydrogenases' have very limited hydrogenase activity with electron acceptors.

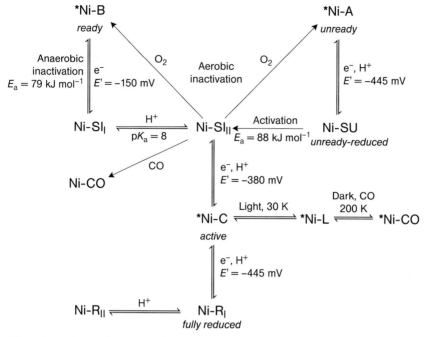

Scheme 8.2 Scheme of reactivity and redox states of the active sites of standard NiFe-hydrogenases. The paramagnetic EPR-active states are marked with an asterisk. The formal redox potentials, E_a, correspond to those measured by FTIR-spectroelectrochemistry of *D. gigas* hydrogenase [169]. The catalytic cycle of the enzyme is proposed to involve cycling between the states Ni-SI, Ni-C and NiR, then back to Ni-SI with the release of H_2. States Ni-A and Ni-B, labeled with an asterisk, are paramagnetic, oxidized forms which become active on reduction (see text). Enzyme in the Ni-C state is light sensitive at cryogenic temperatures, becoming converted to another state, Ni-L. Other states are produced by reversible inhibition with CO. For more details, see [22].

8.2.2
NiFeSe

A subclass of nickel–iron hydrogenases, NiFeSe contains selenium in the form of a selenocysteine residue replacing one of the terminal cysteine ligands (Scheme 8.1b) [31–33]. In the hydrogenase from *Desulfomicrobium baculatum* the [3Fe–4S] cluster found in NiFe hydrogenases is replaced by [4Fe–4S] and a magnesium ion in the large subunit is replaced by Fe(II) [33]. These hydrogenases have high catalytic activity. They do not form the Ni-A (peroxo) state, although they do have an Ni–C oxidation state.

8.2.3
FeFe

FeFe-hydrogenases contain iron, but not nickel. They tend to be used in fermentative metabolism in which H_2 is produced [34, 35] and to be more sensitive to O_2

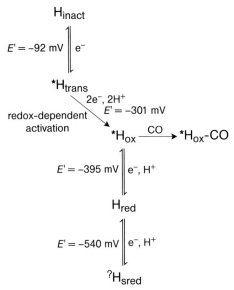

Scheme 8.3 Scheme of reactivity and different redox states of the active site of FeFe-hydrogenases. The EPR-active states are marked with an asterisk. Formal redox potentials (at pH 8.0) measured by FTIR spectroelectrochemistry of *D. desulfuricans* FeFe-hydrogenase [170].

than the NiFe-hydrogenases. The active site responsible for the activation of H_2 in FeFe-hydrogenases, known as the H cluster, consists of a dinuclear iron cluster bonded to the protein backbone through a cysteine sulfur, which bridges to a [4Fe–4S] cluster (Scheme 8.1c). The two iron atoms have CO and CN ligands and are bridged by a unique dithiolate linker. This ligand has never been isolated, but appears from the crystal structure to be either 2-azapropane-1,3-dithiol or propane-1,3-dithiol [33, 36]; the former is an attractive possibility because the nitrogen could act as a base to accept protons during the reaction. The protein, in addition to stabilizing the H cluster structure, maintains the center in an 'entatic state' which facilitates the production of H_2. As with the NiFe-hydrogenases, the protein facilitates the transfer of electrons, H^+ and H_2 and the dimeric FeFe site is stable in several different oxidation states, some of which are paramagnetic (Scheme 8.3).

8.2.4
Fe (non-Fe–S) Hydrogenase (Hmd)

An interesting, and so far unique, hydrogenase is found in methanogenic Archaea, which catalyzes a direct hydrogenation of the substrate using H_2. It does not contain an intermediate electron-transfer pathway of iron–sulfur clusters. This highly evolved mechanism uses an active site that contains iron with carbonyl ligands, but is completely different from the hydrogenases discussed above. In the earlier literature it is described as the 'metal-free' hydrogenase, because the iron present in active preparations was very low. It was not appreciated at first that the activity was due to a small proportion of the active enzyme, which has a very high specific activity; moreover, the enzyme is extremely light sensitive. It is now known as Hmd (H_2-forming methylenetetrahydromethanopterin dehydrogenase) or the iron–sulfur cluster-free hydrogenase. The crystal structures of the Hmd apoproteins from *Methanocaldococcus jannaschii* and *Methanopyrus kandleri* have been reported [37]. Recently the structure of the active protein from *M. jannaschi* has been published [171], which confirms the expected features of the catalytic cycle. As for the catalytic center, Mössbauer spectra show that the iron is low-spin Fe(II) [38] and FTIR spectra indicate the presence of two carbonyl ligands per iron [39]. Upon photolysis, the enzyme releases iron, CO and a unique guanine nucleotide containing a pyridone [40, 41]. The N/O ligands illustrated in Scheme 8.1d were indicated by X-ray absorption spectroscopy [41] and may originate from the pyridone [42]. A possible open site for H_2 binding is modeled at the position *trans* to a CO ligand, considering that the ligand would be expected to increase the acidity of bound H_2.

8.2.5
Biosynthesis of the Active Sites

The assembly of the active sites of the hydrogenases requires complex sequences of reactions, facilitated by specific proteins known variously as chaperones, metallochaperones and scaffold proteins. The genes for these assembly proteins are often found in an operon together with the structural genes for the hydrogenase proteins [20, 43]. In *E. coli*, the synthesis of the complete NiFe center involves a minimum of seven maturation enzymes plus carbamoyl phosphate, GTP and ATP [44]. Carbamoyl phosphate is now known to be the precursor of the cyanide group, but the origin of the carbonyl is still unknown [45]. Starting with the hydrogenase apo-protein, the iron atom of the dinuclear cluster is added first by an unknown mechanism. Specific enzymes are responsible for transport of nickel into the cell and its insertion into the partially formed hydrogenase metallocluster [46].

The biosynthetic pathway for the assembly of the H clusters in the FeFe-hydrogenases is not completely understood, but two novel radical *S*-adenosylmethionine-dependent proteins have been found to be required for the assembly of an active Fe-hydrogenase [47]. By analogy with the biosynthesis of biotin or lipoamide, these enzymes may be responsible for inserting the sulfur in the azapropane-1,3-dithiol ligand.

8.3
Experimental Approaches

Experimentally, hydrogenase activity can be measured by artificial electron mediators such as viologen dyes or the surface of an electrode. Hydrogenases also catalyze reactions with H_2 which involve no net electron transfer, such as the isotopic exchange of $^1H_2$ with $^2H_2$ or ^2H_2O, forming $^2H_2$ and $^1H^2H$ [48].

8.3.1
EPR and Related Methods

EPR spectroscopy provided the first direct evidence that hydrogenases could contain redox-active nickel. Early studies of hydrogenases from sulfate-reducing and methanogenic bacteria revealed three types of EPR spectra, which were identified as due to nickel by substitution with ^{61}Ni [49]. As mentioned above, two signals, designated Ni-A and Ni-B, were observed in the oxidized state and one in the reduced state designated Ni-C [50]. The facile reduction from Ni(III) to Ni(II) (Scheme 8.2) showed that the redox chemistry of the center takes place at the nickel. The iron atom in the NiFe center appears to bear little unpaired electron density, as seen by the very small hyperfine couplings to ^{57}Fe shown by ENDOR of enzyme grown in ^{57}Fe medium [51]. Mediator titrations showed that the redox potentials of the nickel are pH dependent, consistent with the involvement of protons in the reduction of the center [52–54].

The paramagnetic states of the NiFe-hydrogenases have proved to be a valuable tool for observing hydrogen bound to hydrogenase, which cannot be seen by X-ray diffraction or FTIR spectroscopy. Hyperfine interactions with protons have been studied by pulsed EPR, ENDOR and related techniques [23]. Geometric information was obtained by exploiting the orientation selectivity of the EPR spectrum and, more precisely, by studies of single crystals. The results are consistent with the model for the Ni-C oxidation state, in which the center has Ni(III) with a bridging hydride to the iron at the position indicated by X in Scheme 8.1a [55].

Further work with DFT calculations has helped to define the likely intermediates in catalysis. A combination of techniques is narrowing the search for the position where H_2 binds to hydrogenases. Orientation-selective ENDOR spectroscopy, combined with single-crystal EPR and density-functional calculations, has led to the conclusion that the Ni-C state has a bridging hydride ligand [23, 29, 56].

8.3.2
FTIR Spectroscopy

The identity of the unusual diatomic ligands to the iron was established by FTIR spectroscopy of *Allochromatium vinosum* NiFe-hydrogenase. Unusual infrared absorption bands were shown to shift in response to changes in the oxidation state of the nickel [57]. The effect of these ligands, which are very unusual in biology, is to maintain the low-spin state of the ferrous iron ion of the dinuclear center. FTIR can observe these ligands in all oxidation states of the dinuclear center, in contrast to EPR, which only observes paramagnetic states (Scheme 8.2). The redox titration of the Ni-A–Ni-SU couple, monitored by FTIR, has also been reported for the *D. fructosovorans*, *A. vinosum* and *D. vulgaris* NiFe-hydrogenases [58–60]. The pH dependences of the formal measured redox potentials for these standard NiFe-hydrogenases are indicative of a one-electron/one-proton step, in agreement with the small shift in the frequencies of the CO and CN$^-$ ligands, which indicates that there is compensation of the charge density at the active site.

8.3.3
Protein Film Voltammetry (PFV)

In the hands of Armstrong's group, protein film voltammetry (PFV) has provided an incisive method not only to measure redox potentials of the metal centers of hydrogenases, but also to observe the catalytic activity of the enzymes in all their forms and to follow the deactivation/activation processes in real time [61, 62]. The hydrogenase is adsorbed as a film on the surface of an edge-cut graphite rotating electrode, which allows facile electron exchange. The electrode is mounted in an anaerobic chamber into which various gases may be introduced. The rate of diffusion of H$_2$ molecules to the enzyme is controlled by the rotation rate of the electrode. The electric current is a measure of catalytic activity and the driving voltage can be altered at will. Analysis of the non-turnover cyclic voltammograms of the CO-inhibited hydrogenase allows the determination of the redox potentials of the iron-sulfur clusters and the electrode coverage [63]. Rapid changes in applied voltage can be used to measure the instantaneous activity of the enzyme, while chronoamperometry, in which a current is measured during the steady application of a voltage, can be used to follow activation and inactivation [63]. This strategy allows a total control of the redox state of the enzyme by modulating the potential of the electrode [62]. By using cyclic voltammetry and potential step techniques and by controlling experimental conditions as pH, H$_2$ partial pressure and scan rate, the mechanism of the interconversion from the ready to the active states was defined as a two-step mechanism involving a chemical step and an electrochemical step [64]. The exposure of the enzyme to O$_2$ at different potentials and the further analysis of the reductive reactivation showed that the unready form was mainly formed, hence it should contain a product of the partial reduction of O$_2$, probably a peroxo species [65]. The reactivation of this species by reducing agents is so slow *in vitro* that it could pose a problem to the organism. Lamle *et al.* showed that CO and H$_2$ acted not only to drive the irreversible activation process, but also to displace the peroxo or hydroxo bridging species and thus facilitate reactivation of the unready species [66].

Using PFV, the ability of several hydrogenases to tolerate exposure to O$_2$ was tested, showing a different behavior. Whereas the *D. desulfuricans* FeFe-hydrogenase was irreversibly damaged by the O$_2$, the *R. eutropha* membrane-bound NiFe-hydrogenase showed reversible inactivation, followed by fast reactivation. *A. vinosum* and *D. gigas* [NiFe] hydrogenases showed a different reactivity with O$_2$ depending on the reaction conditions, but it was clear that the reversibility of *A. vinosum* was higher than that of the *D. gigas* enzyme [67]. The PFV methodology has also been used by Leger and co-workers for studying the kinetics of wild-type [68] and mutant samples of *D. fructosovorans* NiFe hydrogenase [69]. The electrochemical technique allowed studies of the effects of mutations near the distal iron-sulfur cluster of the hydrogenase on intra- and intermolecular electron transfer [69].

8.4
Catalytic Mechanisms of Hydrogenases

Examination of the structures of NiFe-hydrogenases suggests that the components of the enzyme reaction (Equation 8.1), namely H_2, H^+ and electrons, are brought to the dinuclear center by different paths through the protein (Figure 8.3). It is possible to divide the hydrogenase catalytic cycle into six steps, which are, in the direction of H_2 oxidation:

1. Diffusion of hydrogen molecules from the surface of the protein to the active site. Although there are several ways in which H_2 can diffuse through the protein [70], a particular series of channels were revealed by structural determination of an NiFe-hydrogenase saturated with xenon, which becomes localized in the channels [71].

2. Heterolytic splitting of hydrogen molecule after binding to the bimetallic active site. This may be summarized as

$$M<>Fe+ H_2 \rightarrow M<>FeH+ H^+ \tag{8.3}$$

$$M<>FeH \rightarrow M<>Fe+ H^+ + 2e^- \tag{8.4}$$

where "<>" represents a double-or triple-bridged dinculear center.

It is noticeable that the NiFe-hydrogenases, the FeFe-hydrogenases and Hmd each contain a low-spin Fe(II) with CO and, usually, CN^- ligands at the active site (Scheme 8.1). As the three types of hydrogenases are not phylogenetically related, the strict conservation indicates that these elements are the key to H_2 conversion to hydride.

3. Oxidation of the hydride to H^+. This step requires an acceptor that can take a transient two-electron reduced state, then donate electrons one at a time to the electron-transfer chain. The reduction step for the NiFe-hydrogenases might be $Ni^{III} \rightarrow Ni^I$ or for the FeFe-hydrogenases $Fe^{II}-[4Fe-4S]^{2+} \rightarrow Fe^I-[4Fe-4S]^+$ or even $[4Fe-4S]^{2+} \rightarrow [4Fe-4S]^0$ [22, 72].

4. H^+ transfer from the active site to the water solvent. In the FeFe-hydrogenases, the azapropane-1,3-dithiol bridging ligand has been modeled with its NH proton close to where H_2 is presumed to bind, so it could act as an H^+ acceptor [17]. In the NiFe-hydrogenases, cysteine ligand of the Ni is generally considered as an H^+ acceptor [17, 21, 22]. A conserved histidine residue is in a position to promote a hydrogen bond to one of the NiFe bridging cysteines, which DFT calculations have predicted would favor the protonation of the N^ε of the histidine [73]. Possible pathways for hydrons leading from the active site have been proposed, including a glutamate residue [74] and numerous internal water molecules in the protein structure, including ligands to a magnesium ion [75].

5. Electron transfer from the active site to the distal redox cluster. Most hydrogenases comprise a chain of iron-sulfur clusters, separated by distances of 1.2-1.4 nm, from the catalytic center to a point on the surface where an electron-transfer protein could bind (Figure 8.3). The distances between the clusters seem to be

critical for efficient electron transfer [76], although the redox potentials do not. In *D. gigas* hydrogenase, the clusters are, in sequence, a [4Fe−4S], a [3Fe−4S] and a [4Fe−4S] cluster. The central cluster has a much less negative reduction potential than those of the others and the H^+/H_2 potential, but this does not appear to affect the rate of reaction significantly [77]. An important factor is that the transfer of electrons is accompanied by hydrons, to preserve charge neutrality. In hydrogenases, this requires that there is movement of protons to compensate for transfer of electron density. A corollary of this is that if there is no possibility of charge compensation, electron transport is blocked. Possibly for this reason, the distal [4Fe−4S] cluster has an essential histidine ligand and substitution of this by any other ligand decreases the catalytic rate [69]. The FeFe-hydrogenases from eukaryotic algae such as *Scenedesmus obliquus* are small monomeric proteins, lacking the chain of iron-sulfur clusters [78]. This shows that the H cluster can be self-sufficient for hydrogenase activity. These hydrogenases are believed to exchange electrons directly with the iron-sulfur clusters of ferredoxins.

6. Intermolecular electron transfer from the distal cluster to the redox partner. This part of the hydrogenase structure shows the greatest variability, due to the wide variety of electron carriers used. For *D. gigas* hydrogenase, the acceptor is a positively charged cytochrome c_3. A 'crown' of glutamate residues around the distal [4Fe−4S] cluster [27, 79] attracts and orients cytochrome c3 through the positive charges of lysine residues around the hemes [80–82]. Hydrogenases that reduce membrane-bound quinones such as menaquinone generally do so through an additional heme-containing hydrophobic membrane anchor subunit. As will be discussed later, the membrane-bound hydrogenases are particularly suited for binding to the hydrophobic surfaces of carbon electrodes.

8.5
Progress So Far with Biological Hydrogen Production Systems

8.5.1
Fermentation

Anaerobic bacteria such as *E. coli* and *Clostridium* species produce H_2 as a product of fermentation of organic materials. Hydrogen production serves to maintain the redox balance of the metabolic cycles in their cells. These fermentations may be applied to generate H_2 as a fuel from many types of biomass, including the waste products of agriculture and food production, which have a high biological oxygen demand [83–85]. Usually the fermentation is associated with the production of other compounds such as organic acids and solvents, which can also be useful. The nature of the feedstock means that it is not practicable to use pure cultures of bacteria in order to select for H_2 production. The production of H_2 has principally been optimized by the selection of growth conditions such as pH, temperature, nutrients and dilution rate and by bioreactor design [86]. For example, the growth of methanogens, which would produce methane from H_2 and CO_2, is prevented by operating outside the pH range in which methane production is energetically favorable [83].

8.5.2
Oxygenic Photosynthesis

In principle, the cleanest hydrogen biotechnology is to use photosynthesis to produce reducing equivalents for hydrogen production and release O_2. This requires the isolation and application of metabolic systems from algae or cyanobacteria. [87–90].

Photosynthesis involves the creation of oxidizing and reducing species by photolysis of chlorophylls, coupled to complex arrangements of electron carriers [91]. In oxygenic photosynthesis, Photosystem II generates O_2 and Photosystem I generates compounds with negative reduction potentials that could potentially generate H_2. Light is a destructive factor in most biological systems and in particular the strong oxidizing potentials generated in Photosystem II mean that the D1 protein has a short half-life. This problem, known as photoinhibition, is overcome by continuous rapid synthesis and reassembly of the protein complex [92]. Another problem when nitrogenase or an FeFe-hydrogenase is the source of H_2 is that these enzymes are highly sensitive to O_2 and must be kept separate from oxygen-evolving systems. Therefore, the common practice in biotechnology of using cells as 'bags of enzymes' (see for example [93]) will not work for photosynthetic water splitting. It requires the participation of the full machinery for protein synthesis and therefore a fully-functioning cell. A second problem for the use of photosynthetic organisms such as algae is that they have evolved to replicate themselves as efficiently as possible, conserving their resources such as hydrogenase. Existing organisms are certainly not optimized as hydrogen factories for us. Much genetic engineering of the cellular regulatory systems would be required.

Benemann *et al.* [11] conducted an early demonstration of the feasibility of this approach, using a combination of oxygenic photosynthesis with hydrogenase to produce detectable amounts of hydrogen and oxygen. For various reasons the yields of hydrogen were small. The systems used (spinach chloroplasts and *Clostridium kluyveri* hydrogenase) were not stable; the O_2 produced was inhibitory to the hydrogenase and would reoxidize the reduced ferredoxin, thereby short-circuiting the system. Since then, steps have been taken towards improving the efficiency and longevity of the system, such as the use of thermophilic algae and cyanobacteria, in which the proteins are more robust [94], and embedding the components on surfaces or in gel matrices [95]. These photochemical systems produce both H_2 and O_2 in a mixture and, moreover, in dilute aqueous solution. The mixture of gases is undesirable from the point of view of safety. It cannot be used directly in conventional fuel cells, because the platinum electrodes would catalyze the conversion of the mixture back to H_2O. In order to obtain H_2 as a fuel, it would have to be separated, for example by suitable permeable membranes, with inevitable loss of efficiency.

A further step is to avoid the use of cells or organelles altogether and to generate H_2 and O_2 from genetically engineered molecular systems. A chimeric protein complex has recently been constructed, comprising the PsaE subunit of Photosystem I from the thermophilic cyanobacterium *Thermosynechococcus elongatus* and the membrane-bound NiFe-hydrogenase from *R. eutropha* [96]. The resulting hydrogenase-PsaE fusion protein associated spontaneously with ferredoxin and PsaE-free Photosystem I and was capable of light-driven electron transfer to H^+, forming H_2.

Alternatively, the processes of photochemical O_2 production and anaerobic H_2 production can be made to operate alternately over time, for example by switching the growth conditions. A novel intervention by Melis *et al.* [97, 98] in the eukaryotic alga *Chlamydomonas reinhardtii* was to suppress O_2 formation by growing the algae under sulfur-limited conditions. This is because sulfur-containing amino acids are required for the constant renewal of the D1 subunit of the Photosystem II subunits, to repair damage done by light [92]. The cells then switch to fermentative metabolism and consume their reserves of starch, producing H_2. Restoration of sulfate to the growth medium then allows O_2 production to resume and starch reserves to be recovered [99]. In line with these results, progress has been reported on photosynthetic algae, genetically modified to avoid O_2 evolution under light, which allows them to reach anaerobic conditions compatible with hydrogen evolution [100].

The hydrogenases of *C. reinhardtii* are of the FeFe type [101] and are sensitive to O_2. Moreover, H_2 inhibits the endogenous hydrogenases of these organisms and suppresses the biosynthesis of hydrogenase. By genetic engineering it should be possible to substitute the FeFe hydrogenases by ones that are less sensitive to O_2, such as those from *R. eutropha*, and to modify the regulatory mechanisms for hydrogenase biosynthesis.

Cyanobacteria (prokaryotes) have the capacity to produce both H_2 and O_2 under appropriate conditions. The conditions under which hydrogen production may be optimized have been the subject of much investigation [94]. H_2 production by nitrogenases is a major source of hydrogen in cyanobacterial cultures. Nitrogenase is not as efficient a catalyst as hydrogenase and, moreover, it is highly sensitive to O_2. In filamentous cyanobacteria, nitrogenase is spatially separated from the oxygenic photosynthetic reactions in specialized cells known as heterocysts. In N_2-fixing organisms, the H_2 produced is often recovered by specific membrane-bound hydrogenases, which oxidize it to recover energy in the form of ATP.

8.5.3
Anaerobic Photosynthesis

Higher yields of H_2 from organic wastes can in principle be obtained by the use of anaerobic photosynthetic bacteria, which use organic or inorganic electron donors instead of H_2O to boost the production of H_2 [102, 103] (Figure 8.2). Unlike oxygenic photosynthesis, this type of photosynthesis can use both near-infrared and visible wavelengths and is less sensitive to the destructive effects of light on the photosystems. Photobioreactors, using anaerobic photosynthetic bacteria such as *Rhodobacter* species, can use organic wastes as reductant. H_2 has been produced on a pilot scale. In practice, most of the H_2 appears to derive from nitrogenase rather than hydrogenase [104–106].

8.5.4
Emulation: Hydrogenase Model Compounds

The active site of NiFe- and FeFe-hydrogenases is essentially an organometallic compound, a dinuclear center with CO and/or CN ligands. In the case of the FeFe-hydrogenase, the bimetallic center is bound to the rest of the protein only through a single cysteine residue. It sits in a cavity in the protein, analogous to the Fe-Mo cofactor in nitrogenase [107]. Nevertheless the protein environment is one of the most important features of the hydrogenases, which minimizes the peaks and troughs in free energy during the course of the reaction of the catalytic centers with H_2 [17]. The protein can maintain a vacant site at which H_2 or hydride can bind, while protecting it from inhibitory gas molecules. The position of the CO and CN ligands in the centers also appears to be critical. In the proteins, the ligands are located in pockets in the protein, with hydrogen bond partners to the CN ligands. This would assist the correct assembly of the centers, but would probably also assist the activation of H_2/hydride at the vacant site.

Inorganic chemists have attempted to synthesize compounds resembling structural and catalytic features of the hydrogenase active site and examined their activity in O_2 production. Pickett and co-workers synthesized a close analogue of the iron-only hydrogenase H cluster, linking a diiron complex to a [4Fe−4S] cluster (Scheme 8.4a) [108]. The subsite model was capable of catalyzing proton reduction at a potential of -1.13 V (versus Ag/AgCl). Liu and Darensbourg explored derivatives of the classic symmetrical Fe^IFe^I organometallic compound $(\mu\text{-pdt})[Fe(CO)_3]_2$ (pdt = propanedithiolate) as models of the iron dimer of the H cluster in FeFe hydrogenases. In the X-ray structures of the FeFe-hydrogenases, the ligands to one iron atom of the H cluster are rotated so that a CO ligand bridges to other atom of the pair. In a recent study, using the unique properties of an *N*-heterocyclic carbene, they synthesized an asymmetric mixed-valent $Fe^{II}Fe^I$ compound with a rotated state, which reproduced accurately the EPR signals and the IR spectra of the $H_{as\ isolated}$ and H_{ox} cluster of the native enzyme (Scheme 8.4b) [109]. In the same direction, Rauchfuss and co-workers synthesized another mixed-valence complex with spectroscopic properties and the reactivity resembling those of the H_{ox} cluster [110]. In this case they used diphenylphosphinovinylidene as a ligand on one iron and a trimethylphosphine on the other (Scheme 8.4c). The rotation of the ligands to one iron atom causes an asymmetry between the two iron atoms, which would facilitate binding of H_2 to one iron atom. DFT calculations indicate that H_2 or CO would be expected to bind to the terminal iron of the H cluster [17].

In the FeFe-hydrogenases, the dithiolate linker is a three-atom bridge of which the middle atom might be a nitrogen, which, as already mentioned, would be a pendant base well positioned to facilitate proton transfer out of the enzyme. DuBois and co-workers explored the chemistry of first-row metals with diphosphine ligands incorporating a pendant base as a proton relay in H_2 activation [111]. A complex was synthesized (Scheme 8.4d) which was a very effective catalyst for H_2 production, displaying high rates and long lifetimes. The turnover number found, $350\ s^{-1}$ at $22\ ^\circ C$, is comparable to those reported for NiFe-hydrogenases ($500\text{-}700\ s^{-1}$ at $30\ ^\circ C$ [112]).

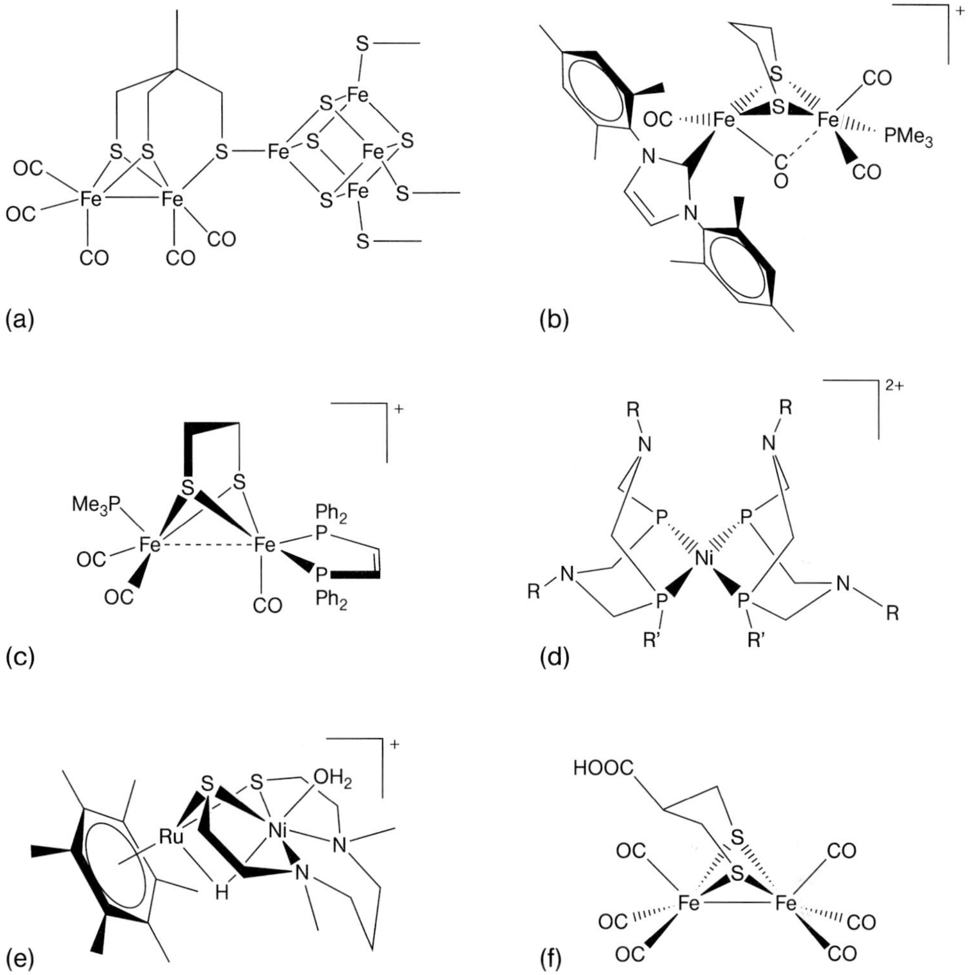

Scheme 8.4 (a) H cluster framework synthesized by Pickett and co-workers [108]. (b) Mixed-valent diiron compound with a rotated state according to Lin and Darensbourg [109]. (c) Unsaturated mixed-valence diiron model of the H_{ox} state synthesized by Rauchfuss and co-workers [110]. (d) Nickel complex with diphosphine ligands incorporating a pendant base [111]. (e) Ni compound with two pendant bases on the phosphines, synthesized by DuBois and co-workers [113]. Depending on the substituents on nitrogen and phosphorus, the compounds were very active for H_2 oxidation or production. (f) Carboxylic acid derivative for attachment to an amine modified electrode synthesized by Darensbourg and co-workers [117].

Addition of H_2 was the limiting step, oxidation of H_2 being fast once it was incorporated to the coordination sphere [113]. The effects of coligands on the pK_a of such complexes was studied, demonstrating that there is considerable electronic communication between the nitrogen atom and the metal center [114].

Another model of the NiFe-hydrogenase was published by Ogo *et al.* Combining two aqueous solutions of Ni(S$_2$N$_2$) and [(C$_6$Me$_6$)Ru(H$_2$O)$_3$]$^{2+}$, the resultant water compound reacted in water with hydrogen to form a hydride and protons (Scheme 8.4e) [115].

In order to investigate the electrocatalytic properties of such complexes, different strategies have been applied to attach them to electrodes. Pickett and co-workers reported the incorporation of iron-only hydrogenase model complexes into functionalized polypyrrole electrodes, obtaining electrocatalytic currents for hydrogen production when a proton source was provided [116]. Electrocatalysis by this compound was measured, but at too high an overpotential for use in a commercial device. Another approach to linking model complexes to electrodes has been proposed by Darensbourg and co-workers, who modified model complexes with carboxylic acids to form amide bonds with amine functionalized electrodes (Scheme 8.4f) [117]. The bond was stable and all the reaction steps were identified. However electrocatalytic proton reduction on the electrodes was not observed due to the irreversible reduction of the iron complex.

A point to bear in mind with these biomimetic systems is that it may be necessary to avoid poisoning of the catalyst by the use of a selective filter membrane to exclude molecules such as O_2, H_2S and CO, as is done in fuel cells [4, 118].

A new approach to building hydrogenase models and modulating the properties of the metal site has been initiated by Jones *et al.*, who reported the incorporation of a diiron compound with an α-helical synthetic peptide [119]. The design of such hydrogenase maquettes offers the prospect of developing robust and cheap water-soluble H_2 production catalysts [119].

8.5.5
Hydrogenases on Electrodes

Yagi's group were pioneers in electro-enzymatic hydrogen generation with hydrogenases isolated from *Desulfovibrio vulgaris* adhering to glassy carbon electrodes [120]. As already discussed, hydrogenases bound to electrodes are outstanding tools for the study of the kinetics, mechanisms of reaction and inactivation and reactivation processes of the enzymes. Other practical applications of hydrogenase electrodes include H_2 sensors [121, 122]. Hydrogenase electrodes have also been used as a source of reducing power in the electro-enzymatic preparation of chiral compounds [123]. Armstrong *et al.* [124] reviewed the factors that allow electron transfer between electrode surfaces and the redox centers in redox proteins or enzymes. For renewable H_2 production (Figure 8.1), several strategies have been employed to obtain a catalytic response from hydrogenase-modified electrodes, to accelerate the electrochemical reduction of hydrons to hydrogen in electrolysis and its reverse reaction, the oxidation of hydrogen in fuel cells [6, 125, 126]:

1. Indirect electron transfer between the redox sites in the protein and the electrode may be mediated by soluble redox carriers with redox potentials close to the H^+/H_2 couple, such as viologens [54, 127]. This approach has been useful in estimating the midpoint potentials of the redox carriers in the protein which can be observed by spectroscopy [128]. Some electron-transfer proteins may also

act as mediators. Bianco and co-workers showed that the tetraheme cytochrome c_3, which has hemes exposed to its surface and which is the natural substrate of *Desulfovibrio desulfuricans* FeFe-hydrogenase, could act as electron mediator in the electrocatalytic hydrogen evolution at pyrolytic graphite electrodes (PGEs) [129–131].

2. Glassy carbon electrodes surface-modified with redox mediators could act as the electron source to soluble *D. vulgaris* hydrogenase for catalytic H_2 evolution [132].

3. Entrapment of hydrogenases in polymeric semiconductive matrices on PGEs was developed by Varfolomeyev and Bachurin [133, 134]. Hydrogenases have been also entrapped in viologen polymers, which acted as electron conductors, generated in the electrode [135, 136], or in an amphiphilic bilayer assembly covering an electrode that also contains a viologen compound [137]. De Lacey *et al.* reported a hydrogenase electrode formed by successive layers of the enzyme and a viologen compound immobilized by avidin–biotin affinity interactions, in which the current density of H_2 oxidation was proportional to the quantity of hydrogenase layers [138]. Another method of co-immobilization of hydrogenase and the redox mediator was based on intercalation of hydrogenase between two layers of a mixture of clay and a viologen polymer deposited on a glassy carbon electrode [121]. A similar clay-based layer-by-layer deposition method has been reported in which the natural redox partner of the hydrogenase, cytochrome c_3, replaced viologen as redox mediator [139]. Redox-mediated electro-enzymic oxidation of H_2 has also been reported for bacterial cells with hydrogenase activity adsorbed on glassy carbon electrodes [140].

4. A mediated response for the hydrogenase can be monitored by immobilizing hydrogenase on a carbon electrode and using methylviologen as a redox mediator. This procedure allowed a qualitative study of the activation/inactivation processes [141]. One step further is building a multi-monolayer hydrogenase electrode based on the avidin-biotin interactions and measuring the mediated current. With this kind of electrode, a quantitative analysis of the hydrogenase mechanism was carried out and the measured kinetics were in good agreement with previously published data [142].

5. Direct, non-mediated, electrocatalysis by hydrogenases covalently immobilized on glassy carbon electrodes: Schlereth *et al.* reported on the activity of *R. eutropha* Z-1 hydrogenase covalently immobilized on a glassy carbon electrode [143]. This complex enzyme has the capacity to reduce NAD^+, affording the possibility of other NAD-dependent reactions of interest to biotechnology. It was able to support the electro-enzymatic reduction of NAD^+ without external promoters or electron mediators. Bergel and co-workers described direct electrocatalysis of NAD^+ reduction with *R. eutropha* H16 hydrogenase adsorbed on platinum electrodes [144, 145]. The product of the electro-enzymatic reduction of NAD^+ was demonstrated to be biologically active [146]. *R. eutropha* hydrogenase has also been entrapped during the electro-polymerization of pyrrole on a Pt electrode and the modified electrode catalyzed the reduction of NAD^+ [147]. A fluidized bed reactor

for the continuous reduction of $NADP^+$ to NADPH with hydrogen has been developed by Greiner *et al.* [148], using hydrogenase I from *Pyrococcus furiosus* directly adsorbed on conducting graphite beds with excellent stability.

6. Promoters, non-redox substances which assist the binding of proteins to surfaces, have been used to bind hydrogenases to electrodes and facilitate their electron exchange. This methodology was introduced by Eddowes and Hill [149] and Veeger and co-workers [150]. Early examples of the alternative strategy to promote the adsorption of hydrogenases consisted of coating poly-L-lysine on either a mercury electrode surface [151] or a PGE [152]. In these modified electrodes, the adsorbed polycationic film bonded the negatively charged molecule of *D. vulgaris* hydrogenase and promoted a direct electron transfer between the electrode and the redox centers of the enzyme, which was not detected in the absence of poly-L-lysine. Under low-overpotential conditions it was shown that the PGE modified with poly-L-lysine and hydrogenase behaved as an H_2 electrode, the electrons rapidly exchanged between the electrode and the active site of the hydrogenase [152]. By comparison, this electrode behaved as the mercury electrode did, but avoided the denaturation of the hydrogenase observed on its adsorption on the mercury drop surface [153]. A similar stabilizing effect on the adsorption of proteins at a PGE was observed by Armstrong *et al.* with polymyxin B, a cyclic decapeptide with several amine groups [154]. Films of *A. vinosum* hydrogenase PGE with polymyxin B showed a higher turnover number for unmediated H_2 oxidation than those measured with soluble organic compounds such as methylene blue or methylviologen as electron acceptors. In these experiments, rapid catalytic hydrogen oxidation, close to mass transfer-controlled rates, were observed [63]. Armstrong and co-workers showed that NiFe-hydrogenase molecules from *A. vinosum* adsorbed on edge-cut PGE, modified with polymyxin B, catalyzed hydrogen oxidation at rates comparable to those with electrodeposited platinum and, more importantly, are much more resistant to CO poisoning than the noble metal [8].

7. Direct electrochemistry, without aid of promoters, has also been reported for NiFe-hydrogenases adsorbed on a carbon electrode [155, 156]. The efficiency of binding of hydrogenases to electrodes, with and without promoters, can be rationalized on the basis of the structural motifs that control electron transfer *in vivo* between the NiFe-hydrogenases such as that from *D. gigas* and cytochrome c_3. An analysis of the electrostatic potential distribution over the surface of this protein showed an asymmetric charge distribution with the most external [4Fe−4S] cluster located in the negative region of the protein surface, that is, at the origin of a strong dipole moment in the enzyme molecule [157]. Using this rationale, *D. gigas* hydrogenase was covalently coupled to PGE modified with a monolayer of 4-aminophenyl groups in an orientation that allowed direct electron transfer to the electrode. As the enzyme was covalently bound to the electrode surface, these electrodes showed exceptionally high operational stability [157]. Hydrogenase electrodes have also been constructed with FeFe-hydrogenases. Hagen and co-workers reported a direct electron exchange of the hydrogenase from *Megasphaera elsdenii*

with glassy carbon electrodes [158]. In this case, polymyxin did not stabilize the response of the electrodes and even attenuated the electrode response. As expected, this FeFe-hydrogenase electrode was more sensitive to CO poisoning that the NiFe-hydrogenase electrodes and their electrochemical response decayed rapidly after CO addition. FeFe hydrogenase from *Desulfovibrio vulgaris* adsorbed on PGE also catalyzed direct H_2 production in the absence of any promoter [159]; however, the catalytic current was lower than those observed with *A. vinosum* NiFe-hydrogenase adsorbed on an electrode that was surface modified with polymyxin B sulfate [63].

8. To gain stability for the development of fuel cells, covalent attachment of the enzyme is needed. De Lacey and co-workers attached *D. gigas* [NiFe] hydrogenase through its glutamate residues to an amine-modified carbon electrode, which increased the electrode stability from hours to weeks [157].

8.5.5.1 Sensitivity and Resistance of Hydrogenases to O_2, CO and Other Inhibitory Gases

O_2, NO, H_2S and CO are strong inhibitors of the active sites of hydrogenases, just as they are poisons for Pt electrodes. Inactivation by O_2 is a serious drawback for the utilization of NiFe- and FeFe-hydrogenases on electrodes [34], although in the case of NiFe-hydrogenases the inhibitory effect can be reversed by a reductive treatment [160]. However organisms that produce these gases metabolically (Figure 8.2) have evolved mechanisms to counteract the inhibitory effects.

In some hydrogenases from H_2-oxidizing bacteria, the active site appears to be modified so as to restrict access to O_2 [161, 162]. The O_2-resistant soluble NiFe-hydrogenase from *R. eutropha* is a complex protein in which there is FTIR spectroscopic evidence for the involvement of an additional cyanide ligand to the nickel [163]. The enzyme from a mutant organism, lacking the *hypX* gene that is required for insertion of the extra CN ligand, is sensitive to O_2 [163]. The *R. eutropha* enzyme gives no Ni-C EPR signal and there is recent evidence that it contains an additional FMN group [164]. It has been suggested that this is located so as to accept hydride directly from the NiFe center, possibly avoiding an O_2-sensitive paramagnetic state [165].

A second oxygen-tolerant NiFe-hydrogenase occurs in *R. eutropha*, which is membrane bound. This hydrogenase maintains a significant amount of H_2 uptake activity in the presence of high levels of O_2 and full activity was recovered upon removal of H_2 [10].

Karyakin *et al.* compared some advantages of hydrogenases in fuel cell electrodes with those of platinum [166]. In agreement with the findings of Armstrong and co-workers [8], they showed that hydrogenases adsorbed on PGE could oxidize hydrogen at rates comparable to those for electrodeposited platinum and with less susceptibility to CO poisoning [6, 8]. Hydrogenases are more selective than platinum for H_2, so the problem of poisoning electrodes by impurities from the fuel gas can be solved using enzymatic electrodes. The enzymes adsorbed on bare or polypyrrole-modified PGEs were tested under different H_2–CO gas mixtures; no CO inhibition was observed up to a CO partial pressure of 0.1%, and even at 1% CO the activity was decreased by just 10%. By contrast, platinum loses 99% of its activity after 10 min

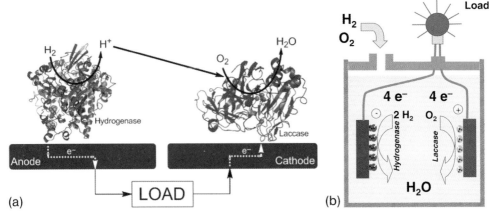

Figure 8.4 (a) Schematic representation of a hydrogenase/ laccase bio-fuel cell. (b) On the anode hydrogenase oxidizes hydrogen to protons. The electrons are used by the laccase on the cathode to reduce oxygen to water. When using a hydrogenase which is not affected by H_2 as *R. metallidurans*, the fuel cell operates without the need for a proton transfer membrane for protecting the hydrogenase from oxygen inactivation [126].

under 0.1% CO. Moreover, the inhibition by CO was reversible and, after exposing the enzyme to pure CO, 100% activity was recovered after flushing the CO with hydrogen, whereas recovery of noble metals poisoned by CO requires special procedures [166]. The results with Na_2S were similar and showed insensitivity of the hydrogenase up to 5 mM sulfide.

An important step was the realization by Vincent *et al.* [126] that if hydrogenases were used that are unreactive with O_2, these effectively contain their own intramolecular barrier, obviating the need for a membrane to separate the O_2-consuming side of a fuel cell from the H_2-consuming side. This made possible the design of an entirely enzymatic fuel cell, using oxidases which are unreactive with H_2. The cell constructed used the membrane-bound hydrogenase of *R. eutropha* at the anode for hydrogen oxidation and the robust copper-containing oxidase laccase from *Trametes versicolor* at the cathode (Figure 8.4a, b). Better performance was shown by *R. metallidurans* hydrogenase, leading to a fuel cell which operated in a non-explosive mixture of air with 4% H_2. The cell was able to generate a usable electric current from 3% H_2 in still air. In the presence of CO it achieved almost 1 V at open-circuit conditions, which represents a noticeable advantage over platinum electrodes [167]. Three cells were connected in series and provided enough power to power a wristwatch for 24 h.

All these devices have been based on enzyme adsorption on the electrodes and surface electrochemistry is a two-dimensional process. Therefore, in order to increase the catalytic rates and produce smaller fuel cells, an increased surface area is required. Multi-walled carbon nanotubes were grown on microscale electrodes. The nanotubes were modified with diazonium salts to cover them with amine functionalities and

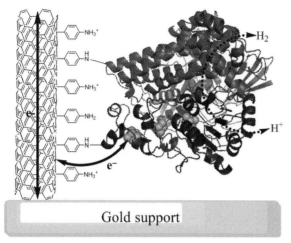

Figure 8.5 Carbon nanotube electrode for *D. gigas* hydrogenase. Carbon nanotubes were grown on a gold microchip. The nanotubes were electrochemically modified with amine functionalities for the orientation and covalent attachment of the hydrogenase [168].

covalently attach the hydrogenase (Figure 8.5). The electrodes prepared in this way had an electroactive area of 130 C cm^{-2}, 43 times higher than the area obtained upon modification by the same method of a polished PGE. The catalytic currents measured were 33 times larger than those at the PGE, meaning that most of the electroactive area was covered by the hydrogenase. This result was similar to the best results reported by other workers for hydrogenase directly adsorbed on carbon electrodes, which almost equaled the performance of Pt-based electrodes. The operational stability of those electrodes was tested under hydrogen electrocatalytic oxidation conditions. The covalent bonding of the enzyme to the electrode was crucial to gain an operational stability of more than 1 month, retaining 90% of initial activity [168].

8.6
Conclusion and Future Directions

We should not underestimate the challenges involved in trying to exploit biological hydrogen production for a major source of energy. Studies are at the stage of proof of principle and clearly a huge effort of research and development would be needed to develop practical renewable systems to replace fossil fuels.

Domestication of Hydrogenases The ideal organism for biological hydrogen production might be one that produces oxygen by photosynthesis and diverts a high proportion of the resulting reducing equivalents to generate hydrogen. Such organisms exist, but the yields of hydrogen are low. In view of the complexity of biosynthesis and regulation of hydrogenases, this would be a major project in cell biology. An analogy may be drawn with the domestication of animals and plants to support an agricultural existence, which took millennia of selective breeding. This process could be greatly accelerated by genetic engineering.

Emulation of Hydrogenases The features of the hydrogenase molecule that allow it to be so efficient and specific have become clearer as a result of recent research. The elucidation of the structures and mechanisms of hydrogenase action has revealed novel organometallic chemistry and novel applications of theoretical chemistry. A major challenge is to create a stable structure with a vacant site with the appropriate geometry for hydrogen binding and activation and controlled access for electrons, hydrons and H_2.

Hydrogenase Electrodes The technology has been demonstrated to work on a small scale. Fuel cells could operate without the membranes, which are a weakness in current designs. Other attractive features are the lack of reliance on precious metals, room temperature operation, the use of non-explosive H_2–air mixtures and resistance to pollutant gases. The challenges include creating greater stability and sufficiently high current densities.

Abbreviations

DFT	density functional theory
ENDOR	electron–nuclear double resonance
EPR	electron paramagnetic resonance
FMN	flavin mononucleotide
FTIR	Fourier transform infrared
Hmd	H_2-forming methylene-H_4MPT dehydrogenase
MPT	methanopterin
NAD	nicotinamide adenine dinucleotide
NADP	nicotinamide adenine dinucleotide phosphate
PFV	protein film voltammetry
PGE	pyrolytic graphite edge electrode

References

1 Navarro, R.M., Pena, M.A. and Fierro, J. L. G. (2007) *Chemical Reviews*, **107**, 3952.

2 Palo, D.R., Dagle, R.A. and Holladay, J.D. (2007) *Chemical Reviews*, **107**, 3992.

3 Esswein, M.J. and Nocera, D.G. (2007) *Chemical Reviews*, **107**, 4022.

4 Borup, R., Meyers, J., Pivovar, B., Kim, Y.S., Mukundan, R., Garland, N., Myers, D., Wilson, M., Garzon, F., Wood, D., Zelenay, P., More, K., Stroh, K., Zawodzinski, T., Boncella, J., McGrath, J.E., Inaba, M., Miyatake, K., Hori, M.,

Ota, K., Ogumi, Z., Miyata, S., Nishikata, A., Siroma, Z., Uchimoto, Y., Yasuda, K., Kimijima, K.I. and Iwashita, N. (2007) *Chemical Reviews*, **107**, 3904.

5 Gordon, R.B., Bertram, M. and Graedel, T.E. (2006) *Proceedings of the National Academy of Sciences of the United States of America*, **103**, 1209.

6 Karyakin, A.A., Morozov, S.V., Karyakina, E.E., Zorin, N.A., Perelygin, V.V. and Cosnier, S. (2005) *Biochemical Society Transactions*, **33**, 73.

7 Markovic, N.M., Lucas, C.A., Grgur, B.N. and Ross, P.N. (1999) *Journal of Physical Chemistry B*, **103**, 9616.

8 Jones, A.K., Sillery, E., Albracht, S. P. J. and Armstrong, F.A. (2002) *Chemical Communications*, 866.

9 Karyakin, A.A., Morozov, S.V., Voronin, O.G., Zorin, N.A., Karyakina, E.E., Fateyev, V.N. and Cosnier, S. (2007) *Angewandte Chemie International Edition*, **46**, 7244.

10 Vincent, K.A., Cracknell, J.A., Lenz, O., Zebger, I., Friedrich, B. and Armstrong, F.A. (2005) *Proceedings of the National Academy of Sciences of the United States of America*, **102**, 16951.

11 Benemann, J., Berenson, K., Kaplan, N.O. and Kamen, M.D. (1973) *Proceedings of the National Academy of Sciences of the United States of America*, **70**, 2317.

12 Gaffron, H. and Rubin, J. (1943) *Journal of General Physiology*, **26**, 219.

13 Belaich, J.P. Bruschi, M. and Garcia J.L. (eds) (1990) *Microbiology and Biochemistry of Strict Anaerobes Involved in Interspecies Hydrogen Transfer*, Plenum Press, New York.

14 Redwood, M.D., Mikheenko, I.P., Sargent, F. and Macaskie, L.E. (2008) *FEMS Microbiology Letters*, **278**, 48.

15 Mohanty, A.K., Misra, M. and Drzal, L.T. (2002) *Journal of Polymers and the Environment*, **10**, 19.

16 Nicholls, D.G. and Ferguson, S.J. (2002) *Bioenergetics 3*, Academic Press, London,

17 Siegbahn, P. E. M., Tye, J.W. and Hall, M.B. (2007) *Chemical Reviews*, **107**, 4414.

18 Menon, S. and Ragsdale, S.W. (1996) *Biochemistry*, **35**, 15814.

19 Cammack, R., Frey, M. and Robson, R. (eds), (2001) *Hydrogen as a Fuel: Learning from Nature*, Taylor and Francis, London.

20 Vignais, P.M. and Billoud, B. (2007) *Chemical Reviews*, **107**, 4206.

21 Fontecilla-Camps, J.C., Volbeda, A., Cavazza, C. and Nicolet, Y. (2007) *Chemical Reviews*, **107**, 4273.

22 De Lacey, A.L., Fernandez, V.M., Rousset, M. and Cammack, R. (2007) *Chemical Reviews*, **107**, 4304.

23 Lubitz, W., Reijerse, E. and van Gastel, M. (2007) *Chemical Reviews*, **107**, 4331.

24 Vincent, K.A., Parkin, A. and Armstrong, F.A. (2007) *Chemical Reviews*, **107**, 4366.

25 Burgdorf, T., Lenz, O., Buhrke, T., van der Linden, E., Jones, A.K., Albracht, S. P. J. and Friedrich, B. (2005) *Journal of Molecular Microbiology and Biotechnology*, **10**, 181.

26 Vignais, P.M., Billoud, B. and Meyer, J. (2001) *FEMS Microbiology Reviews*, **25**, 455.

27 Volbeda, A., Charon, M.H., Piras, C., Hatchikian, E.C., Frey, M. and Fontecillacamps, J.C. (1995) *Nature*, **373**, 580.

28 Higuchi, Y., Yagi, T. and Yasuoka, N. (1997) *Structure*, **5**, 1671.

29 Volbeda, A., Martin, L., Cavazza, C., Matho, M., Faber, B.W., Roseboom, W., Albracht, S. P. J., Garcin, E., Rousset, M. and Fontecilla-Camps, J.C. (2005) *Journal of Biological Inorganic Chemistry*, **10**, 239.

30 Foerster, S., van Gastel, M., Brecht, M. and Lubitz, W. (2005) *Journal of Biological Inorganic Chemistry*, **10**, 51.

31 He, S.H., Teixeira, M., Legall, J., Patil, D.S., Moura, I., Moura, J. J. G., Dervartanian, D.V., Huynh, B.H. and Peck, H.D. (1989) *Journal of Biological Chemistry*, **264**, 2678.

32 Patil, D.S. (1994) *Methods in Enzymology*, **243**, 68.

33 Garcin, E., Vernede, X., Hatchikian, E.C., Volbeda, A., Frey, M. and Fontecilla-Camps, J.C. (1999) *Structure with Folding & Design*, **7**, 557.

34 Adams, M. W. W. (1990) *Biochimica et Biophysica Acta*, **1020**, 115.

35 Nicolet, Y., Cavazza, C. and Fontecilla-Camps, J.C. (2002) *Journal of Inorganic Biochemistry*, **91**, 1.

36 Peters, J.W., Lanzilotta, W.N., Lemon, B.J. and Seefeldt, L.C. (1998) *Science*, **282**, 1853.

37 Pilak, O., Mamat, B., Vogt, S., Hagemeier, C.H., Thauer, R.K., Shima, S., Vonrhein, C., Warkentin, E. and Ermler, U. (2006) *Journal of Molecular Biology*, **358**, 798.

38 Shima, S., Lyon, E.J., Thauer, R.K., Mienert, B. and Bill, E. (2005) *Journal of the American Chemical Society*, **127**, 10430.

39 Lyon, E.J., Shima, S., Boecher, R., Thauer, R.K., Grevels, F.W., Bill, E., Roseboom, W. and Albracht, S. P. J. (2004) *Journal of the American Chemical Society*, **126**, 14239.

40 Shima, S., Lyon, E.J., Sordel-Klippert, M.S., Kauss, M., Kahnt, J., Thauer, R.K., Steinbach, K., Xie, X.L., Verdier, L. and Griesinger, C. (2004) *Angewandte Chemie International Edition*, **43**, 2547.

41 Korbas, M., Vogt, S., Meyer-Klaucke, W., Bill, E., Lyon, E.J., Thauer, R.K. and Shima, S. (2006) *Journal of Biological Chemistry*, **281**, 30804.

42 Lyon, E.J., Shima, S., Buurman, G., Chowdhuri, S., Batschauer, A., Steinbach, K. and Thauer, R.K. (2004) *European Journal of Biochemistry*, **271**, 195.

43 Schwartz, E., Henne, A., Cramm, R., Eitinger, T., Friedrich, B. and Gottschalk, G. (2003) *Journal of Molecular Biology*, **332**, 369.

44 Blokesch, M. and Bock, A. (2006) *FEBS Letters*, **580**, 4065.

45 Forzi, L., Hellwig, P., Thauer, R.K. and Sawers, R.G. (2007) *FEBS Letters*, **581**, 3317.

46 Bock, A., King, P.W., Blokesch, M. and Posewitz, M.C. (2006) *Advances in Microbial Physiology*, **51**, 1.

47 Posewitz, M.C., King, P.W., Smolinski, S.L., Zhang, L.P., Seibert, M. and Ghirardi, M.L. (2004) *Journal of Biological Chemistry*, **279**, 25711.

48 Cammack, R., Fernandez, V.M. and Hatchikian, E.C. (1994) *Methods in Enzymology*, **243**, 43.

49 Albracht, S. P. J., Graf, E.-G. and Thauer, R.K. (1982) *FEBS Letters*, **140**, 311.

50 LeGall, J., Ljungdahl, P.O., Moura, I., Peck, H.D., Xavier, A.V., Moura, J. J. G., Teixeira, M., Huynh, B.H. and DerVartanian, D.V. (1982) *Biochemical and Biophysical Research Communications*, **106**, 610.

51 Huyett, J.E., Carepo, M., Pamplona, A., Franco, R., Moura, I., Moura, J. J. G. and Hoffman, B.M. (1997) *Journal of the American Chemical Society*, **119**, 9291.

52 Cammack, R., Patil, D.S., Aguirre, R. and Hatchikian, E.C. (1982) *FEBS Letters*, **142**, 289.

53 Cammack, R., Fernandez, V.M. and Schneider, K. (1986) *Biochimie*, **68**, 85.

54 Cammack, R., Patil, D.S., Hatchikian, E.C. and Fernandez, V.M. (1987) *Biochimica et Biophysica Acta*, **912**, 98.

55 Stein, M. and Lubitz, W. (2002) *Current Opinion in Chemical Biology*, **6**, 243.

56 Brecht, M., van Gastel, M., Buhrke, T., Friedrich, B. and Lubitz, W. (2003) *Journal of the American Chemical Society*, **125**, 13075.

57 Bagley, K.A., Duin, E.C., Roseboom, W., Albracht, S. P. J. and Woodruff, W.H. (1995) *Biochemistry*, **34**, 5527.

58 DeLacey, A.L., Stadler, C., Fernandez, V.N., Hatchikian, E.C., Fan, H.J., Li, S.H. and Hall, M.B. (2002) *Journal of Biological Inorganic Chemistry*, **7**, 318.

59 Bleijlevens, B., van Broekhuizen, F.A., De Lacey, A.L., Roseboom, W., Fernandez, V.M. and Albracht, S. P. J. (2004) *Journal of Biological Inorganic Chemistry*, **9**, 743.

60 Fichtner, C., Laurich, C., Bothe, E. and Lubitz, W. (2006) *Biochemistry*, **45**, 9706.

61 Leger, C., Jones, A.K., Roseboom, W., Albracht, S. P. J. and Armstrong, F.A. (2002) *Biochemistry*, **41**, 15736.

62 Leger, C., Elliott, S.J., Hoke, K.R., Jeuken, L. J. C., Jones, A.K. and Armstrong, F.A. (2003) *Biochemistry*, **42**, 8653.

63 Pershad, H.R., Duff, J. L. C., Heering, H.A., Duin, E.C., Albracht, S. P. J. and

Armstrong, F.A. (1999) *Biochemistry*, **38**, 8992.

64 Jones, A.K., Lamle, S.E., Pershad, H.R., Vincent, K.A., Albracht, S. P. J. and Armstrong, F.A. (2003) *Journal of the American Chemical Society*, **125**, 8505.

65 Lamle, S.E., Albracht, S. P. J. and Armstrong, F.A. (2004) *Journal of the American Chemical Society*, **126**, 14899.

66 Lamle, S.E., Albracht, S. P. J. and Armstrong, F.A. (2005) *Journal of the American Chemical Society*, **127**, 6595.

67 Vincent, K.A., Parkin, A., Lenz, O., Albracht, S. P. J., Fontecilla-Camps, J.C., Cammack, R., Friedrich, B. and Armstrong, F.A. (2005) *Journal of the American Chemical Society*, **127**, 18179.

68 Leger, C., Dementin, S., Bertrand, P., Rousset, M. and Guigliarelli, B. (2004) *Journal of the American Chemical Society*, **126**, 12162.

69 Dementin, S., Belle, V., Bertrand, P., Guigliarelli, B., Adryanczyk-Perrier, G., De Lacey, A.L., Fernandez, V.M., Rousset, M. and Leger, C. (2006) *Journal of the American Chemical Society*, **128**, 5209.

70 Cohen, J., Kim, K., King, P., Seibert, M. and Schulten, K. (2005) *Structure*, **13**, 1321.

71 Montet, Y., Amara, P., Volbeda, A., Vernede, X., Hatchikian, E.C., Field, M.J., Frey, M. and Fontecilla-Camps, J.C. (1997) *Nature Structural Biology*, **4**, 523.

72 Watt, G.D. and Reddy, K. R. N. (1994) *Journal of Inorganic Biochemistry*, **53**, 281.

73 Buhrke, T., Brecht, M., Lubitz, W. and Friedrich, B. (2002) *Journal of Biological Inorganic Chemistry*, **7**, 897.

74 Dementin, S., Burlat, B., De Lacey, A.L., Pardo, A., Adryanczyk-Perrier, G., Guigliarelli, B., Fernandez, V.M. and Rousset, M. (2004) *Journal of Biological Chemistry*, **279**, 10508.

75 Volbeda, A. and Fontecilla-Camps, J.C. (2005) *Coordination Chemistry Reviews*, **249**, 1609.

76 Page, C.C., Moser, C.C., Chen, X.X. and Dutton, P.L. (1999) *Nature*, **402**, 47.

77 Rousset, M., Montet, Y., Guigliarelli, B., Forget, N., Asso, M., Bertrand, P., Fontecilla-Camps, J.C. and Hatchikian, E.C. (1998) *Proceedings of the National Academy of Sciences of the United States of America*, **95**, 11625.

78 Florin, L., Tsokoglou, A. and Happe, T. (2001) *Journal of Biological Chemistry*, **276**, 6125.

79 Volbeda, A., Garcia, E., Piras, C., deLacey, A.L., Fernandez, V.M., Hatchikian, E.C., Frey, M. and Fontecilla-Camps, J.C. (1996) *Journal of the American Chemical Society*, **118**, 12989.

80 Morelli, X., Czjzek, M., Hatchikian, C.E., Bornet, O., Fontecilla-Camps, J.C., Palma, N.P., Moura, J. J. G. and Guerlesquin, F. (2000) *Journal of Biological Chemistry*, **275**, 23204.

81 Pereira, I. A. C., Romao, C.V., Xavier, A.V., LeGall, J. and Teixeira, M. (1998) *Journal of Biological Inorganic Chemistry*, **3**, 494.

82 Yahata, N., Saitoh, T., Takayama, Y., Ozawa, K., Ogata, H., Higuchi, Y. and Akutsu, H. (2006) *Biochemistry*, **45**, 1653.

83 Hawkes, F.R., Dinsdale, R., Hawkes, D.L. and Hussy, I. (2002) *International Journal of Hydrogen Energy*, **27**, 1339.

84 Van, S.W. Ginkel, Oh, S.E. and Logan, B.E. (2005) *International Journal of Hydrogen Energy*, **30**, 1535.

85 Levin, D.B., Pitt, L. and Love, M. (2004) *International Journal of Hydrogen Energy*, **29**, 173.

86 Nath, K. and Das, D. (2004) *Applied Microbiology and Biotechnology*, **65**, 520.

87 Kruse, O., Rupprecht, J., Mussgnug, J.R., Dismukes, G.C. and Hankamer, B. (2005) *Photochemical & Photobiological Sciences*, **4**, 957.

88 Happe, T., Hemschemeier, A., Winkler, M. and Kaminski, A. (2002) *Trends in Plant Science*, **7**, 246.

89 Zhang, L.P., Happe, T. and Melis, A. (2002) *Planta*, **214**, 552.

90 Appel, J. and Schulz, R. (1998) *Journal of Photochemistry and Photobiology B*, **47**, 1.

91 Blankenship, R.E. and Hartman, H. (1998) *Trends in Biochemical Sciences*, **23**, 94.

92 Adir, N., Zer, H., Shochat, S. and Ohad, I. (2003) *Photosynthesis Research*, **76**, 343.

93 Ball, P., Ziemelis, K. and Allen, L. (2001) *Nature*, **409**, 225.

94 Tamagnini, P., Leitao, E., Oliveira, P., Ferreira, D., Pinto, F., Harris, D.J., Heidorn, T. and Lindblad, P. (2007) *FEMS Microbiology Reviews*, **31**, 692.

95 Rao, K.K. and Hall, D.O. (1984) *Trends in Biotechnology*, **2**, 124.

96 Ihara, M., Nishihara, H., Yoon, K.S., Lenz, O., Friedrich, B., Nakamoto, H., Kojima, K., Honma, D., Kamachi, T. and Okura, I. (2006) *Photochemistry and Photobiology*, **82**, 676.

97 Melis, A., Zhang, L.P., Forestier, M., Ghirardi, M.L. and Seibert, M. (2000) *Plant Physiology*, **122**, 127.

98 Melis, A. (2007) *Planta*, **226**, 1075.

99 Posewitz, M.C., Smolinski, S.L., Kanakagiri, S., Melis, A., Seibert, M. and Ghirardi, M.L. (2004) *Plant Cell*, **16**, 2151.

100 Melis, A. and Happe, T. (2001) *Plant Physiology*, **127**, 740.

101 Forestier, M., King, P., Zhang, L.P., Posewitz, M., Schwarzer, S., Happe, T., Ghirardi, M.L. and Seibert, M. (2003) *European Journal of Biochemistry*, **270**, 2750.

102 Lee, C.M., Chen, P.C., Wang, C.C. and Tung, Y.C. (2002) *International Journal of Hydrogen Energy*, **27**, 1309.

103 Asada, Y. and Miyake, J. (1999) *Journal of Bioscience and Bioengineering*, **88**, 1.

104 Vignais, P.M., Colbeau, A., Willison, J.C. and Jouanneau, Y. (1985) *Advances in Microbial Physiology*, **26**, 155.

105 Koku, H., Eroglu, I., Gunduz, U., Yucel, M. and Turker, L. (2002) *International Journal of Hydrogen Energy*, **27**, 1315.

106 Kovacs, K.L., Maroti, G. and Rakhely, G. (2006) *International Journal of Hydrogen Energy*, **31**, 1460.

107 Howard, J.B. and Rees, D.C. (2006) *Proceedings of the National Academy of Sciences of the United States of America*, **103**, 17088.

108 Tard, C., Liu, X.M., Ibrahim, S.K., Bruschi, M., De Gioia, L. Davies, S.C., Yang, X., Wang, L.S., Sawers, G. and Pickett, C.J. (2005) *Nature*, **433**, 610.

109 Liu, T.B. and Darensbourg, M.Y. (2007) *Journal of the American Chemical Society*, **129**, 7008.

110 Justice, A.K., Rauchfuss, T.B. and Wilson, S.R. (2007) *Angewandte Chemie International Edition*, **46**, 6152.

111 Curtis, C.J., Miedaner, A., Ciancanelli, R., Ellis, W.W., Noll, B.C., DuBois, M.R. and DuBois, D.L. (2003) *Inorganic Chemistry*, **42**, 216.

112 Frey, M. (2002) *Chembiochem*, **3**, 153.

113 Wilson, A.D., Shoemaker, R.K., Miedaner, A., Muckerman, J.T., DuBois, D.L. and DuBois, M.R. (2007) *Proceedings of the National Academy of Sciences of the United States of America*, **104**, 6951.

114 Jacobsen, G.M., Shoemaker, R.K., McNevin, M.J., DuBois, M.R. and DuBois, D.L. (2007) *Organometallics*, **26**, 5003.

115 Ogo, S., Kabe, R., Uehara, K., Kure, B., Nishimura, T., Menon, S.C., Harada, R., Fukuzumi, S., Higuchi, Y., Ohhara, T., Tamada, T. and Kuroki, R. (2007) *Science*, **316**, 585.

116 Ibrahim, S.K., Liu, X.M., Tard, C. and Pickett, C.J. (2007) *Chemical Communications*, 1535.

117 Thomas, C.M., Rudiger, O., Liu, T., Carson, C.E., Hall, M.B. and Darensbourg, M.Y. (2007) *Organometallics*, **26**, 3976.

118 Biyikoglu, A. (2005) *International Journal of Hydrogen Energy*, **30**, 1181.

119 Jones, A.K., Lichtenstein, B.R., Dutta, A., Gordon, G. and Dutton, P.L. (2007) *Journal of the American Chemical Society*, **129**, 14844.

120 Yagi, T. (1976) *Proceedings of the National Academy of Sciences of the United States of America*, **73**, 2947.

121 Qian, D.J., Nakamura, C., Wenk, S.O., Ishikawa, H., Zorin, N. and Miyake, J. (2002) *Biosensors & Bioelectronics*, **17**, 789.

122 Lutz, B.J., Fan, Z.H., Burgdorf, T. and Friedrich, B. (2005) *Analytical Chemistry*, **77**, 4969.

123 Simon, H., Bader, J., Gunther, H., Neumann, S. and Thanos, J. (1984) *Annals*

of the New York Academy of Sciences, **434**, 171.

124 Armstrong, F.A., Hill, H. A. O. and Walton, N.J. (1988) *Accounts of Chemical Research*, **21**, 407.

125 Tsujimura, S., Fujita, M., Tatsumi, H., Kano, K. and Ikeda, T. (2001) *Physical Chemistry Chemical Physics*, **3**, 1331.

126 Vincent, K.A., Cracknell, J.A., Clark, J.R., Ludwig, M., Lenz, O., Friedrich, B. and Armstrong, F.A. (2006) *Chemical Communications*, 5033.

127 Fultz, M.L. and Durst, R.A. (1982) *Analytica Chimica Acta*, **140**, 1.

128 Heineman, W.R., Meckstroth, M.L., Norris, B.J. and Su, C.H. (1979) *Bioelectrochemistry and Bioenergetics*, **6**, 577.

129 Bianco, P. and Haladjian, J. (1991) *Electroanalysis*, **3**, 973.

130 Draoui, K., Bianco, P., Haladjian, J., Guerlesquin, F. and Bruschi, M. (1991) *Journal of Electroanalytical Chemistry*, **313**, 201.

131 Lojou, E., Gludici-Orticoni, M.T. and Bianco, P. (2005) *Journal of Electroanalytical Chemistry*, **579**, 199.

132 Hoogvliet, J.C., Vanos, P., Vandermark, E.J. and Vanbennekom, W.P. (1991) *Biosensors & Bioelectronics*, **6**, 413.

133 Varfolomeyev, S.D. and Bachurin, S.O. (1984) *Journal of Molecular Catalysis*, **27**, 305.

134 Varfolomeyev, S.D. and Bachurin, S.O. (1984) *Journal of Molecular Catalysis*, **27**, 315.

135 Eng, L.H., Elmgren, M., Komlos, P., Nordling, M., Lindquist, S.E. and Neujahr, H.Y. (1994) *Journal of Physical Chemistry*, **98**, 7068.

136 Morozov, S.V., Karyakina, E.E., Zorin, N.A., Varfolomeyev, S.D., Cosnier, S. and Karyakin, A.A. (2002) *Bioelectrochemistry*, **55**, 169.

137 Parpaleix, T., Laval, J.M., Majda, M. and Bourdillon, C. (1992) *Analytical Chemistry*, **64**, 641.

138 De Lacey, A.L., Detcheverry, M., Moiroux, J. and Bourdillon, C. (2000) *Biotechnology and Bioengineering*, **68**, 1.

139 Lojou, E. and Bianco, P. (2006) *Electroanalysis*, **18**, 2426.

140 Tatsumi, H., Takagi, K., Fujita, M., Kano, K. and Ikeda, T. (1999) *Analytical Chemistry*, **71**, 1753.

141 Mege, R.M. and Bourdillon, C. (1985) *Journal of Biological Chemistry*, **260**, 14701.

142 De Lacey, A.L., Moiroux, J. and Bourdillon, C. (2000) *European Journal of Biochemistry*, **267**, 6560.

143 Schlereth, D.D., Fernandez, V.M., Sanchez-Cruz, M. and Popov, V.O. (1992) *Bioelectrochemistry and Bioenergetics*, **28**, 473.

144 Delecouls, K., Saint-Aguet, P., Zaborosch, C. and Bergel, A. (1999) *Journal of Electroanalytical Chemistry*, **468**, 139.

145 Cantet, J., Bergel, A. and Comtat, M. (1996) *Enzyme and Microbial Technology*, **18**, 72.

146 Cantet, J., Bergel, A., Comtat, M. and Seris, J.L. (1992) *Journal of Molecular Catalysis*, **73**, 371.

147 Devaux-Basseguy, R., Gros, P. and Bergel, A. (1997) *Journal of Chemical Technology and Biotechnology*, **68**, 389.

148 Greiner, L., Schroder, I., Muller, D.H. and Liese, A. (2003) *Green Chemistry*, **5**, 697.

149 Eddowes, M.J. and Hill, H.A.O. (1979) *Journal of the American Chemical Society*, **101**, 4461.

150 Van Dijk, C., Van Leeuwen, J.W., Veeger, C., Schreurs, J. P. G. M. and Barendrecht, E. (1982) *Bioelectrochemistry and Bioenergetics*, **9**, 743.

151 Van Berkel-Arts, A., Dekker, M., Vandijk, C., Grande, H., Hagen, W.R., Hilhorst, R., Krusewolters, M., Laane, C. and Veeger, C. (1986) *Biochimie*, **68**, 201.

152 Bianco, P. and Haladjian, J. (1992) *Journal of the Electrochemical Society*, **139**, 2428.

153 Bianco, P., Haladjian, J. and Giannandreaderocles, S. (1994) *Electroanalysis*, **6**, 67.

154 Armstrong, F.A., Butt, J.N. and Sucheta, A. (1993) *Metallobiochemistry, Part D*, **227**, 479.

155 Varfolomeyev, S.D., Yaropolov, A.I. and Karyakin, A.A. (1993) *Journal of Biotechnology*, **27**, 331.

156 Johnston, W., Cooney, M.J., Liaw, B.Y., Sapra, R. and Adams, M.W.W. (2005) *Enzyme and Microbial Technology*, **36**, 540.

157 Rudiger, O., Abad, J.M., Hatchikian, E.C., Fernandez, V.M. and De Lacey, A.L. (2005) *Journal of the American Chemical Society*, **127**, 16008.

158 Butt, J.N., Filipiak, M. and Hagen, W.R. (1997) *European Journal of Biochemistry*, **245**, 116.

159 Guiral-Brugna, M., Giudici-Orticoni, M.T., Bruschi, M. and Bianco, P. (2001) *Journal of Electroanalytical Chemistry*, **510**, 136.

160 De Lacey, A.L., Fernandez, V.M. and Rousset, M. (2005) *Coordination Chemistry Reviews*, **249**, 1596.

161 Duche, O., Elsen, S., Cournac, L. and Colbeau, A. (2005) *FEBS Journal*, **272**, 3899.

162 Buhrke, T., Lenz, O., Krauss, N. and Friedrich, B. (2005) *Journal of Biological Chemistry*, **280**, 23791.

163 Bleijlevens, B., Buhrke, T., van der Linden, E., Friedrich, B. and Albracht, S.P.J. (2004) *Journal of Biological Chemistry*, **279**, 46686.

164 van der Linden, E., Burgdorf, T., De Lacey, A.L., Buhrke, T., Scholte, M., Fernandez, V.M., Friedrich, B. and Albracht, S.P.J. (2006) *Journal of Biological Inorganic Chemistry*, **11**, 247.

165 van der Linden, E., Faber, B.W., Bleijlevens, B., Burgdorf, T., Bernhard, M., Friedrich, B. and Albracht, S. P. J. (2004) *European Journal of Biochemistry*, **271**, 801.

166 Karyakin, A.A., Morozov, S.V., Karyakina, E.E., Varfolomeyev, S.D., Zorin, N.A. and Cosnier, S. (2002) *Electrochemistry Communications*, **4**, 417.

167 Tye, J.W., Hall, M.B. and Darensbourg, M.Y. (2005) *Proceedings of the National Academy of Sciences of the United States of America*, **102**, 16911.

168 Alonso-Lomillo, M.A., Rudiger, O., Maroto-Valiente, A., Velez, M., Rodriguez-Ramos, I., Munoz, F.J., Fernandez, V.M. and De Lacey, A.L. (2007) *Nano Letters*, **7**, 1603.

169 deLacey, A.L., Hatchikian, E.C., Volbeda, A., Frey, M., FontecillaCamps, J.C. and Fernandez, V.M. (1997) *Journal of the American Chemical Society*, **119**, 7181.

170 Roseboom, W., De Lacey, A.L., Fernandez, V.M., Hatchikian, E.C. and Albracht, S. P. J. (2006) *Journal of Biological Inorganic Chemistry*, **11**, 102.

171 Shima, S., Pilak, O., Vogt, S., Schick, M., Stagni, M.S., Meyer-Klaucke, W., Warkentin, E., Thauer, R.K. and Ermler, U. (2008) *Science*, **321**, 572–575.

9

PAH Bioremediation by Microbial Communities and Enzymatic Activities

Vincenza Andreoni and Liliana Gianfreda

9.1
Introduction

Polycyclic aromatic hydrocarbons (PAHs), listed as priority pollutants [1], are hydrophobic and highly persistent contaminants with long half-lives in soils, sediments, air and biota and show long-range transport.

Depending on their molecular structure and complexity, PAHs exhibit low water solubility and in turn limited bioavailability. They may bioaccumulate in human and animal tissues and be biomagnified in food chains. These properties may result in potential significant impacts of PAHs on human health and the environment. PAH contamination is therefore of major concern for toxicological risk assessment of contaminated field sites and for the feasibility of (bio)remediation technologies to restore contaminated soils. Moreover, very often a mixed pollution may occur. For example, toxic metals and hydrocarbons may be simultaneously present in a given environment.

Several strategies have been devised to remediate polluted environments: physical and chemical methods and biological approaches, requiring the involvement of biological agents [2, 3]. Despite the fact that PAHs have a high resistance to chemical and biochemical degradation, a variety of microorganisms can degrade certain PAHs and biological methodologies may be implemented for restoring PAH-polluted environments.

Bioremediation, either as an intrinsic degradation or as an engineered process, usually refers to the use of microorganisms (mainly), but also of plants, enzymes and plant–microorganism associations that degrade or transform pollutants to less toxic or non-toxic products, with mitigation or elimination of environmental contamination and thus with protection of soil quality [2, 4–7].

Among the managed processes, composting of PAH-contaminated soils, using mixtures with manure and/or carbon sources, has been found to be effective in the bioremediation of highly petroleum-contaminated soil when other methods are too expensive or ineffective [8].

Handbook of Green Chemistry, Volume 3: Biocatalysis. Edited by Robert H. Crabtree
Copyright © 2009 WILEY-VCH Verlag GmbH & Co. KGaA, Weinheim
ISBN: 978-3-527-32498-9

Phytoremediation can also be a useful biological treatment. It is usually considered a secondary polishing approach for PAH-contaminated soils previously treated by land farming. The key elements for successful phytoremediation are the use of plant species with the ability to proliferate in the presence of PAHs and strains of plant growth-promoting rhizobacteria (PGPR) that increase plant tolerance to PAHs and accelerate plant growth.

Whereas bioremediation may be utilized *in situ* or *ex situ*, phytoremediation is a typical *in situ* methodology, usually offering numerous advantages over *ex situ* technologies.

The scope of this chapter is to provide a brief survey of many aspects concerning the fate of PAHs in polluted environments, and the properties of pollutant-restoring biological agents, including microorganisms, enzymes and plants. Some suitable strategies to remediate PAH-polluted sites and to improve the efficiency of the selected agents are also briefly discussed.

9.2
Fate of PAHs in the Environment

In soils and groundwater, complex contamination very often occurs. Severe alterations of chemical, biological and biochemical properties may result that in turn influence negatively the soil functions and the survival of living organisms.

Usually, an organic compound in soil can be transferred from one point to another without any modification of its chemical structure. It can be weakly or strongly associated with inorganic and organic colloids through adsorption mechanisms. It can be also absorbed by plant roots or volatilized, depending on the soil zone where it is located and on its chemical and physical properties. If it accumulates on the 'soil surface', it can undergo photodecomposition and the process will be strongly influenced by intrinsic and extrinsic properties of the soil and also by the inherent chemical structure of the compound. The organic chemicals can also undergo chemical or biological decomposition depending on certain soil components (e.g. oxides, microorganisms) and properties (e.g. redox state, pH). A distribution between soil and ground water may also occur upon leaching processes.

The fate of PAHs in soil can be affected to a great extent by the interactions occurring at interfaces between organic and inorganic soil colloids and chemicals through sorption/desorption mechanisms. PAHs tend to sorb strongly to soil colloids, particularly organic matter, although clay minerals, strong soil sorbents for contaminants, may also contribute to affect their behavior in soil. As a result, the movement of PAHs, their availability for plant or microbial uptake and their transformation by abiotic or biotic agents will be strongly affected.

As PAHs have low water solubility, they will be partitioned between air, soil solution and soil matrix, and this will affect strongly the rate of all processes involving them in soil. Moreover, their elevated hydrophobicity renders their persistence in the environment very high. The persistence of an organic contaminant is, however, not only due to its intrinsic molecular property, but is much more a result of environmental microhabitats that affect both the mass transfer of the xenobiotic to the microorganism and its degradative activity [9]. It is generally believed that PAHs must enter the water phase to be biodegraded. The degrading bacteria, therefore, drive the continuous dissolution of PAHs from the soil matrix by constantly removing PAHs from the water phase [10].

The bioavailability of PAHs is further complicated when they interact with non-aqueous phase liquids (NAPLs) and soil colloids, thus making them less available or completely unavailable for microorganisms [11, 12].

To gain access to adsorbed PAHs, degrading bacteria need to attach to surfaces possibly through biofilm formation, favored by chemotaxis, which may guide bacteria to swim using flagellae towards these pollutants adsorbed on a surface, followed by attachment.

The fate of contaminants in soil is also affected by the 'aging' phenomenon (that is, long persistence in soil). The latter strongly decreases the availability of the contaminants because it favors their binding to soil and soil components or their incorporation and sequestration in structural micropores of mineral lattices or in hydrophobic remote areas of the soil organic matrix [13, 14]. Consequently, desorption of the contaminant becomes very slow and a limited amount of it goes to the aqueous phase, and the contaminant becomes not readily bioavailable.

The combination of the above phenomena leads to a different distribution and partitioning of the contaminants in soil, rendering them less readily bioavailable, resistant to biodegradation and thus more persistent in soil [15].

If heavy metals are also present as co-contaminants, the biodegradation of PAHs can be reduced by metal toxicity. Metal bioavailability, determined by numerous processes affecting their distribution and their fate in soil (Figure 9.1), governs the extent to which metals affect biodegradation.

Metals affect organic biodegradation through impacting both the physiology and ecology of degrading microorganisms. Metal contamination may not prevent aromatic catabolism in soils, but may retard activity. Reduced microbial activity may originate from the change in microbial community structure after long-term exposure to a heavy metal. Doelman *et al.* [16] observed that metal-contaminated soil contained more metal-resistant microorganisms, but with a restricted ability to degrade organic pollutants. The presence of multiple contaminants may pose extreme challenges to the maintenance of a phylogenetically and functionally diverse microbial community. In soils contaminated with both heavy metals and hydrocarbons, only those microorganisms that tolerate both the contaminants may survive.

The influence of heavy metals on PAH degradation has recently been emphasized and the effect of various metals on the degradation of phenanthrene has been investigated. The degradation of phenanthrene was found to be retarded by the presence of copper and high levels of the metal caused incomplete mineralization and accumulation of its metabolites [17]. At a soil Zn^{2+} concentration of 140 mg kg^{-1} and a phenanthrene concentration of 40 mg kg^{-1} a marginal stimulation of the phenanthrene biodegradation rate was observed; at Zn^{2+} concentrations at or above the 'action' values (that is, the level of a contaminant at which soil quality is deemed to

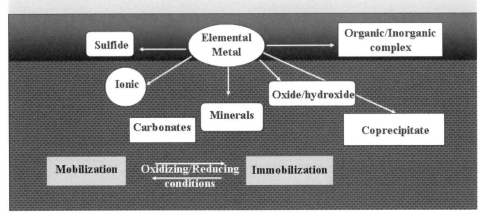

Figure 9.1 Forms of metals in soil.

impair the functional properties of the soil), phenanthrene degradation was instead inhibited [18].

9.3
Population of PAH-Polluted Environments

Organic pollution, although reported to reduce diversity in microbial communities in a variety of environments [19, 20], leads to the enrichment of specific microbial populations. Phylogenetic and degradative diversity of PAH-degrading bacteria has been mostly investigated in terrestrial habitats, leading to the isolation of *Pseudomonas*, *Sphingomonas* spp., *Tsukamurella*, *Rhodococcus* spp., *Arthrobacter* spp., *Nocardia* spp. and *Mycobacterium* spp. [21–24].

Sphingomonads and the fast-growing mycobacteria, a descendant line distinct from that of the slow growers which cluster the mycobacterial pathogens, are considered essential members of the PAH-degrading bacterial community in PAH-contaminated soils. With the development of specific PCR–DGGE methods to monitor rapidly the community composition of sphingomonads and fast-growing mycobacteria, Leys and co-workers [25, 26] recognized the different species present in mixed *Sphingomonas* and *Mycobacterium* communities of several PAH-contaminated soils of diverse origin with different contamination profiles, pollution concentrations and geochemical properties. They found high *Sphingomonas* cell concentrations (10^5–10^6 per gram of soil) confirming the important role of these bacteria in colonizing and degrading

PAH-contaminated soils. Furthermore, they detected fast-growing species such as *M. frederiksbergense* and *M. austroafricanum* only in most PAH-contaminated soils. Surprisingly, they detected 16S rRNA genes related to *M. tusciae* sequences in all contaminated but not in the uncontaminated soils. This *Mycobacterium* species so far has not been reported to be capable of biodegrading PAHs.

Hydrocarbon-degrading bacteria have been found in different temperate and tropical zone ecosystems. *Pseudomonas* spp. were the primary isolates for aerobic PAH degradation from Antarctic soils, in which they degraded naphthalene efficiently at low temperatures, although they harbored degradative genes similar to those of mesophilic strains, thus strongly supporting the contention that the degrading genes derived from the same origin. Lateral transfer of naphthalene dioxygenase (*ndo*) genes may have occurred in Antarctic soil among different species via large plasmids [27].

Catabolic genotypes involved in microbial degradation pathways of representative fractions of petroleum hydrocarbons, including *n*-alkanes, aromatics and PAHs, are widespread also in Arctic soils [28, 29]. Margesin *et al.* [30] determined the prevalence of seven catabolic genotypes involved in the degradation of petroleum hydrocarbon in bacterial populations present in both contaminated and pristine Alpine soils from Tyrol (Austria). They found a significant correlation between the level of contamination and the prevalence of the seven genotypes considered. A significantly higher percentage of genotypes containing genes from Gram-negative bacteria (*P. putida alkB, xylE* and *ndoB* and *Acinetobacter alkM*) in contaminated (50–75%) than in pristine (0–12.5%) soils indicated that these organisms were enriched following contamination. Moreover, genotypes containing genes from Gram-positive bacteria (*Rhodococcus alkB1* and *alkB2* and *Mycobacterium nidA*) were detected at high frequency in both contaminated (41.7–75%) and pristine (37.5–50%) soils, indicating that they were already present before a contamination event.

The pyrene dioxygenase (*pdo1*) gene was the dominant PAH gene in urban soils, suggesting that mycobacteria and related genera were the dominant PAH degraders [31], while pseudomonads and sphingomonads were of minor importance. The degradation of highly bioavailable PAHs in soil is possible done by sphingomonads and other fast-growing bacteria(*r* strategists), whereas later they are outcompeted by relatively slow-growing bacteria (*K* strategists) [26]. *P. putida* and *Acinetobacter*, which rapidly colonize and grow on hydrocarbon contaminants, are *r* strategists; *Rhodococcus* and *Mycobacterium*, which tend to be more successful in nutrient-limited situations, are *K* strategists. Populations of *K* strategists are usually more stable and permanent members of the community [32].

PAH-degrading bacteria have been detected also in contaminated estuarine sediments [33, 34] and in salt marsh rhizosphere [22]. In a total of five different plant samples, the main bacterial groups were Gram-negative pseudomonads, Gram-positive (predominantly nocardioform) and the Gram-positive, spore-forming group *Paenibacillus*. Furthermore, whereas 75% of the pseudomonad isolates hybridized to the classical *nah* gene from *P. putida* NCIB9816-4 and approximately the same number hybridized to the *nag* genes cloned from *C. testosterone* GZ42, no homology of the *Paenibacillus* isolates to all the tested gene probes was found [22].

Naphthalene dioxygenase (*ndoB*) catabolic genotypes were enriched from the interior of plant roots of *Scirpus pungens* in response to pollution in a contaminant-dependent manner [35].

9.4
Microbial Degradation of PAHs

PAHs are biodegradable under both aerobic [36] and anaerobic conditions [37]. In aerobic conditions, bacteria and fungi utilize oxygen for both ring activation and cleavage of the aromatic nucleus and as an electron acceptor for its complete degradation.

The biodegradation of PAHs can serve as carbon and energy sources for the degrading organisms (assimilative biodegradation) or for intracellular detoxification. The intracellular hydroxylation of PAHs in bacteria is an initial step in preparing ring fission and assimilation, whereas in fungi it is an initial step in detoxification [38].

PAHs with low molecular weights are more rapidly degraded than less soluble and higher molecular weight compounds. Indeed, only a very limited number of bacteria can grow in pure cultures on PAHs with five or more aromatic rings. Benzo[*a*] anthracene and benzo[*a*]pyrene are examples of more recalcitrant high molecular weight PAHs that probably cannot sustain microbial growth. In the environment, these compounds are probably degraded by cometabolic processes, if they are biodegraded at all. Environmental bacterial isolates often degrade only a narrow range of PAHs and the patterns of the simultaneous degradation of PAH mixtures are complex. Often, they are cooperative processes that involve a consortium of strains with complementary capacities.

PAH bacterial metabolism usually occurs via the initial incorporation of molecular oxygen by an initial dioxygenase enzymatic system [36], leading to the formation of a *cis*-dihydrodiol. Subsequently, the latter is further metabolized to a few intermediates, including catechol, protocatechuic acid and gentisic acid, which are channeled into TCA cycle intermediates by their dioxygenase-mediated ring cleavage in either the *meta* or *para* position. Scheme 9.1 summarizes the initial attack of low molecular PAHs, leading to the dihydroxylated derivatives.

The biodegradability of PAHs in the absence of oxygen by a diversity of organisms makes possible beneficial remediation under anaerobic conditions. Anaerobic zones frequently develop in soils during compaction, and also in soils and sediments contaminated with hydrocarbons, where the indigenous aerobic microbial population is stimulated and rapid depletion of the dissolved oxygen content of the groundwater may occur. Water entering the contaminated environment carries only a small amount of dissolved oxygen, due to its low solubility and O_2 replenishment from the atmosphere is slow.

The microbial utilization of aromatic compounds in the absence of oxygen occurs without the perturbation of the benzene nucleus (Scheme 9.1). Anaerobic degradation of naphthalene, and in some instances of phenanthrene, has been measured in microbial communities in soil and sediments under conditions of denitrification [39] and sulfate reduction [40, 41] and under methanogenic conditions [42]. In a marine, sulfate-reducing sediment, fluorene, phenanthrene and fluoranthene were mineralized, whereas pyrene and benzo[a]pyrene were not [43].

With the identification of a sequence of metabolites during naphthalene and phenanthrene degradation by consortia, a pathway for these PAHs has been depicted (Scheme 9.1). As confirmed with a pure culture of a sulfate-reducing bacterium, strain NaphS2, phenanthrene is carboxylated by the addition of an external bicarbonate molecule to yield phenanthrenecarboxylic acid [44]. This acid is formed also through the addition of fumarate, catalyzed by naphthyl-2-methylsuccinate synthase (Scheme 9.1) [45]. Several intermediates, including two ring-cleaved products, have been identified [41, 46]. The two intermediates of the naphthalene pathway, 2-naphthoic acid and decalin-2-carboxylic acid, being metabolites unique to the anaerobic degradation of these PAHs, can serve to assess an actual naphthalene degradation occurring in field sites [47].

A wide variety of fungi (including ligninolytic, non-ligninolytic, mitosporic and ectomycorrizhal fungi) have been proved to transform PAHs. They include *Aspergillus ochraceus*, *Cladosporium sphaerospermum*, *Cunninghamella elegans* and *C. echinulata*, *Phanerochaete chrysosporium*, *Bjerkandera* sp., *Trametes versicolor* and *T. harianum*, *Pleorotus ostreatus* and the yeast *Saccharomyces cerevisiae* [5, 48, 49]. Several features of these microorganisms render them attractive organisms to be applied as bioremediating agents. Indeed, they usually tolerate higher concentrations of pollutants than bacteria and have broad substrate specificity and the key enzymes to the process are mainly of a constitutive nature, thus preventing their originating organisms from being adapted to the chemical being degraded.

The most efficient degraders are white rot fungi, and the lignin-degrading enzyme system (LDS), the very powerful extracellular oxidative enzymatic system produced by ligninolytic fungi, has been recognized to be involved in PAH mineralization. These enzymes mediate the transformation of PAHs from an unavailable form into partially degraded or oxidized products of low molecular weight that can be easily absorbed and taken up by cells and successively intracellularly transformed.

The LDS systems are combinations of lignin peroxidase (LiP) and Mn-dependent peroxidase (MnP), two glycosylated heme-containing peroxidases, and of laccase (L), a copper-containing phenoloxidase. The three enzymes together with an H_2O_2-generating system and cellulolytic and hemicellulolytic enzymes are the main effective agents during the decay of wood and may be involved in the transformation of numerous pollutants. The amounts produced and strengths of LSD enzymes are different for each type of white rot fungi, resulting in different oxidative activities. Such enzymes are found naturally in the soil system and much experimental evidence (see below) has demonstrated their efficiency in PAH degradation also *in vitro*, that is, as cell-free enzymes in the absence of their originating cells.

Non-ligninolytic fungi oxidize PAHs via a membrane-bound cytochrome P450 monooxygenase by incorporating one atom of the oxygen molecule into the PAH to form an arene oxide. Most arene oxides are unstable and can undergo either enzymatic hydration to form *trans*-dihydrodiols or non-enzymatic rearrangement to form

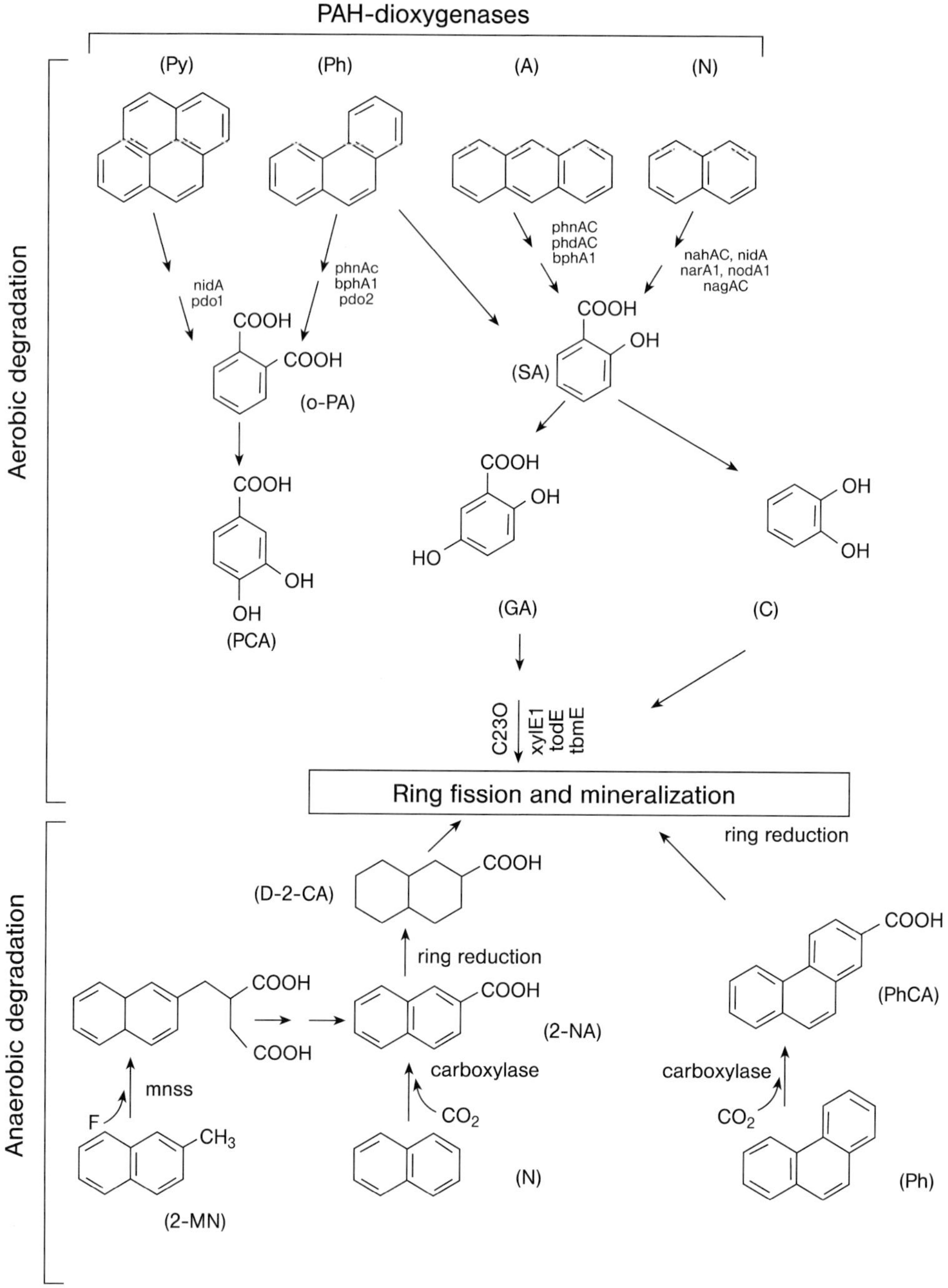
PAH-dioxygenases
(Py)
(Ph)
(A)
(N)
nidA
pdo1
phnAc
bphA1
pdo2
phnAC
phdAC
bphA1
nahAC, nidA
narA1, nodA1
nagAC
COOH
COOH
(o-PA)
(SA)
OH
(PCA)
(GA)
HO
(C)
C23O
xylE1
todE
tbmE
Ring fission and mineralization
ring reduction
(D-2-CA)
(2-NA)
carboxylase
CO_2
mnss
F
CH_3
(2-MN)
(PhCA)
Aerobic degradation
Anaerobic degradation

phenols, which can be conjugated with sulfate, glucose, xylose or glucuronic acid. Non-ligninolytic fungi are able to oxidize PAHs to *trans*-dihydrodiols, phenols, tetralones, quinones, dihydrodiol epoxides and various conjugates of the hydroxylated intermediates, but only a few have the ability to degrade PAHs to CO_2.

9.5
Dioxygenases as Key Enzymes in the Aerobic Degradation of PAHs and Markers of Bacterial Degradation

A broad range of oxygenases, differing in structure, mechanism and cofactor requirement, are distributed among the microorganisms that grow on PAHs. Aromatic ring-hydroxylating dioxygenases generally consist of a terminal dioxygenase and a reductase chain, which transfers electrons from NAD(P)H to the terminal dioxygenase. The reduced terminal dioxygenase catalyzes the direct insertion of O_2 into the substrate. Some terminal dioxygenases are homomultimers, whereas others are heteromultimers consisting of a large subunit (α) and a small subunit (β) [50].

The large subunits with the same reported substrate specificity are in general closely related and their DNA sequences are conserved. Phylogenetic studies of amino acid sequences of the proteins involved in the initial oxidative attack of PAHs and in their ring cleavage show significant sequence homology, indicating a common ancestry and allowing the design of group-specific primer sets for detection by polymerase chain reaction (PCR) [51, 52].

Among the Gram-negative PAH-degrading bacteria, several groups of genes for the initial naphthalene oxygenation are known: *ndo* genes from *P. putida* NCIB9816, *nah* genes from *P. putida* G7 and NCIB 9816-4, *dox* genes derived from *Pseudomonas* sp. C18 and *pah* genes cloned from *P. putida* OUS82 and *P. aeruginosa* PaK1. Each gene name is referred to the substrate used by the strains:

◄ ──────────────────────────────────────

Scheme 9.1 Schematic presentation of initial aerobic and anaerobic bacterial attack of PAHs and enzymes and genes involved in the respective reactions. Reaction under aerobic conditions: naphthalene (N) is initially oxygenated by a naphthalene dioxygenase to yield 1,2-dihydroxynaphthalene, which is further degraded to salicylate (SA). SA is then metabolized via catechol (C) or gentisic acid (GA). Anthracene (A) is dioxygenated at the 1,2-position to yield 1,2-dihydroxyanthracene, which is later metabolized to SA and C. Phenanthrene (Ph) is mostly oxygenated by a phenanthrene dioxygenase to yield 3,4-dihydroxyphenanthrene, which is further metabolized through two pathways: one follows the N biodegradation pathway with the formation of SA and the other leads to protocatechuic acid (PCA). Pyrene (Py) is mostly dioxygenated at the 1,2-position. A successive derivative, 4-phenanthroic acid, undergoes a second dioxygenase reaction and is further metabolized to o-phthalic acid (o-PA) and further metabolism proceeds via catabolic pathways similar to those of Ph (for detailed reactions, see [36]). Reactions under anaerobic conditions: for N, addition of CO_2 to yield 2-naphthoic acid (2-NA), which is further degraded by sequential reduction steps to decalin-2-carboxylic acid (D-2-CA); Ph is also directly carboxylated to phenanthrenecarboxylic acid (PhCA) as the initial step towards mineralization; for 2-methylnaphthalene (2-MN), addition of fumarate (F), catalyzed by naphthyl-2-methylsuccinate synthase. Source: adapted from [44, 47].

nah for naphthalene degradation, *ndo* for naphthalene dioxygenation (equivalent to *nah*), *dox* for dibenzothiophene oxidation and *pah* for phenanthrene degradation. These genes are grouped as the 'classical *nah*-like genes' for their gene organization and sequence similarity (about 90%) to that of the *nah* genes of strain G7. In some cases, the enzymes encoded by NAH plasmids have broad specificities, allowing the host to grow on several three-ring PAHs as the sole carbon and energy sources [53].

Other phenanthrene and naphthalene catabolic genes are instead evolutionarily different from the *nah*-like genes. They have been characterized from various Gram-negative and Gram-positive bacteria. They are *phd* genes from *Comamonas testosteroni* strain GZ39, *nag* genes from *Ralstonia* sp. U2 and *phn* genes from *Burkholderia* sp. RP007, *Alcaligenes faecalis* AFK2 and *Sphingomonas* and its related species [36]. Moreover, *phn* genes of the last two microorganisms present a novel structure with a similarity of 31–39% and less then 60% to that of other aromatic ring dioxygenases, respectively. Strain AFK2 utilizes phenanthrene but not naphthalene [54]; in contrast, *Sphingomonas* F199 can grow on various mono- and polycyclic aromatic hydrocarbons, biphenyl and dibenzothiophene [55].

Dioxygenases involved in PAH degradation by Gram-positive bacteria are distantly related to PAH dioxygenases of Gram-negative strains. The *nar* and *phd* genes have been identified in *Rhodococcus* sp. strain NCIMB12038 [56] and in *Nocardioides* sp. KP7 [57], respectively. Similar dioxygenase system genes, *nid* and *pdo* genes, which encode a novel polycyclic aromatic ring dioxygenase, have been reported from *Mycobacterium* sp. PYR-1 [58] and *Mycobacterium* sp. 6PY1 [59], respectively. Whereas PYR-1 can mineralize pyrene, 1-nitropyrene, phenanthrene, anthracene, fluoranthene and benzo[*a*]pyrene and has either mono- or dioxygenases to catalyze the initial attack of PAHs, 6PY1 mineralizes pyrene and phenanthrene and the initial attack of each PAH involves a specialized enzyme, *pdo1* for pyrene and phenanthrene and *pdo2* preferentially for phenanthrene. Whereas the classical *nah*-like genes and the *phn* genes are involved in the degradation of two- and three-ring PAHs, the *pdo1* genes facilitate growth on the four-ring PAH pyrene [59].

The catechol 2,3-dioxygenase genes comprise a diverse family of genes [60] that code for a group of enzymes with aromatic ring fission activity towards a wide range of aromatic pollutants. The gene phylogeny of these very closely related sequences does not follow strictly a taxonomic relationship with the bacterial hosts, since these genes are mainly found on plasmids and their evolution and conservation rates are heavily affected by traits such as selection pressures, horizontal transfer and mobile genetic elements [61].

Collectively, these features make the α-subunits of dioxygenases amenable to retrieval by PCR and allow for the identification of retrieved fragments in the context of existing databases. On account of the key position of the initial and ring oxygenases during aerobic PAH degradation, the genes encoding these enzymes have been used as targets to detect the presence of hydrocarbon degraders at the DNA level and to construct specific gene probes. The probes may be employed to determine the overall genetic diversity and the density and frequency of specific gene lines required to degrade a target compound at a site.

Once a probe has been developed, it can be expanded for many uses, including quantitative PCR, real-time PCR and functional gene probe arrays.

The development over the years of several primers has allowed the detection and quantification in environmental samples of the presence of specific genotypes encoding the key steps in PAH biodegradation and in their mRNA [51, 52, 62]. For example, *nid* gene sequences encoding for the large (*nidA*) and the small (*nidB*) subunits of dioxygenase system have been used to create a gene probe for the detection of PAH-degrading mycobacteria in soils undergoing bioremediation [62].

The intrinsic biodegradative potential of harbor sediments contaminated with PAHs was examined in bioslurry microcosms, during a 4 month treatment, by correlating the microbial community structure, through PLFA analyses, with catabolic gene presence and with PAH loss. A threefold increase in total microbial biomass and a dynamic microbial community composition were revealed and both strongly correlated with changes in the PAH chemistry. The copies of genes associated with PAH degradation increased by four orders of magnitude. Declines in the concentrations of phenanthrene correlated with PLFA indicative of *Rhodococcus* spp. and/or actinomycetes and genes encoding for naphthalene and C23O degradative enzymes [63].

Therefore, the greater numbers of catabolic gene copies within a contaminated area, in comparison with those in uncontaminated soils, can be used as evidence of natural attenuation or of the effectiveness of engineered bioremediation.

Both PCR-amplified fragments and nucleotides derived from functional genes can be used to fabricate functional gene arrays. Rhee *et al.* [64] developed a comprehensive 50-mer-based oligonucleotide microarray containing probes from all known genes involved in PAH biodegradation and metal reduction and used it to monitor the biodegradation potential and activity of enrichment cultures and soil microcosms supplemented with naphthalene. With this gene array, they successfully detected changes in the microbial community structures in enrichment and soil microcosms and found that the naphthalene-degrading genes from *Rhodococcus*-type bacteria were dominant in naphthalene enrichments, whereas the genes involved in naphthalene and PAH degradation from Gram-negative bacteria were most abundant in soil microcosms.

Numerous environmental factors can activate or repress gene expression and thereby modulate microbial activity. The influence of agents increasing PAH bioavailability on the expression of the catabolic gene *nahAC* has been studied by using an RT-PCR assay and to evaluate the regulation of the *nahAC* gene during phenanthrene degradation [65]. Gene expression was successfully determined by extraction of bacterial mRNA followed by RT-PCR amplification of the *nahAC* gene and of an internal housekeeping gene (*rpoD*). The *rpoD* gene, which encodes a housekeeping sigma factor for transcription initiation, served as a baseline control for evaluation/interpretation of catabolic gene expression. With this gene expression assay, temporal changes in *nahAC* expression were shown when phenanthrene was degraded. Hence monitoring gene expression may allow the detection of subtle changes in the expression of degradative genes, due to temporal changes in cell physiology or changes in toxicity associated with the accumulation of intermediates.

9.6
PAH Transformation by Extracellular Fungal Enzymes

Enzymes mainly involved in PAH transformation as cell-free or extracellular enzymes are oxidoreductases including phenoloxidases and peroxidases. The cleavage of the aromatic ring and the formation of unstable substrate cation radicals with subsequent non-enzymatic transformation (for example, C–C or ether cleavage or oxidative coupling) and polymerization are reactions that may occur. Phenoloxidases, including tyrosinases and laccases, require molecular oxygen for activity, whereas peroxidases that comprise horseradish peroxidase (HPR), ligninases (LiP and MnP) and chloroperoxidases (CPO) utilize hydrogen peroxide. In some cases, for example manganese peroxidases, the reaction depends on the presence of other components such as divalent manganese.

Phenoloxidases and peroxidases are produced by a large number of living cells (microorganisms, plants and animals), but their main producers are white rot fungi, suggesting the major role of these enzymes in lignin transformation. The biochemical properties, the catalytic features and the potential use of these enzymes for practical applications have been exhaustively studied and numerous reviews have been published [66, 67].

Both groups catalyze by different mechanisms the oxidation of phenolic and non-phenolic aromatic compounds through an oxidative coupling reaction that results in the formation of polymeric products of increasing complexity. Cross-coupling reactions may occur between substrates of different nature. Oxidation of relatively inert substrates by the co-presence of more reactive molecules also may occur.

As previously reported, the lignin-degrading system (LDS), or some of its enzymatic constituents, plays the primary role in the degradation of PAHs (Table 9.1) [68]. Peroxidases such as CPOs and HPRs or enzymatic and non-enzymatic hemoproteins such as cytochrome c (Cyt c) and hemoglobin (HMG) and also tyrosinases showed different capabilities to oxidize PAHs (Table 9.1) [7].

The enzymes reported in Table 9.1 had different activities and substrate spectra. Among them, the best catalysts were lignin peroxidase and chloroperoxidase. Much lower specific activities and less catalytic capability toward different PAHs were shown by HMG, Cyt c and Cyt P450. The LiPs requiring H_2O_2 (usually endogenously generated) are particularly attractive as pollutant degraders because they are relatively non-specific, have a very high oxidation-reduction potential and can potentially oxidize xenobiotics that are usually difficult to be affected by other peroxidases [69, 70]. The LiP from the white rot fungus *Nematoloma frowardii* transformed 58.6% of anthracene and 34.2% of pyrene, whereas only 31.5% of anthracene and 11.2% of pyrene were oxidized by MnP from the same fungus. CPO from *C. fumago* displayed high activity on several PAHs and also a broad substrate spectrum (Table 9.1). CPO is a versatile enzyme capable of catalyzing different types of reactions such as catalase, halogenase, peroxidase and cytochrome P450-like substrate oxidations. Its catalytic flexibility renders it an attractive catalyst for environmental application.

The different reactivities of the considered proteins and enzymes involved in PAH degradation are a result of different factors related to the structure of both the catalyst

Table 9.1 Oxidoreductive enzymes, enzymatic and non-enzymatic hemoproteins capable of degrading PAHs.

Enzyme	Sources	PAHs
Chloroperoxidase	*Caldariomyces fumago*	Acenaphthene, anthracene, azulene, benzothiophene, dibenzothiophene, fluoranthene, fluorene, 2-methylanthracene, 9-methylanthracene, phenanthrene, pyrene, thianthrene
Cytochrome *c*	*Saccharomyces cerevisiae*	Anthracene, azulene, benzo[a]pyrene, benzothiophene, carbazole, dibenzothiophene, 9-methylanthracene, pyrene, thianthrene
Cytochrome P450 Hemoglobin	*Pseudomonas putida*	Crysene, fluoranthene, fluorene, pyrene Acenaphthene, anthracene, azulene, benzo[a]pyrene, benzothiophene, fluoranthene, fluorene, phenanthrene, pyrene, thianthrene
Laccase	*Trametes versicolor, Coriolopsis gallica, Nematoloma frowardii*	Anthracene, benzo[a]pyrene, acenaphthene, 2-methylanthracene, 9-methylanthracene, phenanthrene
Lignin peroxidase	*Phanerochaete chrysosporium, Nematoloma frowardii*	Acenaphthene, anthracene benzothiophene, benzo[a]pyrene carbazole, dibenzothiophene, fluoranthene, 2-methylanthracene, 9-methylanthracene, phenanthrene, pyrene
Manganese peroxidase	*Phanerochaete laevis, Nematoloma frowardii*	Anthracene, benzo[a]anthracene, benzo[a]pyrene, fluoranthene, fluorene, phenanthrene
Tyrosinase	*Mushroom*	Anthracene, fluoranthene, phenanthrene, pyrene

Source: adapted from [7] and [68].

and the substrate involved in the reaction and also to their biological functions. For instance, CPO has a cysteine residue as axial ligand of the heme iron. This residue is an electron donor and favors the cleavage of the O–O bond of hydrogen peroxide. By contrast, in the other peroxidases and hemoproteins, the residue that coordinates the heme iron is a histidine residue, which is an electron acceptor. Moreover, the ionization potential of each compound involved strongly influences the ability of the enzyme in the oxidation process, thus confirming that the mechanism involves a free radical. Furthermore, studies performed with PAH derivatives, known as intermediates or potential dead-end products of microbial PAH metabolism, demonstrated that the hydroxylated PAH metabolites were oxidized by all the oxidoreductases, whereas PAH quinones and oxo metabolites were not transformed. Biodegradation of PAHs and their metabolites increases with their oxidation stage, suggesting that enzymatic oxidation followed by the action of indigenous microorganisms could be an effective PAH remediation strategy.

Although the catalytic efficiency has to be considered when selecting an enzyme catalyst, factors such as stability, ability to act in harsh environment (e.g. strong organic media), cost and availability should be considered. Indeed, the use of enzymes as detoxifying agents may present several bottlenecks, such as (a) low catalytic efficiency with low reaction yields, (b) presence of complex pollutant mixtures, (c) low stability, mainly in organic media such as those possibly present in PAH-contaminated sites, and (d) low ability to adapt themselves to survive in harsh environments or to reproduce themselves with high enzyme costs. This implies that for enzymes to be effective bioremediation agents they should have a broad substrate specificity and high substrate affinity and catalytic velocity [values of Michaelis–Menten constant (K_m) in the μmol L^{-1} range and of catalytic constant (k_{cat}) >100 min^{-1}], should not require cofactors for activity as cofactors such as NDPH and glutathione are prohibitively expensive, should be resistant to conditions such as proteolysis, changes in pH and temperature and the presence of inhibitors, and production and formulation processes for the enzymes should be cheap, that is, a bacterial expression of >5% yield of enzymes and shelf-life of months.

Several strategies have been addressed to improve the effectiveness of oxidor-eductive enzymes towards the transformation of organic pollutants and PAHs. One of the most interesting aspects of these enzymes is their capability to transform relatively recalcitrant compounds if additional co-substrates or proper redox, mediators are present. Chemical mediators are believed to be oxidized by the enzyme and then undergo oxidative coupling with enzyme substrates. For instance, this process seems to be essential to the laccase-mediated transformation of recalcitrant compounds and effective mediators for laccase were 2,2'-azinobis (3-ethylbenzthiazoline-6-sulfonic acid) (ABTS), 3-hydroxyanthranilate, 1-hydroxy-benztriazole (HTB) [66] and also natural mediators such as veratryl alcohol, phenylacetic acid, hydroxybenzoic acid, hydroxybenzyl alcohol, vanillin, acetova-nillone, acetosyringone, syringaldehyde, 2,4,6- trimethylphenol, *p*-coumaric acid, ferulic and sinapic acids and tyrosine [7, 71]. For instance, in the presence of HBT, the transformation of acenaphthylene, acenaphthene and pyrene by laccase from *T. versicolor* increased from less than 10% up to ~100% [7]. Similarly, vanillin, acetovanillone, 2,4,6-trimethylphenol and, above all, *p*-coumaric acid strongly promoted the removal of PAHs by laccase from *P. cinnabarinus* ss3. The *p*-coumaric acid resulted in a better laccase mediator than ABTS and close similarity to HBT, attaining 95% removal of anthracene and benzo[*a*]pyrene and around 50% of pyrene within 24 h [71].

The successful *in vitro* transformation of PAHs by laccase in the presence of natural phenolic mediators supports the potential role of these systems for the *in vivo* degradation of PAHs in soil. Laccase-aided decontamination of these pollutants could be directed not only towards their immobilization (humification) by oxidative coupling but also towards their oxidative transformation (promoted by phenolic mediators) to products easily taken up by soil microflora.

Another attractive strategy is the generation of improved and optimized catabolic enzymes by directed evolution methods or by site-specific mutagenesis. Enzymes with increased stability and activity under selected conditions, for example, in solvents or at high temperatures, and enzymes with new and/or improved activities can be generated and used to develop biocatalytic processes. The laccase gene from

Myceliophthora thermophia (MtL) was transformed into *Saccharomyces cerevisiae* and subjected to directed evolution. After 10 rounds of directed evolution, a mutant with an eightfold increase in laccase expression and a 22-fold increase in the k_{cat} for ABTS was created. The final mutant had a total activity 170-fold higher than the wild type. The MtL enzyme holds great potential for the bioremediation of PAHs due to its high thermal stability that enables it to work at the elevated temperatures needed to increase the solubility of highly recalcitrant PAHs. Moreover, engineering of a laccase with activity in the absence of mediators would be a suitable target for directed evolution [72].

9.7
In Situ Strategies to Remediate Polluted Soils

Bioremediation techniques require the optimization of the conditions for microbial growth and biodegradation as they are dependent on the catabolic activities of the indigenous microflora.

The implementation of a complete remediation program usually requires more than one step: (a) knowledge of the past history of the polluted area and activities leading to the contamination of the site; (b) examination and quantification of the severity of the contamination problem; (c) development of the remediation action program to target the specific contaminant or group of contaminants; (d) development of treatments and a treatment sequence suited for the wastes and the site; and (e) an effective monitoring program to evaluate the effectiveness of the adopted bioremediation plan.

The positive acceptance of a bioremediation process in many cases depends, however, on the overall cost, which has to be no greater than that of existing chemical and physical treatments.

9.7.1
Intrinsic or Natural Attenuation

Natural attenuation of contaminated sites is becoming more and more interesting, because it seems to permit bioremediation with minimal cost, avoiding land disruption and human exposure. Although usable at numerous sites, it can rarely be used as a sole treatment process since natural engineered biodegradation processes are very slow.

Biodegradation, chemical transformation, stabilization (that is, binding and sequestration by clays and humus), volatilization, dispersion, dissolution and dilution are all natural attenuation processes of organic contaminants in soil and groundwater. However, owing to the slow rate of such natural processes, certain chemicals may persist for years. For instance, their susceptibility to biodegradation may change drastically, depending on several factors related to the chemical and physical properties of both the chemical and the environment in which they are present.

Methods to assess microbial natural attenuation of organic pollutants include analysis of the subsurface geology and hydrology of the site, qualitative and quantitative pollutant profiles, pollutant bioavailability, composition and activity of the microflora and microcosm studies. An appropriate evaluation of natural attenuation, however, requires the demonstration that the transformation processes are taking place at a rate that is protective of human health and the environment.

Natural attenuation is mainly used for BTEX and more recently for chlorinated hydrocarbons; the natural attenuation of many other contaminants such as PAHs, PCBs and pesticides has not been investigated extensively.

PAHs are common in freshwater ecosystems and particularly in river sediments where they accumulate. As microorganisms in sediments have a high functional diversity and are exposed to humic or fulvic compounds containing aromatic structures, sediments commonly contain some PAH degraders. Freshwater sediments can be oxic (that is, under rapid aerated flow streams), but are frequently characterized by extensive anoxic zones. The availability of electron acceptors would then be crucial in controlling the biodegradation of PAHs. At low dissolved oxygen concentrations, degradation may even occur by alternative electron acceptors, including nitrate, Fe and Mn oxides, sulfate and carbon dioxide [73]. Moreover, river sediments continuously receive organic residues acting as valuable sources of energy and carbon for microorganisms and stimulating biological activity and O_2 consumption. Laboratory batch experiments were performed with PAH-polluted sediments to quantify the extent of PAH degradation under oxic and anoxic conditions and evaluate the feasibility of natural attenuation to remediate the river sediments [74]. The results indicated that PAH attenuation in polluted sediments under anaerobic conditions, even favorable for anaerobic metabolism, is exceedingly slow. In fact, three- and four-ring PAHs were degraded by the indigenous microbial community under aerobic conditions, but anaerobic degradation based on iron and sulfate reduction was not coupled with even phenanthrene degradation. The addition of cellulose stimulated aerobic and anaerobic respiration, but had no effect on PAH degradation [74]. Alternative approaches including dredging, deposition on land and O_2 diffusion through bioventing or composting/landfarming appear to be a better solution.

9.7.2
Biostimulation and Bioaugmentation

Normally soils contain bioavailable microbial substrates that become limiting and soil bacteria for the most part exist under starvation conditions. The presence of organic contaminants or metals can impose additional stress on microbial communities, resulting in decreased viable bacterial populations and/or activities. The absence of natural biodegradation in a contaminated site indicates that both localized environmental conditions and physical and biological constraints may limit partially or totally the degradation of pollutants.

Several strategies may be adopted to help natural biodegradation processes to work faster and overcome the above constraints: bioaugmentation, that is, the addition of exogenous specific strains to polluted sites for enhancing the biological activity of the existing populations, and biostimulation, that is, the addition of supplementing carbon sources or other nutrients to stimulate the activity of indigenous or inoculated degrading strains.

The utility of bioaugmentation is supported by the apparent enhanced bioremediation rate after the addition of competent microorganisms [75].

The use of pure cultures into multi-substrate-polluted systems, such as soil or groundwater, has had variable results [76]. The addition of known PAH-degrading strains to soils with indigenous bacteria with adequate PAH-degrading capacity slightly enhanced PAH bioremediation [77, 78] as the indigenous bacteria in soils with a long history of PAH contamination are probably very well adapted to PAH degradation and the addition of additional PAH-degrading bacteria seems to be of little advantage. These results, similar to those reported for other cultures [79, 80], are in contrast to an observed enhanced PAH biodegradation by augmentation with *Burkholderia cepacia* EA100 [81]. However, even if added strains do not enhance the rate of PAH degradation, their survival is improved by the presence of PAHs [78]. Although bioaugmentation is still a source of controversy within environmental microbiology, it is considered as a way of enhancing the genetic capacity of a given site, thus corresponding to an increase in the gene pool and in the genetic diversity of the site [82]. Gene bioaugmentation is now defined as the process of obtaining enhanced activity after gene transfer from an introduced donor organism to a member of the indigenous soil population. In this context, the long-term survival of the introduced donor strain does not represent the major bottleneck in bioaugmentation processes.

Thomson *et al.* [83] pointed out technical advances to perform a strain selection based on an *in situ* understanding of organism abundance, functional activity and population dynamics in the habitat from which they are derived. A strain, derived from a population, temporally and spatially prevalent in a specific habitat, more likely persists as an inoculum when reintroduced than one that is transient or alien to such a habitat. Once an abundant population has been identified, the second phase of the selection procedure is to identify strains that are capable of degrading the contaminant. In the bioremediation of co-contaminated soils, until recently the bioaugmentation strategy was focused on the introduction of a microorganism that was both metal resistant and capable of organic degradation. As such approach is frequently unsuccessful, a dual-bioaugmentation strategy in the remediation of co-contaminated systems has been proposed on the basis of field study experiments [84]. The strategy involved the co-inoculation of a metal-resistant population with an organic degrading population, the primary mode of action being metal detoxification, such that organic degradation was no longer inhibited.

Effective strategies to enhance organic degradation in the presence of toxic metals include also reducing the bioavailable metal concentration. They include amendment with kaolinite and montmorillonite clays, with chelating resins and more recently with the delivery of a rhamnolipidic biosurfactant produced by *Pseudomonas aeruginosa* to alter the microbial metal uptake, thus resulting in an enhanced rate of bioremediation [85].

Determination of the potential success of bioaugmentation, however, requires an understanding of the bioavailability of the pollutant, the survival and activity of the added microorganism(s) or its genetic material and the general environmental conditions that control soil bioremediation rates.

The understanding of the survival and fate of added microorganisms and of the critical parameters for the design of bioaugmentation processes can be now routinely addressed through the use of gene probes, PCR technology and immunoassay technology.

To overcome chemical and environmental constraints, delivery systems may be designed to provide nutrients (nitrogen and phosphorus), oxygen and other electron acceptors to stimulate and maintain the activity of microbial degraders. Similarly, environmental conditions can be improved to obtain optimal values of pH, moisture and so forth for the microbial degradation. Substances more amenable to biodegradation than the target contaminant can be added to soil to stimulate the microbial cometabolic transformation of the pollutants, otherwise not degraded.

The addition of nutrients can lead to an increased development of indigenous microorganisms, which themselves either biostimulate the process or hinder the process by consuming the added nutrient or carbon source. In cases where co-metabolism is desired, the consumption of added substrates by indigenous microorganisms, incapable of co-metabolizing the pollutant, can hinder the growth of the added microorganisms.

The addition of mature composted materials is a means for the bioaugmentation of a contaminated soil to be remediated. Composts have an enormous potential for bioremediation as they are capable of sustaining diverse populations of microorganisms, such as bacteria and ligninolytic fungi, all with the potential to degrade a variety of aromatic pollutants. Two possible hypotheses might explain the effect of compost: the first is that the degradation is due to the activity of the microorganisms inherent to the compost and the second is that the degradation is due to the simple function of the compost acting as an organic substrate. According to the first, consecutive biotransformation of pollutants and their metabolites can occur because of the synergistic interactions of the microorganisms in the system. Five-ring PAHs were mineralized up to 17% in composts, thus supporting this hypothesis [86]. According to the second, composts can entrap pollutants within the organic matrix, thereby reducing pollutant bioavailability. Furthermore, the compost, as a soil ameliorant, is capable of changing pH, moisture content and soil structure and may act as a nutrient source, thereby improving the contaminated soil environment for indigenous or inoculated microorganisms. Thus, the depletion of the PAH compounds in soil might be due to real microbial degradation or to sorption on the organic matter. The fact that PAHs sorbed on the soil matrix were also degraded in the soil–compost mixture may account for the attack by esoenzymes such as peroxidases and laccases. To date, the use of compost has not been widely applied as a bioremediation strategy, also because of the concern deriving from the mixing of non-contaminated material with contaminated soil and resulting in a far greater quantity of contaminated material in the case of an unsuccessful bioremediation [87].

Composting has been demonstrated to be effective in biodegrading PAHs present in contaminated soil at both the laboratory and field scales [88, 89]. In the course of compost maturation, the pollutants are degraded by the active microflora within the mixture. For example, pyrene was found to be degraded in the composting of soil–sludge mixtures [90]. The disappearance of benzo[a] pyrene in a soil under composting in the presence and absence of *Phanerochaete chrysosporium* was monitored during 95 composting days. No appreciable difference in benzo[a]pyrene removal between the uninoculated (65.6%) and inoculated (62.8%) systems was observed and nearly 100% of the benzo[a]pyrene removed was in the form of bound residue, indicating that the bioaugmentation of the soil composting system with *P. chrysosporium* was ineffective in degrading benzo[a] pyrene. Joyce *et al.* [91], investigating the fate of a mixture of three- and four-ring PAHs under composting conditions with solid municipal waste, over 60 days of composting, found that anthracene, phenanthrene and pyrene were really lost during the active phase of composting (30 days) by a combination of biotic and abiotic mechanisms, principally dominated by the biotic processes. More recently, using laboratory-scale in-vessel composting reactors to investigate the degradation of 16 PAHs from a coal tar-contaminated soil, Antizar-Ladislao *et al.* [92] confirmed that in-vessel composting can reduce PAH concentrations in the soil and that optimal removal occurred at 38 °C where the highest microbial activity was also observed. The main mechanism of PAH removal was biological, although abiotic mechanisms also played a role. With the same composting approach, the biodegradation of an aged coal tar-contaminated soil amended with green waste over 56 days was also demonstrated [93].

9.7.3
Phytoremediation

Plants may perform and enhance the reduction of several organic contaminants, including petroleum hydrocarbons, solvents, pesticides, heavy metals, radionuclides and landfill leachates in soil.

Phytoremediation is the *in situ* use of plants and their associated microorganisms to degrade, contain or render harmless contaminants in soil or groundwater (Figure 9.2).

As phytoremediation enhances the natural attenuation of contaminated sites, it can be considered intermediate between engineering and natural attenuation. Because phytoremediation depends on natural, synergistic relationships among plants, microorganisms and the environment, it does not require intensive engineering techniques or excavation. Human intervention may, however, be required to establish an appropriate plant–microbe community at the site or apply agronomic techniques (such as tillage and fertilizer application) to enhance natural degradation or containment processes.

Plants may utilize different mechanisms to remove hydrocarbons efficiently from a polluted soil. Evapotranspiration effective only towards volatile hydrocarbons, uptake and breakdown of hydrocarbons into smaller products, accumulation in the

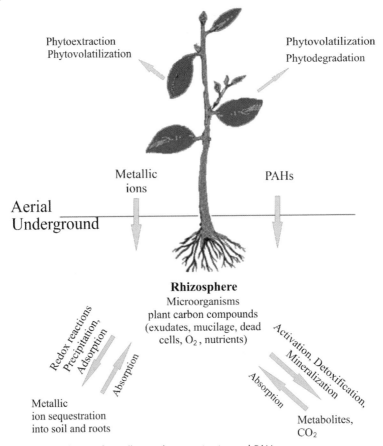

Figure 9.2 Scheme of metallic ion decontamination and PAH decontamination/degradation in phytoremediation processes.

root zone and/or translocation to aerial parts are the main mechanisms of a phytoremediation action (Figure 9.2).

An interesting phenomenon is the synergistic interaction between plants and microorganisms that specifically occurs in the soil environment influenced by plant roots or rhizosphere. The latter represents a complex ecosystem with the potential to accelerate biodegradation of PAHs from soil. This remediation process is referred to as rhizodegradation/rhizosphere bioremediation [4, 94] and often it results in more efficient removal of pollutants from the contaminated soil deriving from an increased microbial activity. Plants and microbes have co-evolved a mutually beneficial strategy, where microorganisms benefit from the plant exudates whereas the plants benefit from the ability of microorganisms to break down toxic chemicals.

Microbial growth may be stimulated by organic plant exudates, impacting PAH bioaccessibility by two contrasting mechanisms. One can improve PAH bioaccessibility, whereas the other can retard it [95]. Some phenolic compounds have been postulated to serve as the primary substrate for PAH co-oxidation and/or as potential inducers of microbial oxygenases that initiate the PAH degradation. The concurrent release of easily degradable organic molecules at higher concentration than the phenolics, however, could repress the expression of pertinent catabolic genes. Hence the evaluation of the microbial exposure to root-derived substrates on catabolic gene expression will become of relevant importance in understanding the mechanisms of enhanced PAH rhizodegradation. The bioluminescent reporter strain *Pseudomonas fluorescens* HK44 with a *nah–lux* fusion, was used to investigate the effect of root material from various plants and potential plant root-derived substrates (sugars, carboxylic acids, amino acids and phenolics) on the expression of a gene (*nahG*) for naphthalene dioxygenase transcription [96]. The *nahG* gene was induced by some phenolics, releasable by plants, but not by root extracts. Higher bacterial growth in reactors with root extracts and naphthalene resulted in a higher level of *nahG* expression by the resulting larger microbial population and in faster naphthalene degradation rates. A large heterotrophic community is conducive to faster degradation rates of labile-C-substrate concentrations and to facilitation of the induction and PAH degradation processes.

In pot experiments, soil contaminated with petroleum hydrocarbons during the Gulf War [97] was successfully treated over 27 months with both *Medicago sativa* and *Phragmites australis* and vegetated soils had significantly higher numbers of PAH degraders than the unvegetated control [98]. *Cannaba sativa* also reduces significantly the concentration of benzo[a]pyrene and chrysene in pot experiments [99].

In field phytoremediation of a soil contaminated with aged hydrocarbons, the decrease in hexadecane, naphthalene and phenanthrene levels was related to the increase in bacteria containing hydrocarbon catabolic genes (*ndoB*, *alkB* and *xylE*) in either the bulk or rhizosphere soil [100]. With *Festuca aurundinacea*, the catabolic activity of rhizosphere was stimulated and the increase in catabolic activity corresponded to an increase in specific catabolic genes; in contrast, with *Trifolium hirtum* the reverse was observed, thus confirming the importance of plant-specific interactions during phytoremediation [101, 102].

The plants used in phytodegradation are generally selected on the basis of their growth rate and biomass, the depth of their root zone, the tolerance and the bioaccumulating capability towards particular contaminants. Various plants have been identified for their potential to facilitate the phytoremediation of sites contaminated with petroleum hydrocarbons and most representative are grass cultivars. Indeed, the highly branched and deep root systems render them attractive candidates for phytoremediation. In addition, the monocots wheat and grass can carry approximately 10-fold more bacteria on the root system, especially on the lower parts, compared with some dicots [103]. Moreover, grasses are also considered suitable for inorganic and organic heavy metal remediation [94].

Full-scale applications of this technology are currently limited to only a small number of cases [104]. Phytoremediation, in addition to being used for treating contaminants in low concentrations, can be seen as a final polishing step following the initial treatment of the high-level contamination.

Pradhan *et al.* [105], by using phytoremediation as either a primary remediation technology or a final polishing step for treatment of soil contaminated with PAHs, observed a 57% reduction in total PAH concentration in a 6 month treatment with *Medicago sativa, Panicum virgatum* and *Schizachyrium scoparium*. In a 10 month experiment performed under field conditions in a loamy soil in South Carolina, *Lolium multiflorum* decreased the pyrene concentration below the detection limit [106].

With a multicomponent phytoremediation system of the soil combining land farming, bioaugmentation with PAH-degrading bacteria and growth of plants (*Festuca aurundinacea*) with PGPR, an improved effective removal of 16 persistent and soil-bound PAHs was obtained under laboratory conditions when compared with the results of the single treatment [107]. Phytoremediation was successful because plant species were able to grow in the presence of high levels of contaminants and the strains of PGPR increased plant tolerance to PAHs and accelerated plant growth in heavily contaminated soils.

The specific advantages, limitations and economics of phytoremediation compared with other approaches have been extensively reviewed [108–110]. Several environmental factors affect or alter the mechanisms of phytoremediation. Soil type and organic matter content can limit the bioavailability of petroleum contaminants. The water content in soil and wetlands affects plant/microbial growth and the availability of oxygen required for aerobic respiration. Temperature affects the rates at which various processes take place. Nutrient availability can influence the rate and extent of degradation in oil-contaminated soil. Finally, sunlight can transform parent compounds into other compounds, which may have different toxicities and bioavailabilities than the original compounds. These various environmental factors cause *weathering* – the loss of certain fractions of the contaminant mixture – with the end result being that only the more resistant compounds remain in the soil.

Successful phytoremediation of petroleum hydrocarbons will require the establishment of appropriate plants and microorganisms at the contaminated site. The stimulation of rhizodegradation processes can be also pursued. Factors to consider include (1) the influence of contaminants on the germination of plants or survival of transplanted vegetation, (2) the effectiveness of inoculating contaminated soils with microorganisms, the increase of their activity, (3) the use of native versus non-native plants and (4) the changes in the physical–chemical properties of the contaminated soil.

9.7.4
Feasibility of Bioremediation Technologies

Since the scope of bioremediation is to decrease the concentration of organic pollutants to levels undetectable or, if measurable, lower than the limits established as safe or tolerable by regulatory agencies, several criteria concerning the agent selected to perform the bioremediation process, the toxicity of the end-products and the conditions at the site, favorable or not to the process, must be met to be bioremediation seriously considered as a practical method for treatment.

Once a bioremediation program has been designed, its feasibility can be evaluated by considering: (a) *the applicability,* strictly related to the properties of the contaminants and the environmental, biological and hydrogeochemical features of the contaminated site; (b) *the treatability studies* to determine the potential for bioremediation, to define required operating and management practices, to design and implement the bioremediation plan and to establish a suitable and effective monitoring program; (c) *the possible limitations* and *drawbacks*; and (d) *the advantages* linked to its potential of harnessing naturally occurring biogeochemical processes, immobilizing or destroying, partially or completely, contaminants.

Only when all these conditions are met will a bioremediation process occur that is successful, productive, non-deleterious of soil quality and of acceptable cost.

References

1 Samanta, S.K., Singh, O.V. and Jain, R.K. (2002) *Trends in Biotechnology*, **20**, 243–248.

2 Adriano, D.C., Bollag, J.-M., Frankenberger, W.T.R., Jr and Sims, R.C. (1999) *Biodegradation of Contaminated Soils*, Agronomy Monograph 37, American Society of Agronomy, Crop Science of America, Soil Science Society of America, p. 772.

3 Iwamoto, T. and Nasu, M. (2001) *Journal of Bioscience and Bioengineering*, **92**, 1–8.

4 Siciliano, S.D. and Germida, J.J. (1998) *Environmental Reviews*, **6**, 65–79.

5 Pointing, S.B. (2001) *Applied Microbiology and Biotechnology*, **57**, 20–32.

6 Gianfreda, L. and Bollag, J.-M. (2002) in *Enzymes in the Environment. Activity, Ecology and Applications* (eds R.G. Burns and R. Dick), Marcel Dekker, New York, pp. 491–538.

7 Torres, E., Bustos-Jaimes, I. and La Borgne, S. (2003) *Applied Catalysis B- Environmental*, **46**, 1–7.

8 Beaudin, N., Caron, R.F., Legros, R., Ramsay, J., Lawlor, L. and Ramsay, B. (1996) *Compost Science & Utilization*, **4**, 37–45.

9 Huesemann, M.H., Hausman, T.S. and Fortman, T.J. (2004) *Biodegradation*, **15**, 261–274.

10 Johnsen, A.R. and Karlson, U. (2004) *Applied Microbiology and Biotechnology*, **63**, 452–459.

11 Birman, I. and Alexander, M. (1996) *Environmental Toxicology and Chemistry/ SETAC*, **15**, 1683–1686.

12 Cavalca, L., Rao, M.A., Bernasconi, S., Colombo, M., Andreoni, V. and Gianfreda, L. (2008) *Biodegradation*, **19**, 1–13.

13 Alexander, M. (2000) *Environmental Science & Technology*, **34**, 4259–4265.

14 Gevao, B., Semple, K.T. and Jones, K.C. (2000) *Environmental Pollution*, **108**, 3–12.

15 Barraclough, D., Kearney, T. and Croxford, A. (2005) *Environmental Pollution*, **133**, 85–90.

16 Doelman, P., Jansen, E., Michels, M. and Van Til, M. (1994) *Biology and Fertility of Soils*, **17**, 177–184.

17 Sokhn, J., De Lej, F.A.A.M., Hart, T.D. and Lynch, J.M. (2001) *Letters in Applied Microbiology*, **33**, 164–168.

18 Wong, K.W., Toh, B.A., Ting, Y.P. and Obbard, J.P. (2005) *Letters in Applied Microbiology*, **40**, 50–55.

19 Juck, D., Charles, T., White, L.G. and Greer, G.W. (2000) *FEMS Microbiology Ecology*, **33**, 241–249.

20 Roling, W.F.M., Milner, M.G., Jones, D.E., Lee, K., Daniel, F. and Swannell, R. J. P (2002) *Applied and Environmental Microbiology*, **68**, 5537–5548.

21 Kanaly, R.A., Bartha, R., Watanabe, K. and Harayama, S. (2000) *Applied and Environmental Microbiology*, **66**, 4205–4211.

22 Daane, L.L., Harjonio, I., Zylstra, G.J. and Hagblom, M.M. (2001) *Applied and Environmental Microbiology*, **67**, 2683–2691.

23 Willmusen, P.A., Karlson, U., Stakebrandt, E. and Kroppenstedt, R.M. (2001)

International Journal of Systematic and Evolutionary Microbiology, **51**, 1715–1722.

24 Saul, D.J., Aislabie, J.M., Brown, C., Harris, L. and Foght, J.M. (2005) *FEMS Microbiology Ecology*, **53**, 141–155.

25 Leys, N.M.J., Ryngaert, A., Bastiaens, L., Wattiau, P., Top, E.M., Verstraete, W. and Springael, D. (2005) *Microbial Ecology*, **51**, 375–388.

26 Leys, N.M.J., Ryngaert, A., Bastiaens, L., Verstraete, W., Top, E.M. and Springael, D. (2004) *Applied and Environmental Microbiology*, **70**, 1944–1955.

27 Ma, Y., Wang, L. and Shao, Z. (2006) *Environmental Microbiology*, **8**, 455–465.

28 Whyte, L.G., Bourbonnière, L., Bellerose, C. and Greer, C.W. (1999) *Bioremediation Journal*, **3**, 69–79.

29 Sotsky, J.B., Greer, C.W. and Atlas, R.M. (1994) *Canadian Journal of Microbiology*, **40**, 981–985.

30 Margesin, R., Labbé, D., Schinner, F., Greer, C.W. and Whyte, L.G. (2003) *Applied and Environmental Microbiology*, **69**, 3985–3092.

31 Johnsen, A.R., de Lipthay, J.R., Sorensen, S.J., Ekelund, F., Christensen, P., Andersen, O., Karlson, U. and Jacobsen, C.S. (2006) *Environmental Microbiology*, **8**, 535–545.

32 Atlas, R.M. and Bartha, R. (1998). *Microbial Ecology: Fundamentals and Applications*, 4th edn, Benjamin/ Cummings, Menlo Park, CA,

33 Gilewicz, M., Ni'matuzahroh, T., Nadalig, T., Budzinski, H., Doumenenq, P., Mitchotey, V. and Bertrand, J.C. (1997) *Applied and Environmental Microbiology*, **48**, 528–533.

34 Hedlund, B.P., Geiselbrecht, A.D., Bair, T.J. and Staley, J.T. (1999) *Applied and Environmental Microbiology*, **65**, 251–259.

35 Siciliano, S.D., Fortin, N., Mihoc, A., Wisse, G., Labelle, S., Beaumier, D., Ouellette, D., Roy, R., Whyte, L.G., Banks, M.K., Schwab, P., Lee, K. and Greer, C.W. (2001) *Applied and Environmental Microbiology*, **67**, 2469–2475.

36 Habe, H. and Omori, T. (2003) *Bioscience, Biotechnology, and Biochemistry*, **67**, 225–243.

37 Gibson, J. and Harwood, C. (2002) *Annual Review of Microbiology*, **56**, 345–369.

38 Cerniglia, C.E. (1993) *Current Opinion in Biotechnology*, **4**, 331–338.

39 Al-Bashir, B., Cseh, T., Leduc, R. and Samson, R. (1990) *Applied Microbiology and Biotechnology*, **34**, 414–419.

40 Zhang, X., Sullivan, E.R. and Young, L.Y. (2000) *Biodegradation*, **11**, 117–124.

41 Meckenstock, R.U., Annweiler, E., Michaelis, W., Richnow, H.H. and Schink, B. (2000) *Applied and Environmental Microbiology*, **66**, 2743–2747.

42 Yuan, S.Y. and Chang, B.V. (2007) *Journal of Environmental Health*, Part B **42**, 63–69.

43 Coates, J.D., Woodward, J., Allen, J., Philip, P. and Lovley, D.L. (1997) *Applied and Environmental Microbiology*, **63**, 3589–3593.

44 Galushko, A., Minz, D., Schino, B. and Widdel, F. (1999) *Environmental Microbiology*, **1**, 415–420.

45 Sullivan, E.R., Zhang, X., Phelps, C. and Young, L.Y. (2001) *Applied and Environmental Microbiology*, **67**, 4353–4357.

46 Annweiler, E., Michaelis, W. and Meckenstock, R.U. (2002) *Applied and Environmental Microbiology*, **68**, 852–858.

47 Young, L.Y. and Phelps, C.D. (2005) *Environmental Health Perspectives*, **113**, 62–67.

48 Muncnerova, D. and Augustin, J. (1994) *Bioresource Technology*, **48**, 97–106.

49 Cameron, M.D. and Aust, S.D. (1999) *Archives of Biochemistry and Biophysics*, **367**, 115–121.

50 Furukawa, K., Hirose, J., Suyama, A., Zaiki, T. and Hayashida, S. (1993) *Journal of Bacteriology*, **175**, 5224–5232

51 Meyer, S., Moser, R., Neef, A., Stahl, V. and Kampfer, P. (1999) *Microbiology*, **145**, 1731–1741.

52 Baldwin, B.R., Nakatsu, C.H. and Nies, L. (2003) *Applied and Environmental Microbiology*, **69**, 3350–3358.

53 Foght, J.M. and Westlake, D.W.S. (1996) *Biodegradation*, **7**, 353–366.

54 Kiyohara, H., Nagao, K., Kouno, K. and Yano, K. (1982) *Applied and Environmental Microbiology*, **43**, 458–461.

55 Fredrickson, J.K., Balkwill, D.L., Romine, M.F. and Shi, T. (1999) *Journal of Industrial Microbiology & Biotechnology*, **23**, 273–283.

56 Larkin, M.J., Allen, C.C.R., Kulakov, L.A. and Lipscomb, D.A. (1999) *Journal of Bacteriology*, **181**, 6200–6204.

57 Saito, A., Iwabuchi, T. and Harayama, S. (1999) *Chemosphere*, **38**, 1331–1337.

58 Khan, A.A., Wang, R.F., Cao, W.W., Doerge, D.R., Wennerstrom, D. and Cerniglia, C.E. (2001) *Applied and Environmental Microbiology*, **67**, 3577–3585.

59 Krivobok, S., Kuony, S., Meyer, C., Louwagie, M., Willison, J.C. and Jouanneau, Y. (2003) *Journal of Microbiology (Seoul, Korea)*, **185**, 3828–3841.

60 Eltis, L.D. and Bolin, J.T. (1996) *Journal of Bacteriology*, **178**, 5930–5937.

61 Williams, P.A., Jones, R.M. and Shaw, L.E. (2002) *Journal of Bacteriology*, **184**, 6572–6580.

62 Hall, K., Miller, C.D., Sorensen, D.L. and Anderson, A.J.R.C. (2005) *Biodegradation*, **16**, 475–484.

63 Ringelberg, D.B., Talley, J.W., Perkins, E.J., Tucker, S.G., Luthy, R.G., Bouwer, E.T. and Frederickson, H.L. (2001) *Applied and Environmental Microbiology*, **67**, 1542–1550.

64 Rhee, S.-K., Liu, X., Wu, L., Chong, S.C., Wan, X. and Zhou, J. (2004) *Applied and Environmental Microbiology*, **70**, 4303–4317.

65 Marlowe, E.M., Wang, J.M., Pepper, I.L. and Mayer, R.M. (2002) *Biodegradation*, **13**, 251–260.

66 Gianfreda, L., Xu, F. and Bollag, J.-M. (1999) *Bioremed. J.*, **3**, 1–25.

67 Gianfreda, L. and Rao, M.A. (2004) *Enzyme and Microbial Technology*, **35**, 339–354.

68 Ruggaber, T.P. and Talley, J.W. (2006) *Practice Periodical of Hazardous, Toxic, and Radioactive Waste Management*, **4**, 3–85.

69 Reddy, C.A. and D'Souza, T.M. (1994) *FEMS Microbiology Reviews*, **13**, 137–152.

70 Higuchi, T. (1993) *Journal of Biotechnology*, **30**, 1–8.

71 Canas, A., Alcade, M., Martinez, M.J., Martinez, A. and Camarero, S. (2007) *Environmental Science & Technology*, **41**, 2964–2971.

72 Ang, E.L., Zhao, H. and Obbard, J.P. (2005) *Enzyme and Microbial Technology*, **37**, 487–496.

73 Bouwer, E.J. and Zehnder, A.J.B. (1993) *Trends in Biotechnology*, **11**, 360–367.

74 Quantin, C., Joner, E.J., Portal, J.M. and Berthelin, J. (2005) *Environmental Pollution*, **134**, 315–322.

75 Edgehill, R.U. (1999) in *Bioremediation of Contaminated Soils* (eds D.C. Adriano, J.-M. Bollag, W.T. Frankenberger, Jr and R.C. Sims), Agronomy Monograph 37, American Society of Agronomy, Crop Science of America, Soil Science Society of America, pp. 289–310.

76 Bouchez, T., Patureau, D., Dabert, P., Juretschko, S., Dore, J., Delegenes, P., Moletta, R. and Wagner, M. (2000) *Environmental Microbiology*, **2**, 179–190.

77 Cavalca, L., Colombo, M., Larcher, S., Gigliotti, C., Collina, E. and Andreoni, V. (2002) *Journal of Applied Microbiology*, **92**, 1058–1065.

78 Van Herwijnen, R., Joffe, B., Ryngaert, A., Hausner, M., Springael, D., Govers, H.A.J., Wuertz, S. and Parson, J.R. (2006) *FEMS Microbiology Ecology*, **55**, 122–135.

79 Bouchez, M., Blanchet, D. and Vandecasteele, J.P. (1995) *Applied and Environmental Microbiology*, **43**, 156–164.

80 Vanbroekhoven, K., Ryngaert, A., Bastiaens, L., Wattiau, P., Vancanneyt, M., Swings, J., de Mot, R. and Springael, D. (2004) *Environmental Microbiology*, **6**, 1123–1136.

81 Alsaleh, E.S., Smith, S. and Mason, J. (1999) in *Bioremediation Technologies for Polycyclic Aromatic Hydrocarbon*

Compounds – The Fifth International In Situ and On-site Bioremediation Symposium (eds A. Leeson and B.C. Alleman), Battelle Press, Cleveland, OH, pp 111–115.

82 Dejonghe, W., Boon, N., Seghers, D., Top, E.M. and Verstraete, W. (2001) *Environmental Microbiology*, **3**, 649–657.

83 Thomson, I.P., van der Gast, C.J., Ciric, L. and Singer, A.C. (2005) *Environmental Microbiology*, **7**, 909–915.

84 Roane, T.M., Josephson, K.L. and Pepper, I.L. (2001) *Applied and Environmental Microbiology*, **67**, 3208–3215.

85 Sandrin, T.R., Chech, A.M. and Maier, R.M. (2000) *Applied and Environmental Microbiology*, **66**, 4585–4588.

86 Martens, R. (1982) *Chemosphere*, **11**, 761–770.

87 Semple, K.T., Reid, B.J. and Fermor, T.R. (2001) *Environmental Pollution*, **112**, 269–283.

88 McFarland, M.J. and Qiu, X.J. (1995) *Journal of Hazardous Materials*, **42**, 61–70.

89 Canet, R., Birnstingl, J.G., Malcom, D.G., Lopez-Real, J.M. and Beck, A.J. (2001) *Bioresource Technology*, **76**, 113–117.

90 Adenuga, A.O., Jr, Johnson, J.H., Cannon, J.N. and Wan, L. (1992) *Water Science and Technology*, **26**, 2331–2334.

91 Joyce, J.F., Sato, C., Cardeans, R. and Surampalli, R.Y. (1998) *Water Environment Research: A Research Publication of the Water Environment Federation*, **70**, 356–361

92 Antizar-Ladislao, B., Lopez-Real, J. and Beck, A.J. (2005) *Waste Management*, **25**, 281–289.

93 Antizar-Ladislao, B., Lopez-Real, J. and Beck, A.J. (2006) *Environmental Pollution*, **141**, 459–468.

94 Singh, O.V., Labana, S., Pandey, G., Budhiraja, R. and Jain, R.K. (2003) *Applied Microbiology and Biotechnology*, **61**, 405–412.

95 Gregory, S.T., Shea, D. and Guthrie-Nichols, E. (2005) *Environmental Science & Technology*, **39**, 5285–5292.

96 Kamath, R., Scnoor, J.L. and Alvarez, P.J.J. (2004) *Environmental Science & Technology*, **38**, 1740–1745.

97 Yateem, A., Balba, M.T., El-Nawawy, A.S. and Al-Awadhi, N. (1999) *Soil Groundwater Cleanup*, **2**, 31–33.

98 Muratova, A., Hubner, T., Narula, N., Wand, H., Turkovskaya, O., Kuschk, P., Jahn, R. and Merbach, W. (2003) *Microbiological Research*, **158**, 151–161.

99 Campbell, S., Paquin, D., Awaya, J.D. and Li, Q.X. (2002) *International Journal of Phytoremediation*, **4**, 157–168.

100 Siciliano, S.D., Germida, J.J., Banks, K. and Greer, C.W. (2003) *Applied and Environmental Microbiology*, **69**, 483–489.

101 Liste, H.H. and Alexander, M. (2000) *Chemosphere*, **40**, 7–10.

102 Wiltse, C.C., Rooney, W.L., Chen, Z., Schwab, A.P. and Banks, M.K. (1998) *Journal of Environmental Quality*, **27**, 169–173.

103 Kuiper, I., Bloemberg, G.V. and Lugtenberg, B.J.J. (2001) *Molecular Plant-Microbe Interactions*, **14**, 1197–1205.

104 Suthersan, S.S. (2002). *Natural and Enhanced Remediation Systems*, Arcadis/Lewis, London,

105 Pradhan, S.P., Conrand, J.R., Paterek, J.R. and Srivastav, V.J. (1998) *Journal Of Soil Contamination*, **7**, 467–480.

106 Lalande, T.L., Skipper, H.D., Wolf, D.C., Reynolds, C.M., Freedman, D.L., Pinkerton, B.W., Hartel, P.G. and Grimes, L.W. (2003) *International Journal of Phytoremediation*, **5**, 1–12.

107 Huang, X-D., El-Alawi, Y, Penrose, D.M., Glick, B.R. and Greenberg, B.M. (2004) *Environmental Pollution*, **130**, 465–476

108 Cunningham, S.D., Berti, W.R. and Hung, J.W. (1995) *Trends in Biotechnology*, **13**, 393–397.

109 Macek, T., Mackova, M. and Kas, J. (2000) *Biotechnology Advances*, **18**, 23–34.

110 Susarla, S., Medina, V.F. and McCutcheon, C. (2002) *Ecological Engineering*, **8**, 647–658.

Index

Handbook of Green Chemistry, Volume 3: Biocatalysis. Edited by Robert H. Crabtree
Copyright © 2009 WILEY-VCH Verlag GmbH & Co. KGaA, Weinheim
ISBN: 978-3-527-32498-9